TOPOLOGICAL SPACES

First published in the German language under the title Topologische Räume
and copyrighted in 1961 by Birkhäuser Verlag, Basel and Stuttgart.

TOPOLOGICAL SPACES

BY

H. J. KOWALSKY

UNIVERSITÄT ERLANGEN
GERMANY

TRANSLATED BY

JAY E. STRUM

GRADUATE SCHOOL OF BUSINESS
NEW YORK UNIVERSITY, NEW YORK, NEW YORK

1964

ACADEMIC PRESS • New York and London

ACADEMIC PRESS INC.
111 Fifth Avenue, New York, New York 10003

United Kingdom Edition published by
ACADEMIC PRESS INC. (LONDON) LTD.
Berkeley Square House, London, W. 1

LIBRARY OF CONGRESS CATALOG CARD NUMBER: 63-23341

PRINTED IN THE UNITED STATES OF AMERICA

PREFACE

The theory of topological spaces or, as it is called, point-set or general topology, has become one of the elementary building-blocks underlying diverse branches of mathematics. Its concepts and methods have enriched numerous other fields of mathematics and given considerable impulse to their further development. For these reasons, topology counts among those few basic structures which alone give access to modern mathematical research.

The totality of point-set topology forms an exceptionally extensive and ramified field of mathematics, containing within itself several autonomous specialities. In order to write a textbook on topology, one whose physical size must necessarily be limited, it is evidently imperative that the author adhere to a very restricted choice of topics. Accordingly, we have confined ourselves to the fundamental notions of the theory, without in any way striving for completeness. Many important topics had to be either omitted or else only touched upon in short notes.

As exposition, this text differs from others mainly in its extensive use of the notion of filter. With a view to applications in analysis, at the very outset of our exposition we make the definition of topological spaces depend upon the correct notion of convergent filter. Sometimes filters make conceptual simplifications more easily possible, and often they allow of symbolic proofs akin to the familiar inequalities of algebra. In any case, our method of exposition necessitates on the part of the reader a short period of adaptation and assimilation.

I should like to give heartfelt thanks to all those who, by their advice and suggestions, have helped me in my work on this book. I am especially thankful to H. Bauer, F. Hofmann, and E. Luft, three gentlemen who took upon themselves the exacting task of proofreading and who, by their constructive criticism, furnished me with most decisive aid.

I wish to offer my most sincere thanks to the publishers for the care which they have given to the preparation of this volume, and for all the many other instances of their solicitude.

Erlangen H. J. Kowalsky

CONTENTS

V. Extensions and Characterization of Topological Spaces

VI. Metric and Uniform Spaces

VII. Topological Groups. Applications

Introduction

The theory of topological spaces is akin to analytic geometry in that it serves as a link between geometry and analysis. It can accordingly be developed from diverse points of view. If we start, for instance, from the intuitive notion of the distance between two points in space, we are at first led to the abstract notion of metric space, and after further generalization, ultimately to the notion of topological spaces. But, on the other hand, we can take analysis as our starting point and raise the general question of how to define the concept of convergence in arbitrary sets. It is from this latter viewpoint that we shall begin our exposition. If we further stipulate that the sought-for notion of convergence is to obey a few simple and natural conditions we shall necessarily be led to the concept of topological space.

With the exception of the last chapter, where we presuppose knowledge of the fundamental concepts of modern algebra, the prerequisites for reading this book are few, taken solely from set theory. Set theory has become so basic to all of mathematics that we can safely assume that the reader is familiar with the main results of this subject; and, accordingly, we have refrained from developing these results in detail. The concepts and results which we have collected in Section 1 should serve, however, to fix notation and clearly establish exactly what it is that we do assume of the reader.

Section 3 should be studied very carefully. It is devoted to the notion of filter, an essential concept for all that follows. Filters generalize ordinary sequences in exactly that way needed to answer the previously raised question of how to define topological convergence in full generality. They lend themselves to short, clear statements; and their use often allows us to write out proofs in formal symbolism.

We shall make extensive use of the material of Section 3, and familiarity with this material is a must for the reader. Ordered structures, in particular the lattice structure of filters, will be especially important. Our definition of order relation is not the usual one, but rather its inverse. This deviation from common usage necessitates a change in the usual habits of thought; but as soon as it is assimilated by the reader, it affords him an especially handy symbolism. The elements of lattice theory are exposed in Section 2 and serve merely as an introduction to what follows. The exposition is classical and can be omitted by readers familiar with that material.

The earlier chapters are primarily devoted to the study of the inner properties of a single topological space. Not until Chapter IV do we introduce mappings, which serve to depict the relations between two different topological spaces. In Chapter V we characterize topological spaces by means of their possible embeddings and their representations as images under continuous maps. These problems are intimately connected with the problem of the extension of topological spaces. We treat the problem of extensions in great generality, establishing a principle which, on the one hand, encompasses as special cases all the well-known types of extension, and on the other hand, also allows of a simple treatment of the problems of completing metric and uniform spaces. We conclude with a sixth chapter on metric and uniform spaces. There is also a seventh chapter, which is more of an appendix, touching upon a few topics from the theory of topological groups and rings, etc., intended to give an idea of some of the applications of the general theory.

The nomenclature we use is standard, except where we have introduced a new concept or symbol. There are several instances where nomenclature has not yet become standardized and where even in the original literature there is danger of confusion. This is especially true for the different separation axioms and the many definitions of compactness. The reader must accordingly pay close attention to the terminology we use when treating such questions. In order to avoid confusion of language in the treatment of compactness, we have followed the use of G. Nobeling and employed the rather ponderous term "totally compact."

The following system of numbering has been employed. Sections are numbered consecutively, independently of chapters. Theorems, definitions, etc. are indicated first by the number of the corresponding section, followed immediately by a special symbol indicative of where they appear within the section. The symbol used for theorems is an arabic numeral; for definitions it is a lower case roman letter of the alphabet; for illustrative examples it is a capital roman numeral; while for the notes or exercises appearing at the end of certain paragraphs, we use capital roman letters. For instance, 12.3 denotes the third theorem, 12c the third definition, 12.III the third example, and 12C the third note or exercise, all from Section 12. In addition, the section number appears at the top of every right-hand page. References to material found within the text appear in round brackets. The symbol ▮ signals the end of a proof.

Numbers in square brackets refer to the bibliography appearing at the end of this book. We have in general made no attempt to list original contributions, nor to give historical credit. We should mention that certain theorems are referred to by name. Our list of references contains only a selection from among the most important textbooks.

I

Elementary Principles

We assume that the reader is familiar with the fundamentals of set theory. The first section of this introductory chapter therefore only lists the most important definitions and results from set theory, without giving any indication of proofs, and is intended mainly to orient the reader and fix notation. For a more detailed treatment, we refer the reader to the texts mentioned in the bibliography. Section 2 contains a brief exposition of those results from lattice theory which are essential for our purposes. In Section 3 we go into the subject of filters in greater detail. The concepts and theorems developed in that section are essential to an understanding of all that follows in this book.

1. Fundamental Concepts of Set Theory

The fundamental relation of set theory is that of an element to a set: the fact that p is an *element* of (lies in, belongs to) the set M will be denoted by "$p \in M$"; while "$p \notin M$" will signify the negation of this statement.[1] The *empty set*, which is characterized by the fact that it contains no elements, will be denoted by the symbol $\emptyset$.

A set M is called a *subset* of the set N if $p \in M$ implies $p \in N$. We also say that "M is contained in N" and denote this relation by $M \subseteq N$. We can equivalently say that "N contains M" and denote this by $N \supseteq M$. The empty set is a subset of every set. The relations $M \subseteq N$ and $N \subseteq M$ imply $M = N$. The equality sign is here used to denote the identity of M and N. If $M \subseteq N$

[1] If one constructs the theory of sets axiomatically, the relation "being an element of" should at first be defined only for so-called *families* or *classes*. Sets can then be defined as special classes, namely, those which are elements of any class. By making such a distinction between sets and classes, the well-known antinomies of set theory can be avoided. However, in order to read this text, all that is needed is an intuitive understanding of the concept of set.

but $M \neq N$, then M is called a proper subset of N; we denote this relation by $M \subset N$.

The set composed of the elements $p_1, \ldots, p_n$ is denoted by $\{p_1, \ldots, p_n\}$. The set containing just the one element p is accordingly denoted by $\{p\}$ and should be carefully distinguished from the element p itself. Thus the symbol $\{\emptyset\}$ denotes the set whose sole element is the empty set. Let M be an arbitrary but fixed set (a reference set for the discussion at hand). The symbol $\mathscr{E}(p)$ will signify that the property $\mathscr{E}$ pertains to the element p, while the symbol $\{p : \mathscr{E}(p)\}$ will denote the set of all elements p for which the property $\mathscr{E}$ actually holds true. This notation is unambiguous only when the set M containing the elements p has been stipulated in advance—$\mathscr{E}$, taken out of context, does not refer to any definite set.[2] Nonetheless, we shall not always bother to explicitly name the set M whenever we refer to the property $\mathscr{E}$, for it will always be clear, from the context or from the definition of $\mathscr{E}$, which reference set M is being used. The following situation occurs frequently and illustrates the above remarks on reference sets: Let M and I denote two nonempty sets. To each $\iota \in I$ let us associate a unique element $p_\iota \in M$. We call this an *indexing* of the set M by the indexing set I. Under this indexing $\{p_\iota : \iota \in I\}$ denotes the set of all indexed elements of M. We shall employ the capital letter I exclusively for indexing sets. Whenever we use the notation $\{p_\iota : \iota \in I\}$ without having previously mentioned the sets M and I, it is to be understood that there exists a fixed reference set M and that a definite rule for indexing M by the set I has been stipulated.

Let us agree to use the letter R to represent a set fixed for the purpose of the discussion at hand, whatever it may be. (We can think of R as symbolizing the words referential, or reference set, or the German "Raum.") We shall call the elements of R *points*. Subsets of R will be denoted by capital Roman letters. Sets, whose elements themselves are subsets of R, will also be called *families* or *classes* of sets. We use capital German letters to denote such families or sets. The family of all subsets of R will be denoted by $P(R)$. Whenever R has been stipulated in advance and we speak of sets, we shall always mean subsets of R.

Let R be fixed. By the *complement* of M we mean the set of all $p \in R$ such that $p \notin M$. This complement is denoted by $\complement M$. Should M consist of only a single point, say $M = \{p\}$, then instead of writing $\complement\{p\}$ we simply write $\complement p$. When we wish to emphasize that R is the referential set with respect to which we are forming complements we write $\complement_R M$ rather than $\complement M$. The following relations are valid:

$$\complement R = \emptyset, \qquad \complement \emptyset = R, \qquad \complement\complement M = M$$
$$M \subseteq N \quad \text{if and only if} \quad \complement M \supseteq \complement N \tag{1.1}$$

[2] If we define $\mathscr{E}(p)$ by the condition $p \notin p$ and do not restrict ourselves to p's belonging to a fixed reference set, then the set $\{p : p \notin p\}$ leads to the famous Russell paradox on "the set of all sets that do not contain themselves as elements."

Let $\mathfrak{S} = \{M_\iota, \iota \in I\}$ be a family of sets. By the *intersection* D of this family of sets we mean the set of all elements common to each member of the family, i.e., $x \in D$ if and only if $x \in M_\iota$ for all ι. Each of the following symbols is sometimes used to denote this intersection:

$$\bigcap_{M \in \mathfrak{S}} M, \qquad \bigcap\{M : M \in \mathfrak{S}\}, \qquad \bigcap_{\iota \in I} M_\iota, \qquad \bigcap\{M_\iota : \iota \in I\}$$

Occasionally, when no possibility of confusion exists, we shall omit the indexing set I and simply write $\bigcap M_\iota$. If I is finite, say $I = \{1, ..., n\}$, we shall also use the notation $M_1 \cap ... \cap M_n$. Two sets, whose intersection is empty, are said to be *disjoint*. The *union* of the family $\mathfrak{S}$ is the set of all elements which belong to at least one of the members of $\mathfrak{S}$. The notation for unions is analogous to that of intersections, $\bigcap$ and $\cap$ being replaced by $\bigcup$ and $\cup$, respectively. Our definitions allow of the family $\mathfrak{S}$ being empty. This is convenient for certain applications—note that $\bigcap \{M : M \in \emptyset\} = R$ and $\bigcup \{M : M \in \emptyset\} = \emptyset$. The formation of unions and intersections is a commutative and associative operation. The following relationships hold:

$$M \subseteq N, \qquad M \cap N = M, \qquad M \cup N = N$$

are all equivalent to each other.

$$M_1 \cup (M_2 \cap M_3) = (M_1 \cup M_2) \cap (M_1 \cup M_3) \tag{1.2}$$

$$M_1 \cap (M_2 \cup M_3) = (M_1 \cap M_2) \cup (M_1 \cap M_3) \tag{1.3}$$

(Distributive Laws)

$$\complement(\bigcup M_\iota) = \bigcap (\complement M_\iota) \quad \text{and} \quad \complement(\bigcap M_\iota) = \bigcup (\complement M_\iota) \tag{1.4}$$

(de Morgan's Laws)

The *product* $A_1 \times ... \times A_n$ of the sets $A_1, ..., A_n$ is defined to be the set of all ordered n-tuples $(a_1, ..., a_n)$, where $a_1 \in A_1, ..., a_n \in A_n$.[3] If one of the A_i is empty, the above product is also considered empty. If all the sets $A_1, ..., A_n$ are equal to the set A, we simply write A^n instead of $A_1 \times ... \times A_n$.

An arbitrary subset ρ of the product $A \times B$ is called a *relation* (in $A \times B$). Rather than say $(a, b) \in \rho$ we usually state that the relation ρ holds between a and b. If ρ is as above, then

$$\rho^{-1} = \{(b, a) : (a, b) \in \rho\}$$

is a relation defined in $B \times A$, called the *inverse relation* to ρ. The set of all elements $a \in A$, for which there exists a $b \in B$ such that $(a, b) \in \rho$ is called the

[3] For the most general definition of the product of any family of sets, see Section 22.

domain of ρ. The *range* of ρ is defined to be the set of all $b \in B$ for which there exists an $a \in A$ such that $(a, b) \in \rho$. If M is a subset of $A \times B$, we denote by ρM the set of all $b \in B$, for which there exists an $a \in M$ such that $(a, b) \in \rho$. Hence ρA equals the *range* of ρ and $\rho^{-1}B$ equals the *domain* of ρ. If M and the domain of ρ are disjoint, ρM is the empty set. The following relationships hold:

$$M \subseteq \rho^{-1}(\rho M) \tag{1.5}$$

if and only if M is contained in the domain of ρ.

$$M_1 \subseteq M_2 \quad \text{implies} \quad \rho M_1 \subseteq \rho M_2 \tag{1.6}$$

$$\rho(\cup M_\iota) = \cup \rho M_\iota \tag{1.7}$$

$$\rho(\cap M_\iota) \subseteq \cap \rho M_\iota \tag{1.8}$$

In (1.8) we generally do not have equality, even when the indexing set I is finite.

Suppose ρ is a relation in $A \times B$ and σ a relation in $B \times C$. The set of all couples $(a, c) \in A \times C$, for which there exists a $b \in B$ such that $(a, b) \in \rho$ and $(b, c) \in \sigma$, is itself a relation defined in $A \times C$. This relation is called the *product* of ρ and σ and denoted by $\sigma\rho$ (in this order). The formation of products of relations is associative, but generally not commutative. We have

$$(\sigma\rho)^{-1} = \rho^{-1}\sigma^{-1} \tag{1.9}$$

$$(\sigma\rho)\, M = \sigma(\rho M) \tag{1.10}$$

Let φ be a relation in $A \times B$. Then φ is called a *map* or a *mapping* if, for every a in the domain of φ, the set $\varphi\{a\}$ contains only one element. If φ is a mapping, $\varphi\{a\}$ is called the *image* of a and simply denoted by φa. What was the domain D of φ viewed as a relation is now also called the *domain of definition* of the mapping φ; likewise, the range is now also called the *range of images*. We say that "φ maps D into B" and abbreviate this statement by "$\varphi : D \to B$." If the domain of images of φ happens to be all of B, we say that φ maps D *onto* B. The specification of the mapping φ is independent of the set A, depending only upon the domain of definition D. Whenever we write φa it is understood that a belongs to the domain of definition of φ.

If M is a subset of D, then any mapping $\varphi : D \to B$ induces a mapping $\varphi_M : M \to B$, defined by $\varphi_M p = \varphi p$ for all $p \in M$. The map φ_M is called the *restriction* of φ to M.

For any two maps φ and ψ, their product $\psi\varphi$ as relations has been defined above. This product turns out to be a mapping and is called the product of the maps φ and ψ. Let $\varphi : D \to B$ and $\psi : D' \to B'$. The domain of definition of the product-map $\psi\varphi$ is the subset $\psi^{-1}D'$ of D, and we have:

$$\psi\varphi : \varphi^{-1}D' \to B'$$

The inverse relation φ^{-1} to the mapping φ need not itself be a mapping. If, however, φ^{-1} is also a mapping, then φ is called a *one-one mapping* or a *1-1 mapping*. The inverse relation φ^{-1} is then also called the *inverse mapping* to φ. The product of two 1-1 mappings is again a 1-1 mapping. The mapping φ is 1-1 if and only if $\varphi p = \varphi q$ implies $p = q$. That mapping which associates with each element x of the set R the element x again, is manifestly a 1-1 mapping of R onto itself. It is called the *identity mapping*, or simply the *identity* of R.

If M is a subset of the range of images of the mapping φ, then the set $\varphi^{-1}M$ is called the *inverse-image* of M. Note that the set $\varphi^{-1}M$ is always well-defined even when M is not a subset of B; in all cases, $\varphi^{-1}M = \varphi^{-1}(M \cap B)$.

We have seen that $\rho(\bigcap M_i) \subseteq \bigcap \rho M_i$ (1.8) for all relations ρ. If ρ is the inverse relation φ^{-1} of the map φ, we have equality:

$$\varphi^{-1} \bigcap M_\iota = \bigcap \varphi^{-1} M_\iota \tag{1.11}$$

and if φ is a 1-1 mapping we even have

$$\varphi(\bigcap M_\iota) = \bigcap \varphi M_\iota \tag{1.12}$$

The concept of mapping allows us to generalize the distributive laws (1.2). Let $\{K_\iota : \iota \in I\}$ be a system of indexing-sets K_ι and let Φ be the set of all mappings φ where $\varphi : I \rightarrow UK_\iota$ and $\varphi\iota \in K_\iota$ for all $\iota \in I$. Lastly, let there be associated with each $\iota \in I$ a system of sets $\{M_{\iota,\kappa_\iota} : \kappa_\iota \in K_\iota\}$. Then we have

$$\bigcup_{\iota \in I} \Big(\bigcap_{\kappa_\iota \in K_\iota} M_{\iota,\kappa_\iota} \Big) = \bigcap_{\varphi \in \Phi} \Big(\bigcup_{\iota \in I} M_{\iota,\varphi\iota} \Big) \tag{1.13}$$

$$\bigcap_{\iota \in I} \Big(\bigcup_{\kappa_\iota \in K_\iota} M_{\iota,\kappa_\iota} \Big) = \bigcup_{\varphi \in \Phi} \Big(\bigcap_{\iota \in I} M_{\iota,\varphi\iota} \Big)$$

Let ρ be a relation defined in $A \times A = A^2$. Then ρ is called *reflexive* if $(a, a) \in \rho$ for all $a \in A$. The relation ρ is called *transitive* if $(a, b) \in \rho$ and $(b, c) \in \rho$, taken together, imply that $(a, c) \in \rho$; i.e., if the product $\rho\rho \subseteq \rho$. Finally, ρ is *symmetric* if $(a, b) \in \rho$ implies $(b, a) \in \rho$. A relation defined in A^2 is called an *equivalence relation in A* if it is reflexive, symmetric, and transitive.

If ρ is an equivalence relation in A, then the set of all elements equivalent to a, or $\rho\{a\}$, is called the equivalence class generated by a. Any element $b \in \rho\{a\}$ is called a *representative* of this class. For any $a, b \in A$ we either have $\rho\{a\} = \rho\{b\}$, or else $\rho\{a\}$ and $\rho\{b\}$ are disjoint. Hence the equivalence relation ρ decomposes the set A into pairwise disjoint equivalence classes. The set of these classes is called the quotient set of A by the equivalence ρ and is designated by A/ρ.

If ρ is a relation, defined in A^2, which is reflexive and transitive (but not necessarily symmetric) we can form an equivalence relation in A by taking $\rho \cap \rho^{-1}$. This equivalence is called the equivalence induced by ρ. Let us also note that any map $\varphi : A \rightarrow B$ induces an equivalence $\tilde{\varphi}$ in A, defined as follows: Any two elements $p, q \in A$ are equivalent if and only if $\varphi p = \varphi q$.

Any reflexive and transitive relation ρ in A^2, having the identity relation as its induced equivalence relation $\rho \cap \rho^{-1}$, is called an *order* or *order relation* in A.[4]

If ρ is an order, instead of $(p, q) \in \rho$ we usually write $p \leqslant q$ (or $q \geqslant p$). The definition of order relation can then be stated as

$$\begin{array}{llllll} & p \leqslant p & \text{for all} & p & & \\ \text{If} & p \leqslant q & \text{and} & q \leqslant r, & \text{then} & p \leqslant r. \\ \text{If} & p \leqslant q & \text{and} & q \leqslant p, & \text{then} & p = q. \end{array} \tag{1.14}$$

Whenever $p \leqslant q$ and $p \neq q$ one can simply write $p < q$ or $q > p$.

If $\leqslant$ is an order in the set A, the couple $(A, \leqslant)$ is called an *ordered set.* Whenever the order relation has been fixed and no confusion can arise, we can also call A itself an ordered set. Elements p and q of the ordered set A are called *comparable* if either $p \leqslant q$ or $q \leqslant p$. In general, two elements of an ordered set need not be comparable. If, however, any two elements of an ordered set are always comparable, we say that the set is *totally* or *linearly* ordered, or else we say that the set is a *chain.* Every subset of the ordered set A is also an ordered set under the restriction of the order of A. Every subset of a chain is again a chain.

We call an element m of the ordered set A *minimal* (*maximal*) in A if $a \in A$ and $a \leqslant m$ ($a \geqslant m$) implies $a = m$. A totally ordered set can possess at most one minimal and one maximal element. The unique minimal element, if it exists, is called the *first element* or *initial element*; likewise the unique maximal element is called the *last* or *terminal element.* A linearly ordered set A is called *well-ordered* if every nonempty subset of A contains a first element. The order relation in such a case is called a *well-ordering.*

The principal of mathematical induction, as applied to sequences of natural numbers, can be generalized to the principle of transfinite induction for well-ordered sets. Let A be a well-ordered set and $\mathscr{E}$ a property defined for elements of A. Then we can enunciate the following

Principle of Transfinite Induction

Suppose that property $\mathscr{E}$ holds true for the first or initial element of A. Suppose further that if $\mathscr{E}$ holds for all $a' < a$ it must also hold for a itself. Then under these two suppositions, it follows that $\mathscr{E}$ holds for every $a \in A$.

In order to be able to apply the principle of transfinite induction to a set we must assure ourselves that the set is or can be well-ordered. The following well-ordering theorem (or axiom) tells us that this is always possible. Every

[4] What we call order is called by some authors *partial-order* or *semiorder*. These authors then call ordered what we shall call totally ordered.

set can be well-ordered, i.e., in every set we can define an order relation with respect to which the set becomes well-ordered.

The well-ordering theorem is better called an axiom. It is a theorem only in the sense that, in an axiomatic construction of set theory, it is usually proved to be a consequence of the Axiom of Choice (with which it is equivalent). The axiom of choice can be stated as follows: Let $\mathfrak{S}$ be a nonempty system of nonempty sets. Then we can be sure that there exists a mapping $\varphi: \mathfrak{S} \to \cup\{M : M \in \mathfrak{S}\}$ such that $\varphi M \in M$ for all $M \in \mathfrak{S}$.

Equally equivalent to the Well-Ordering Theorem is the following lemma, Zorn's lemma, a handier tool for proofs than the axiom of choice or the well-ordering theorem. Let B be a subset of the ordered set A. We call an element $s \in A$ a *lower* (*upper*) *bound* of B in A, if for all $b \in B$ we always have $s \leqslant b$ ($b \leqslant s$). With this concept we can now formulate Zorn's lemma.

Let A be a nonempty set. If every chain in A has a lower (*upper*) *bound, then A must necessarily have at least one minimal* (*maximal*) *element.*

In fact, given any $a \in A$, there exists a minimal (*maximal*) *element $m \in A$ such that $m \leqslant a$ ($a \leqslant m$).*[5]

Two sets A and B are called *equipotent* if there exists a 1-1 mapping of A onto B. The equipotencey of sets is an equivalence relation[6]; equipotent sets belong to the same class, their "potency class," so to speak. A class of equipotent sets is called a *cardinal number*.[6] The equipotency class having A as a representative is designated by $|A|$ and called the *cardinality* of A.

It is possible to define an order relation among cardinal numbers as follows: Let k_1 and k_2 be two cardinal numbers, having A_1 and A_2 as representative sets, i.e., $|A_1| = k_1$ and $|A_2| = k_2$. If the set A_1 is equipotent to a subset of A_2 we shall say $k_1 \leqslant k_2$, by definition. It is easy to show that this definition is independent of the choice of representative sets and that it leads to an ordering of all the cardinal numbers. In fact, this order is a well-ordering and a total ordering—however, in order to prove totality, we need the theory of ordinal numbers, a theory which we cannot go into here.

All finite sets having the same number of elements are equipotent, while two finite sets having unequal numbers of elements have different cardinalities. The cardinality of a set having n elements can therefore be represented by, or identified with, the natural number n. The smallest, or first, cardinal number (with respect to the well-ordering defined above) is the cardinality of the empty set, which we can identify with the number 0. The well-ordering of the cardinals then proceeds with the natural numbers following 0, in their usual order. If a

[5] An elegant proof of the equivalence of the Axiom of Choice, Well-Ordering Theorem, and Zorn's lemma can be found in [21].

[6] The totality of all sets and the totality of all cardinal numbers do not constitute sets but rather classes. The concept of equipotency, and the attendant well-ordering of the cardinal numbers, should be thought of as based upon relations between classes.

set contains an infinitude of elements, its cardinality is called a *transfinite number*. The smallest transfinite number is the cardinality of the set of all natural numbers. It is denoted by $\aleph_0$ (read aleph-zero, where $\aleph$ = aleph is the first letter of the Hebrew alphabet). A set having $\aleph_0$ as its cardinality is said to be countably or denumberably infinite, while any set with cardinality $\leqslant \aleph_0$ is simply said to be countable or denumerable. The cardinals that follow $\aleph_0$ in succession are designated by $\aleph_1$, $\aleph_2$, $\aleph_3$, etc., and called uncountable.

The natural numbers are denumerable, as are the following: the set of signed integers; the set of rational numbers; the set of all polynomials in the unknown x, with integral or rational coefficients; the set of all algebraic numbers. However, the set of real numbers is uncountable—the cardinality of this set is called the cardinality of the continuum and designated by the letter c. It is easy to show that $c > \aleph_0$; and hence $c \geqslant \aleph_1$. The question of whether $c = \aleph_1$ is as yet unresolved. It is known as the (special) continuum problem.

It is possible to define addition and multiplication of cardinal numbers as follows: Let k_1 and k_2 be two cardinal numbers with representative A_1 and A_2, and suppose further that A_1 and A_2 are disjoint. The *sum* of k_1 and k_2 is then defined to be the cardinal number of $A_1 \cup A_2 : k_1 + k_2 = |A_1 \cup A_2|$. The product of k_1 and k_2 is defined as the cardinality of the product-set of the sets A_1 and $A_2 : k_1 k_2 = |A_1 \times A_2|$. The above definitions are independent of the choice of representative sets. For finite cardinals their sums and products are the same as the sums and products of the corresponding natural numbers. However, it happens that for transfinite cardinals k_1 and k_2, their sum $k_1 + k_2$ and their product $k_1 k_2$ are identical. More precisely, we have:

The sum and the product of two transfinite cardinals are both equal to the larger of the two. (1.15)

A consequence of theorem (1.15) is the following result, which we shall often have occasion to use.

Let M be an infinite set and E be the set of all finite subsets of M. Then $|E| = |M|$. (1.16)

One can also define exponentiation for cardinals, i.e., the raising of a given cardinal to a cardinal power. This definition is rather involved and we shall not give it here, but we wish to mention the following facts. Met M be an arbitrary set and $P(M)$ the set of all subsets of M. It is then true that $|P(M)| = 2^{|M|}$. Also $2^{\aleph_0} = c$, while for any cardinal number k, we always have $k < 2^k$. If k' is the direct successor to k in the ordering of all the cardinal numbers, the question of whether $k' = 2^k$ for $k \geqslant \aleph_0$ is called the *general continuum problem*.

We shall employ the following abbreviations:

N = set of all natural numbers
Q = set of all rational numbers
R = set of all real numbers
$\mathbf{R}^n$ = n-dimensional space (for any natural number n, we have already seen that $\mathbf{R}^n$ is the set of all n-tuples of real numbers).

(*Translator's Note*: The reader should note carefully that the above abbreviations are given in boldface type. Whenever the eye sees **R** this shall mean the set of real numbers, as opposed to the referential set R (not in boldface), which will denote the set underlying the general topological space concerning which the author proves most of his theorems. For an instance of this distinction, cf. 2.II and 2.III.)

2. Ordered Sets and Lattices

In Section 1 we saw that an ordered set is a set A in which an order relation $\leqslant$ has been defined. The order relation obeyed the following:

Order Axioms (cf. 1.14)

($\mathbf{O}_1$) $a \leqslant a$.

($\mathbf{O}_2$) If $a \leqslant b$ and $b \leqslant c$, then $a \leqslant c$.

($\mathbf{O}_3$) If $a \leqslant b$ and $b \leqslant a$, then $a = b$.

Definition 2a: *Let A and B be ordered sets and φ a map of A into B. Then we say that φ is*

order-preserving *or* monotonic *if $a_1 \leqslant a_2$ implies $\varphi a_1 \leqslant \varphi a_2$,*
strictly monotonic *if $a_1 < a_2$ implies $\varphi a_1 < \varphi a_2$,*
an order-isomorphism *if φ is a 1-1 map of A onto B and both φ and φ^{-1} are monotonic.*

The ordered sets A and B are said to be order-isomorphic *if there exists an order isomorphism of A onto B.*

If φ is an order isomorphism then, of course, so is φ^{-1}. Also the product of two order isomorphisms is again an order isomorphism and the identity map of a set into itself is an order isomorphism. Hence we see that the relation of being order-isomorphic is an equivalence in the class of all ordered sets.[7] Order isomorphisms preserve minimal and maximal elements, as well as chains.

[7] Order isomorphism is an equivalence relation in the class of all ordered sets.

2.1 *Let A be a totally ordered set, B an ordered set, and φ a strictly monotonic map of A into B. Then φ must necessarily be 1-1 and if φ is also a map of A* onto *B, then φ is an order isomorphism (and hence B must be totally ordered also).*

Proof: Suppose $\varphi a_1 = \varphi a_2$. By hypothesis, a_1 and a_2 are comparable. Suppose $a_1 < a_2$. Since φ is strictly monotonic, $\varphi a_1 < \varphi a_2$, but this is contrary to hypothesis. Likewise, the supposition that $a_2 < a_1$ leads to a contradiction and hence $a_1 = a_2$, i.e., φ is 1-1. Let us further assume that φ maps A onto B. Let $b_1, b_2 \in B$ with $b_1 \leqslant b_2$. The inverse images, $a_1 = \varphi^{-1}b_1$ and $a_2 = \varphi^{-1}b_2$, are comparable. If $a_2 < a_1$, then $\varphi a_2 < \varphi a_1$, but $\varphi a_2 = b_2$, $\varphi a_1 = b_1$ and this would contradict our assumption that $b_1 \leqslant b_2$. Hence $a_1 \leqslant a_2$ and φ^{-1} is monotonic. ∎

Examples

2.I Whenever we speak of **N**, **Q**, and **R** (cf. the last paragraph of Section 1) as being ordered, we mean ordered in their usual way. Hence **N**, **Q**, and **R**, as ordered sets, are totally ordered. The ordered set **R** is also known as the *real line* or *real axis*.

Let a and b be two rational or real numbers with $a < b$. **Q** is order-isomorphic to any open interval of the form $]a, b[= \{x: a < x < b\}$ where a, b, and x belong to **Q**. Likewise for **R** when $a, b, x \in \mathbf{R}$. These statements follow from the fact that the compression mapping: $x \to x/(1 + |x|)$ establishes an order-isomorphism of **Q** or **R** onto $]-1, +1[$, while the mapping

$$x \to \frac{b+a}{2} + \left(\frac{b-a}{2}\right) x$$

establishes an order-isomorphism of $]-1, +1[$ onto $]a, b[$. In general, any two open intervals of **Q** are seen to be order-isomorphic to each other, and likewise for **R**.

2.II In **N** we can define an order different from the usual one, basing it upon divisibility relations. Given $a, b \in \mathbf{N}$ we define $a \leqslant b$ if and only if a is an exact divisor of b. With respect to this new order, **N** is not totally ordered. The integer 1 is a unique minimal element. If we agree to include 0 as a natural number, then it becomes a unique maximal element in our divisibility order.

2.III Let R be any set. Then, as we have seen $P(R)$ denotes the set of all subsets of R. Set-theoretic inclusion defines an order in $P(R)$: $A \leqslant B$ if and only if $A \subseteq B$. With respect to inclusion, $P(R)$ is not totally ordered (except in the trivial case where R contains only one element). The empty set is a unique minimal element, the reference set R is a unique maximal element.

2.IV Let A be a totally ordered set having neither a first nor a last element. If we extend A by adjoining to it two new elements, called $-\infty$ and $+\infty$, and subject to the relations $-\infty \leqslant a \leqslant +\infty$ for all $a \in A$, then we form an extension $\bar{A}$ of A. $\bar{A}$ is totally ordered and has $-\infty$ as its first element and $+\infty$ as its last element. If we apply this procedure to the real line $\mathbf{R}$ we form $\overline{\mathbf{R}}$, the *extended* real line.

We can carry over certain definitions from the real line to any ordered set:

The set $[a, b] = \{x : a \leqslant x \leqslant b\}$ will be called the closed interval having a as initial point and b as terminal point. It is nonempty if $a \leqslant b$. Similarly the sets

$$]a, b[= \{x : a < x < b\}$$

$$[a, b[= \{x : a \leqslant x < b\}, \qquad]a, b[= \{x : a < x < b\}$$

are called respectively, *open*, *right open*, and *left open* intervals. Moreover, the sets

$$[\leftarrow, a] = \{x : x \leqslant a\}, \qquad [\leftarrow, a[= \{x : x < a\}$$

$$[a, \rightarrow] = \{x : a \leqslant x\}, \qquad]a, \rightarrow] = \{x : a < x\}$$

will be considered to be intervals. We evidently have

$$[a, b] = [a, \rightarrow] \cap [\leftarrow, b], \qquad]a, b[=]a, \rightarrow] \cap [\leftarrow, b[, \quad \text{etc.}$$

Definition 2b: *An ordered set A is said to be* without gaps, *or* gap-free *if given any two elements $a, b \in A$ with $a < b$, the open interval $]a, b[$ is always nonempty.*

Between any two comparable elements of a gap-free ordered set there always exists another element. The sets $\mathbf{Q}$ and $\mathbf{R}$ are gap-free; $\mathbf{N}$, however, is not gap-free. Every interval of a gap-free set is itself gap-free.

2.2 *An ordered set A is order-isomorphic to the intervals of rationals if and only if A is denumerable, totally ordered, gap-free, and has a first and last element. If A lacks a first or a last element, or both, then it is order-isomorphic to the corresponding half-open or open intervals.*

Proof: The rational interval $[0, 1]$ certainly possesses the above-named properties, which are all invariant under order-isomorphism. To prove the converse, let us suppose A possesses these properties. Let $a_1, a_2, a_3, \ldots$ be a fixed listing of A, wherein a_1 is the initial element and a_2 the terminal element in the ordering of A. Likewise, let $r_1, r_2, r_3, \ldots$ be a fixed listing of the rational interval $[0, 1]$ with $r_1 = 0$ and $r_2 = 1$. We give below an inductive definition of a mapping φ which we shall show to be strictly monotonic:

$$\varphi : A \rightarrow [0, 1], \qquad \varphi a_1 = r_1, \qquad \varphi a_2 = r_2$$

If φ has already been defined as a strictly monotonic mapping for the elements

$a_1, ..., a_n$ $(n \geqslant 2)$ let p be the largest element among $a_1, ..., a_n$ such that $p < a_{n+1}$ (with respect to the order of A, of course) and let q be the smallest of $a_1, ..., a_n$ such that $q > a_{n+1}$. Since the rationals are gap-free there exists an $r \in [0, 1]$ such that $\varphi p < r < \varphi q$. Let φa_{n+1} be the first rational number r in the fixed listing previously given such that $\varphi p < r < \varphi q$.

We prove that φ, as above constructed, maps A onto $[0, 1]$: If it does not map A onto $[0, 1]$, let n be the smallest index for which r_n does not appear as an image under φ. Then $n > 2$. Among the numbers $r_1, ..., r_{n-1}$ let s be the largest such that $s < r_n$, and t the smallest such that $r_n < t$. Since A is gap-free, there exists at least one $a \in A$ with $\varphi^{-1}s < a < \varphi^{-1}t$. Let k denote the first index such that $s < \varphi a_k < t$. Then $\varphi a_k = r_n$, which is a contradiction. Hence φ is a strictly monotonic map of A onto $[0,1]$ and by Theorem 2.1 it must be an order-isomorphism. ∎

By an *order-theoretic* statement we mean a statement which, save for "purely logical components" and the $\in$ relation of belonging between elements and sets, contains only mention of an order relation.[8] *Order-theoretic concepts* are those which can be defined in terms of order-theoretic statements. The previously given order axioms furnish us with an example of order-theoretic statements. Some order-theoretic concepts are "minimal (maximal) element" and "lower (upper) bound." An order-theoretic statement A which is a logical consequence of the order axioms will be called an *order-theoretic theorem*: i.e., A can be deduced by formal logic from the order axioms, no account ever being taken during the proof of the special nature of any sets entering into the question.

If in an order theoretic statement A, we replace the relation $\leqslant$ wherever it occurs by the inverse relation $\geqslant$, we form a new order-theoretic statement A^*, called the *dual statement* to A. We call this procedure *dualization*. If we dualize an order-theoretic concept, we form the *dual concept*. Iterated twice, dualization leads back to the original statement or concept. If A and A^* are logically identical or equivalent, the order-theoretic statement A is termed *self-dual*. The order axioms furnish us with an example of self-dual statements. The concepts of "minimal element" and "upper bound," when dualized, give us, respectively, the concepts "maximal element" and "lower bound." If $\mathfrak{A}$ is a set of order-theoretic statements, then $\mathfrak{A}^* = \{A^* : A \in \mathfrak{A}\}$ is called the *dual set to* $\mathfrak{A}$.

Suppose $\mathfrak{A}$ is a set of order-theoretic statements and that the order-theoretic statement B is a logical consequence of the statements in $\mathfrak{A}$. Then the dual statement B^* is a logical consequence of the dual set $\mathfrak{A}^*$: We can obtain a demonstration of B^*, using statements in $\mathfrak{A}^*$, by dualizing all the statements from

[8] Informally speaking, purely logical components are: negation, "and," "or," "if ... then" (logical implication), "if and only if" (logical equivalence), identification, "for all," and "there exists." In order to formulate a more precise definition of logical components, one has to develop the machinery of formal logic.

$\mathfrak{A}$ occurring in the proof of B.[9] The set of three order axioms, $\mathfrak{O} = \{(O_1), (O_2), (O_3)\}$ and its dual set $\mathfrak{O}^*$ are composed of pairwise logically equivalent statements, for the set of order axioms is self-dual. Each statement which is a consequence of $\mathfrak{O}^*$ is therefore also a consequence of $\mathfrak{O}$ and hence an order-theoretic theorem. This is the justification of the following.

2.3 Order-Theoretic Duality Principle. *If an order-theoretic statement A is an order-theoretic theorem, then its dual statement A^* is also an order-theoretic theorem.*

This principle allows us to prove two theorems in one, for, according to it, once we have proved an order-theoretic theorem, we have automatically proved its dual. The principle of duality is itself not an order-theoretic theorem, but rather a meta-mathematical theorem about the theory of ordered sets. From the above considerations we realize immediately that we can count among the order-theoretic statements those statements which have as components order-theoretic concepts as well as the already mentioned allowable components (purely logical components, the $\mathscr{E}$-relation, and an order relation). Under dualization, these order-theoretic concepts should be replaced by the corresponding dual concepts.

Definition 2c: *Let M be a subset of the ordered set A. We call an element $a \in A$ an* infinum (*supremum*) *of M in A, if a possesses the following properties:*

(1) $m \in M$ implies $a \leqslant m$ $(m \leqslant a)$

(2) $b \in A$ and $b \leqslant m$ $(m \leqslant b)$ for all $m \in M$, imply $b \leqslant a$ $(a \leqslant b)$.

We shall use the following abbreviations, inf and sup for infinum and supremum, respectively.

Property (1) tells us that an inf is a lower bound, while a sup is an upper bound. Property (2) characterizes the inf as maximal among lower bounds and the sup as minimal among upper bounds. Inf and sup are dual concepts.

2.4 *Every subset M of A can possess at most one inf and one sup.*

Proof: Let a and b be two potential infs of M in A. By property (2) above, $a \leqslant b$ and $b \leqslant a$ must both hold. Hence by $(\mathbf{O}_3)$, $a = b$. The uniqueness of the sup then follows by duality. ∎

Because of this uniqueness theorem, we can speak of *the* inf and *the* sup of M in A. Note that, in our definition of these concepts, the possibility that M is empty has not been excluded. The inf of the empty set in A is the unique maximal element e of A, if it exists. It is characterized by the property $a \leqslant e$

[9] The assertion that we can obtain a proof of B^*, using statements from $\mathfrak{A}^*$, by dualizing a proof of B which uses statements from $\mathfrak{A}$, belongs to the domain of formal logic. For a rigorous proof we need the tools of formal logic (the calculus of predicates).

for all $a \in A$. The element e is called the *unit* or *unit element* of A. Likewise the sup of the empty set is the unique minimal element of A, designated by 0 and called the *zero* or *zero element* of A. We have $0 \leqslant a$ for all $a \in A$. The notions of zero and unit are dual to each other.

In general, subsets of an ordered set need possess neither inf nor sup. The ordered set **Q** possesses neither inf or sup. The set M of all rationals r such that $0 < r < \sqrt{2}$ possesses an inf in Q (namely, 0); this same set, however, possesses no sup in Q since $\sqrt{2}$ is irrational. Moreover, if we conceive of M as being a subset of the real line **R**, then M possesses a sup in **R** (namely, $\sqrt{2}$). Thus inf and sup are relative notions.

Definition 2d: *An ordered set A is said to be* conditionally complete *if every nonempty subset which possesses a lower (upper) bound in A also possesses an inf (sup) in A.*

For example, the real axis **R** is conditionally complete, while **Q** is not.

Let R be an arbitrary set and $P(R)$ the set of all subsets of R, ordered by inclusion. Every family of elements of $P(R)$ possesses an inf and a sup: The inf of the family of elements $\mathfrak{M}$ of $P(R)$ is the intersection $\cap \{M : M \in \mathfrak{M}\}$, while the sup is the union $\cup \{M : M \in \mathfrak{M}\}$. Keeping this set-theoretic example in the back of our minds, we shall use the following symbols for the case of a subset M of any ordered set A:

$$\bigwedge_{a \in M} a \qquad \text{or} \qquad \bigwedge \{a : a \in M\}$$

will designate the inf of the set M with respect to A,

$$\bigvee_{a \in M} a \qquad \text{or} \qquad \bigvee \{a : a \in M\}$$

will designate the sup of the set M, with respect to A.

We read the symbols $\wedge$ and $\vee$ either as inf and sup, or else (order-theoretic) intersection and (order-theoretic) union, respectively. If M is finite, e.g., $M = \{a_1, ..., a_n\}$, we can also write $a_1 \wedge ... \wedge a_n$ and $a_1 \vee ... \vee a_n$ for $\bigwedge_{a \in M}$, and $\bigvee_{a \in M}$, respectively. Since inf and sup are order-theoretic concepts, their use is permitted in order-theoretic statements; under dualization $\wedge$ and $\vee$ are to be interchanged.

2.5 *Assuming that all* $\wedge$ *and* $\vee$ *below do exist, the following results are valid*[10]

(1) $a \geqslant a \wedge b$ and $a \leqslant a \vee b$.

(2) $a \wedge b = b \wedge a$ and $a \vee b = b \vee a$.
(commututivity)

[10] Some of these results can be sharpened, e.g., in (4) the existence of $a \vee b$ already implies that of $a \wedge (a \vee b)$, etc.

(3) $a \wedge (b \wedge c) = (a \wedge b) \wedge c$ and $a \vee (b \vee c) = (a \vee b) \vee c$. (associativity)

(4) $a \wedge (a \vee b) = a$ *and* $a \vee (a \wedge b) = a$.

(5) *If* $a \leqslant b$ *and* $c \leqslant d$, *then* $a \wedge c \leqslant b \wedge d$ *and* $a \vee c \leqslant b \vee d$.

(6) *The statements* $a \leqslant b$, $a \wedge b = a$, $a \vee b = b$ *are pairwise equivalent.*

Proof: In (1)-(5) the second assertion is dual to the first and hence we need only prove the first assertion.

(1) Since $a \wedge b$ is the inf of the set $\{a, b\}$, this assertion is true by property (1) characterizing the infimum.

(2) The elements $a \wedge b$ anf $b \wedge a$ are both the inf of the set $\{a, b)$ and hence must be equal.

(3) Let $d = a \wedge (b \wedge c)$. Then (1) implies that $d \leqslant a$ on the one hand and $d \leqslant b \wedge c$ on the other. Since $b \wedge c \leqslant b$ and $b \wedge c \leqslant c$ it follows that $d \leqslant a$, $d \leqslant b$, $d \leqslant c$ and hence, by the second defining property of infs, we see that $d \leqslant a \wedge b \wedge c$. Conversely, $a \wedge b \wedge c \leqslant a$, $\leqslant b$, and $\leqslant c$, hence $a \wedge b \wedge c \leqslant a$ and $a \wedge b \wedge c \leqslant b \wedge c$; finally $a \wedge b \wedge c \leqslant a \wedge (b \wedge c)$ and thus $(a \wedge b) \wedge c = a \wedge b \wedge c$.

(4) It follows from (1) that $a \wedge (a \vee b) \leqslant a$. But $a \leqslant a$ and by (1) again, $a \leqslant a \vee b$. Hence $a \leqslant a \wedge (a \vee b)$.

(5) From $a \leqslant b$ and $c \leqslant d$ it follows, by (1), that $a \wedge c \leqslant a \leqslant b$ and $a \wedge c \leqslant c \leqslant d$. Hence $a \wedge c \leqslant b \wedge d$.

(6) From $a \leqslant b$ it follows, by (5), that $a = a \wedge a \leqslant a \wedge b$; from (1) we $a \wedge b \leqslant a$. Hence $a \leqslant b$ implies $a = a \wedge b$. Conversely, if $a = a \wedge b$ and $a \wedge b \leqslant b$, then $a \leqslant b$. The second assertion, that $a \leqslant b$ and $a \vee b = b$ are equivalent, now follows by duality. ∎

Definition 2e: *An ordered set L is called a* lattice *if it obeys the following two lattice axioms:*

($\mathbf{L}_1$) Every finite subset of elements of L possesses an inf in L.

($\mathbf{L}_2$) Every finite subset of elements of L possesses a sup in L.

A subset K of the lattice L is called a sublattice *if whenever it contains a finite number of elements it also contains their inf (taken in L) and their sup (taken in L).*

To verify that an ordered set L is a lattice it suffices to show that, given any two elements of L, their sup and inf exist and are in L. Axioms ($\mathbf{L}_1$) and ($\mathbf{L}_2$) are order-theoretic dual statements, but they are not order-theoretic theorems because of the existence of ordered sets which are not lattices. It would therefore be erroneous to try to evoke duality to prove, say, axiom ($\mathbf{L}_2$) from axiom ($\mathbf{L}_1$). A special case of sublattices are the *ideals*. A sublattice J of L is called an ideal if $a \in J$, $b \in L$ and $b \leqslant a$ imply $b \in J$. For example, if a is any element of L, then the interval $[\leftarrow, a]$ is an ideal; it is called the *principal ideal generated by a*. We designate by $\mathfrak{L}$ the set consisting of the order axioms and the lattice axioms, $\mathfrak{L} = \{(\mathbf{O}_1), (\mathbf{O}_2), (\mathbf{O}_3), (\mathbf{L}_1), (\mathbf{L}_2)\}$. An order-theoretic statement which is a logical consequence of $\mathfrak{L}$ will be called a *lattice-theoretic statement*. The lattice axioms $(\mathbf{L}_1)$ and $(\mathbf{L}_2)$ are dual to each other and hence $\mathfrak{L}$ and $\mathfrak{L}^*$ are composed of equivalent statements. By an argument analogous to one already given, we have

2.6 The Lattice-Theoretic Principle of Duality. *If an order-theoretic statement A is a lattice-theoretic theorem, then its dual A^* is also a lattice-theoretic theorem.*

Definition 2f: *A lattice L is said to be* complete *if whenever it contains an arbitrary family of elements (not necessarily finite) it also contains their inf and sup.*

Evidently every complete lattice contains a zero element and a unit element.

Examples

2.V Every totally ordered set is a lattice, e.g., **N**, **Q**, and **R** are lattices, each a sublattice of the succeeding one. None, however, is complete. If we extend the real line **R** by adjoining $-\infty$ and $+\infty$ to form $\overline{\mathbf{R}}$, then $\overline{\mathbf{R}}$ is a complete lattice.

2.VI An ordered set is a complete lattice if and only if it contains a unit element and is conditionally complete.

2.VII **N**, with respect to the divisibility-ordering previously mentioned (cf. 2.III), is a lattice, wherein the inf of any finite set of natural numbers is their greatest common divisor while their sup is their least common multiple.

2.VIII Let $\mathfrak{J}$ be the set of all open intervals of the real line **R** (the empty set included). For any two intervals $J_1, J_2 \in \mathfrak{J}$ let us define the order $J_1 \leqslant J_2$ by set-theoretic inclusion, i.e. $J_1 \leqslant J_2$ if and only if $J_1 \subseteq J_2$. Then $\mathfrak{J}$ becomes a lattice. The inf $J_1 \wedge J_2$ is identical with the set-theoretical intersection $J_1 \cap J_2$ but the sup $J_1 \vee J_2$, being by definition the smallest interval containing J_1 and J_2, need not be identical with $J_1 \cup J_2$, which union need not even be an interval. $P(\mathbf{R})$, the set of subsets of the line **R**, is a complete lattice with respect to set-theoretic inclusion (cf. 2.III). In $P(\mathbf{R})$, lattice infs and sups and set-theoretic intersections and unions are identical, respectively. We have $\mathfrak{J} \leqslant P(\mathbf{R})$, and the order of $P(\mathbf{R})$ induces in $\mathfrak{J}$ just the order that we have already defined on $\mathfrak{J}$. Since, however, the operations of taking sups and of taking set-theoretic unions are not identical in $\mathfrak{J}$, we cannot affirm that $\mathfrak{J}$ is a sublattice of $P(\mathbf{R})$.

2.7 *If an ordered set A contains the inf (sup) of any family of its elements, then it is a complete lattice.*

Proof: Suppose that, given any family of elements of A, their inf is always in A. It remains to prove that their sup always exists in A. Let M be any subset of A and S the set of upper bounds of M in A (S could even be empty). By assumption the inf $d = \bigwedge \{s : s \in S\}$ exists. Now $m \in M$ implies $m \leqslant s$ for all $s \in S$, hence $m \leqslant d$. Conversely let b be an element of A with $m \leqslant b$ for all $m \in M$. Then $b \in S$ and hence $d \leqslant b$. Consequently $d = \bigvee \{m : m \in M\}$ The second assertion of this theorem, pertaining to sups, now follows by duality. ∎

Definition 2g: *Let L be a lattice with zero element* 0. *An element $a \in L$ is called an atom of L if*

(1) $a \neq 0$.

(2) Whenever $0 < b \leqslant a$, then $b = a$.

L is called an atomic lattice *if, given any $b \neq 0$, we can always find an atom a such that $a \leqslant b$.*

The above properties (1) and (2) characterize atoms as the minimal elements of the intervals $]0, \rightarrow] = \{x : x > 0\}$. With respect to set-theoretic inclusion, the set $P(\mathbf{R})$ is an atomic lattice. Its atoms are exactly those subsets of $\mathbf{R}$ containing one element. The lattice $\mathfrak{I}$ of open intervals of 2.VIII, however, contains no atoms.

2.8 *In a lattice L with zero element* 0, *the element a is an atom if $a \neq 0$ and if $a \wedge b \neq 0$ implies $a \leqslant b$, and conversely.*

Proof: Let a be an atom of L. Then, by (1), $a \neq 0$. Since $a \wedge b \neq 0$, we have $0 < a \wedge b \leqslant a$. By (2), this implies $a \wedge b = a$ and hence $a \leqslant b$ by 2.5(6). Conversely, suppose $0 < b \leqslant a$. Then $a \wedge b = b$ and $b \neq 0$ imply $a \wedge b \neq 0$. By hypothesis we then have $a \leqslant b$, i.e., $a = b$. ∎

2.9 *In any lattice,*

$$a \wedge (b \vee c) \geqslant (a \wedge b) \vee (a \wedge c)$$

and

$$a \vee (b \wedge c) \leqslant (a \vee b) \wedge (a \vee c)$$

Proof: We always have $a \wedge b \leqslant a$ and $a \wedge b \leqslant b \leqslant b \vee c$, whence $a \wedge b \leqslant a \wedge (b \vee c)$. Likewise $a \wedge c \leqslant a \wedge (b \vee c)$. These relations establish the first assertion and the second follows by duality. ∎

Definition 2h: *A lattice L is called* distributive *if the following two distributivity laws hold within it:*

$(\mathbf{D}_1)$ $a \wedge (b \vee c) = (a \wedge b) \vee (a \wedge c)$,

$(\mathbf{D}_2)$ $a \vee (b \wedge c) = (a \vee b) \wedge (a \vee c)$.

$P(R)$, the set of subsets of R, is a distributive lattice with respect to set-theoretic inclusion, cf. 1.3. The lattice $\mathfrak{J}$ of intervals, cf. 2.VIII, is not distributive, however: Consider the intervals $J_1 =]0, 1[$, $J_2 =]1, 2[$, $J_3 =]2, 3[$; then $J_1 \vee (J_2 \wedge J_3) = J_1$ but $(J_1 \vee J_2) \wedge (J_1 \vee J_3) =]0, 2[$.

2.10 *For any lattice,* $(\mathbf{D}_1)$ *and* $(\mathbf{D}_2)$ *are equivalent conditions.*

Proof: If we replace in $(\mathbf{D}_1)$ the element a by $a \vee b$ and the element b by a, we get

$$(a \vee b) \wedge (a \vee c) = ((a \vee b) \wedge a) \vee ((a \vee b) \wedge c)$$

By (2.5), $(a \vee b) \wedge a = a$ and from $(\mathbf{D}_1)$ we see that $(a \vee b) \wedge c = (a \wedge c) \vee (b \wedge c)$ Using (2.5) again we have

$$(a \vee b) \wedge (a \vee c) = a \vee ((a \wedge c) \vee (b \wedge c)) = (a \vee (a \wedge c)) \vee (b \wedge c) = a \vee (b \wedge c).$$

Hence, $(\mathbf{D}_1)$ implies $(\mathbf{D}_2)$. The converse follows by duality. ∎

The distributivity conditions $(\mathbf{D}_1)$ and $(\mathbf{D}_2)$ are dual to each other. However, they are not lattice-theoretic theorems, for the existence of nondistributive lattices tells us that $(\mathbf{D}_1)$ and $(\mathbf{D}_2)$ can not be consequences of the lattice axioms. It is not duality alone that implies the validity of one of these assertions when the validity of the other has already been established, but rather the combination of duality and one-half of Theorem 2.10 above.

Notes and Exercises.

2A Every totally ordered set is a distributive lattice.

2B Let $(A_1, \leqslant_1), \ldots, (A_n, \leqslant_n)$ be ordered sets. We can define an order in the product set $A = A_1 \times \ldots \times A_n$ as follows: Let $a = (a_1, \ldots, a_n)$ and $b = (b_1, \ldots, b_n)$ be any two n-tuples in A. We say $a \leqslant b$ if $a_1 \leqslant_1 b_1, \ldots, a_n \leqslant_n b_n$. We call a property *multiplicative* if it holds for the product set A (ordered as above) whenever it holds for all the factors of the product.

Exercise: Show that the following properties are multiplicative:

1. The property of being a (complete) lattice.
2. The property of possessing a zero element (unit element).
3. The property of being an atomic lattice.
4. The property of being a distributive lattice.

2C Let L be a lattice with zero element 0 and unit element e. If a and b are any two elements of L such that $a \wedge b = 0$ and $a \vee b = e$, we say that a and b are *complements* of each other. The lattice L is called *complemented* when each of its elements possesses at least one complement. Lattices with zero and unit which are complemented and also distributive are called *Boolean lattices*.

Exercise: In a Boolean lattice every element possesses a unique complement.

2D Ordered sets which contain only a finite number of elements can be *represented* by diagrams (Hasse diagrams): the elements are represented by small circles. Two of these circles are connected by a straight line if an only if the elements they represent are comparable. Also, we place the circle representing a below that of b in the diagram if and only if $a < b$.

Exercise: Show that the diagrams below represent lattices. Study their special properties as lattices.

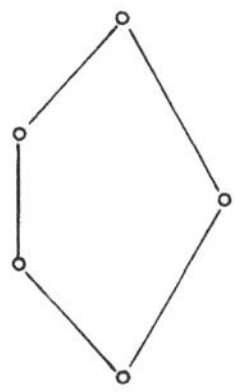

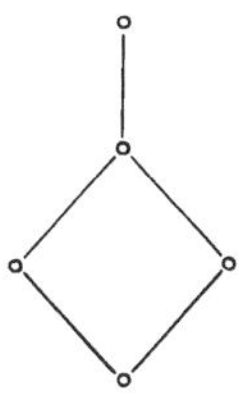

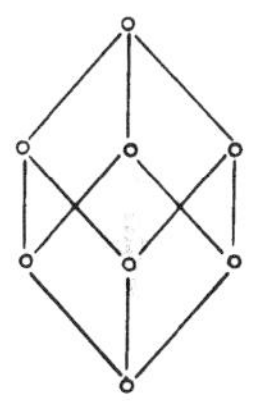

3. Filters

The classical concepts of limit and convergence are based upon the prior concept of a (denumerable) sequence of points. Sequences of points, however, do not prove adequate for further generalization of "convergence to a limit." Instead, we need a more general type of point set—a system of sets called a filter. The following considerations all refer to a fixed but arbitrary set R.

Definition 3a: *A nonempty system of sets* $\mathfrak{a}$ *is called a* filter *in* R *if it possesses the following properties:*

($\mathbf{F}_1$) *If* $A_1, A_2 \in \mathfrak{a}$, *then* $A_1 \cap A_2 \in \mathfrak{a}$

($\mathbf{F}_2$) *If* $A \in \mathfrak{a}$, *and* $A \subseteq B \subseteq R$, *then* $B \in \mathfrak{a}$ [11]

Filters will always be designated by small letters taken from the German alphabet. Property ($\mathbf{F}_1$) implies that if a filter contains a finite numbers of sets it also contains their intersection.

Examples

3.I Let the underlying reference set R be infinite and let k be a cardinal number such that $\aleph_0 \leqslant k \leqslant |R|$. Then the system of all sets $M \subseteq R$ such that

[11] Some authors stipulate, in their definition of filter, that the empty set belongs to no filter. For our purposes, it will be better not to make this limiting stipulation.

$|\complement M| < k$ forms a filter. In particular, when $k = \aleph_0$, this filter is equal to the system of subsets of R having finite complements.

3.II Let M be a fixed subset of R. Then the system $\{A : M \subseteq A\}$ is a filter. This filter is called the *principal filter* generated by M. M can be the empty set: the principal filter generated by the empty set is $P(R)$, the set of all subsets of R. This filter is also designated by $\mathfrak{o}$ and called the *null filter*.

Any nonempty system of sets $\mathfrak{A}$ can always be extended to a system verifying condition $(\mathbf{F}_2)$: Let us designate by $\overline{\mathfrak{A}}$ the system of all sets M for which their exist sets $A_M \in \mathfrak{A}$ with $A_M \subseteq M$. Then $\overline{\mathfrak{A}}$ is uniquely determined by $\mathfrak{A}$ and $\overline{\mathfrak{A}}$ obeys $(\mathbf{F}_2)$.

Definition 3b: *A nonempty system of sets $\mathfrak{A}$ is called a* filter basis, *or simply a* basis, *if $\overline{\mathfrak{A}}$ is a filter. $\overline{\mathfrak{A}}$ is then called the filter generated by $\mathfrak{A}$ and $\mathfrak{A}$ is called a basis of the filter $\overline{\mathfrak{A}}$.*

Every filter is a basis of itself.

3.1 *A nonempty system of sets $\mathfrak{A}$ is a basis if and only if it possesses the following basis property*

(**B**) *Given any two sets $A_1, A_2 \in \mathfrak{A}$ there exists an $A_3 \in \mathfrak{A}$ with $A_3 \subseteq A_1 \cap A_2$.*

Proof: Let $\mathfrak{A}$ be a basis and $A_1, A_2 \in \mathfrak{A}$. Then, by $(\mathbf{F}_1)$, $A_1 \cap A_2 \in \overline{\mathfrak{A}}$ and hence there exists an $A_3 \in \mathfrak{A}$ such that $A_3 \subseteq A_1 \cap A_2$. Conversely, let $\mathfrak{A}$ have the property (**B**). Let $A_1', A_2' \in \overline{\mathfrak{A}}$. Then there exist two sets $A_1, A_2 \in \mathfrak{A}$ with $A_1 \subseteq A_1'$ and $A_2 \subseteq A_2'$. Hence there exists an $A_3 \in \mathfrak{A}$ with

$$A_3 \subseteq A_1 \cap A_2 \subseteq A_1' \cap A_2'.$$

Thus $A_1' \cap A_2' \in \overline{\mathfrak{A}}$ and $\overline{\mathfrak{A}}$ verifies both $(\mathbf{F}_2)$ and $(\mathbf{F}_1)$. ∎

Examples

3.III Every system of sets consisting of only one set M is a basis. The filter generated by this basis is none other than the principal filter generated by M (cf. 3.III). It should be represented by $\overline{\{M\}}$, but instead we shall simply write $\overline{M}$. If M contains only one element—say, $M = \{p\}$—then again, instead of $\overline{\{p\}}$ we shall simply write $\overline{p}$. In particular, $\overline{\varnothing} = \mathfrak{o}$.

3.IV Let $(p_\nu)_{\nu \in N}$ be a sequence of points. The set $A_n = \{p_\nu : \nu \geqslant n\}$ is called the nth terminating section of this sequence. Note that A_n stands for a set and not a subsequence. A point which may appear in several places in the sequence, nevertheless defines only one element of A_n. The system $\mathfrak{A} = \{A_n : n \in N\}$ of all terminating sections of a sequence forms a filter basis; it is called the basis of terminating sections of the sequence or else the *Fréchet basis* of the sequence. Every filter which possesses a Fréchet basis is called a *Fréchet filter*. Each sequence determines a unique Fréchet filter but different sequences may determine the same Fréchet filter.

Among all the sets belonging to the filter $\mathfrak{a}$ consider those of smallest cardinality. We shall call their common cardinality the *potency* of $\mathfrak{a}$ and designate it by $Pz(\mathfrak{a})$. Hence $Pz(\mathfrak{a}) = \min\{|A| : A \in \mathfrak{a}\}$. Evidently $0 \leqslant Pz(\mathfrak{a}) \leqslant |R|$ and $Pz(\mathfrak{a}) = 0$ is equivalent to $\mathfrak{a} = \mathfrak{o}$.

Two distinct bases of the same filter can possess entirely different cardinalities. For instance, the principal filter $\ulcorner p \urcorner$ generated by the one point $p \in R$ possesses the basis $\{p\}$ consisting of the one-point set $\{p\}$—but $\ulcorner p \urcorner$ is always a basis for itself, so that if the underlying set R is infinite then so is the basis $\ulcorner p \urcorner$.

Among all the cardinalities of the possible bases of a filter $\mathfrak{a}$, there is a smallest one which we call the ***rank*** of the filter $\mathfrak{a}$ and designate by Rk $(\mathfrak{a})$. Any basis $\mathfrak{A}$ of the filter $\mathfrak{a}$, for which $|\mathfrak{A}|$ = rank $(\mathfrak{a})$, is called a *minimal basis.*

3.2 *For any filter* $\mathfrak{a}$ *either*

$$\text{Rk}(\mathfrak{a}) = 1, \qquad \textit{or else} \qquad \aleph_0 \leqslant \text{Rk}(\mathfrak{a}) \leqslant 2^{Pz(\mathfrak{a})}$$

Rk $(\mathfrak{a}) = 1$ *characterizes principal filters.*

Proof: Since filter bases are never empty, Rk $(\mathfrak{a}) \geqslant 1$. We now show that Rk $(\mathfrak{a}) = 1$ is equivalent to $\mathfrak{a}$ being a principal filter. Certainly, if $\mathfrak{a}$ has a basis consisting of one set, Rk $(\mathfrak{a}) = 1$. Suppose then that $\mathfrak{a}$ has a finite basis $\{A_1, ..., A_n\}$. Consider $A = A_1 \cap ... \cap A_n$. This set is contained in all sets of $\mathfrak{a}$. Hence by (F_1), $A \in \mathfrak{a}$; i.e., $\{A\}$ is a basis for $\mathfrak{a}$ and Rk $\mathfrak{a} = 1$. Hence Rk $\mathfrak{a} > 1$ implies Rk $\mathfrak{a} \geqslant \aleph_0$. Finally, let M be a set in $\mathfrak{a}$ with $|M| = Pz(\mathfrak{a})$. Then $A = \{A \cap M : A \in \mathfrak{a}\}$ is a basis of $\mathfrak{a}$ and so Rk $\mathfrak{a} \leqslant |\mathfrak{A}| \leqslant |P(M)| = 2^{Pz(\mathfrak{a})}$. ∎

The concept of filter is important because it contains summarized within itself many of the most common properties of systems of sets. Its use therefore leads to greater clarity and simplicity of exposition. One other feature of filters is important: In the set of all filters we can introduce an order relation with respect to which the set of filters will be shown to form a lattice. This special lattice structure of the set of all filters will make for convenient algebraic manipulation.

Now filters are systems of sets and we can surely order them like any system of sets, by inclusion. Nevertheless, in order to have more symmetric formulas, it is better for our purposes to use the inverse relation to set-theoretic inclusion.[12]

Definition 3c: *For any two filters* $\mathfrak{a}$ *and* $\mathfrak{b}$ *of* R *we define* $\mathfrak{a} \leqslant \mathfrak{b}$ *if and only if* $B \in \mathfrak{b}$ *implies* $B \in \mathfrak{a}$.

Thus $\mathfrak{a} \leqslant \mathfrak{b}$ is equivalent to the set-theoretic inclusion, "set $\mathfrak{b} \subseteq$ set $\mathfrak{a}$." This definition of $\leqslant$ is an order, the proof following immediately from the corresponding properties of $\subseteq$. In this book we shall always employ the above

[12] Our exposition differs from the usual ones, where inclusion itself, rather than its inverse, is used to order filters and hence where much that we do is done in exactly the dual way.

defined order for filters. The symbol $F(R)$ will be used to designate the set of all filters of R, ordered as above. Note carefully that subsets of $F(R)$ and systems of such subsets will in general by marked off by an asterisk. The relation $\mathfrak{a} \leqslant \mathfrak{b}$ is usually read "$\mathfrak{a}$ is finer than $\mathfrak{b}$," or "$\mathfrak{b}$ is less fine than $\mathfrak{a}$."

Since every basis generates a unique filter, we can stipulate the order of filters by examining their bases.

3.3 (*Comparison Test*) *Let* $\mathfrak{A}$ *and* $\mathfrak{B}$ *be bases which generate the filters* $\overline{\mathfrak{A}}$ *and* $\overline{\mathfrak{B}}$. *Then* $\overline{\mathfrak{A}} \leqslant \overline{\mathfrak{B}}$ *if and only if for every* $B \in \mathfrak{B}$ *there exists an* $A \in \mathfrak{A}$ *such that* $A \subseteq B$.

Proof: Let $\overline{\mathfrak{A}} \leqslant \overline{\mathfrak{B}}$ and $B \in \mathfrak{B}$ (hence $B \in \overline{\mathfrak{B}}$). It follows that $B \in \overline{\mathfrak{A}}$ and thus there exists an $A \in \mathfrak{A}$ with $A \subseteq B$. Conversely, if $B' \in \overline{\mathfrak{B}}$ then there exists a $B \in \mathfrak{B}$ with $B \subseteq B'$. If $\mathfrak{A}$ and $\mathfrak{B}$ fulfill the conditions of the theorem, then there is an $A \in \mathfrak{A}$ such that $A \subseteq B \subseteq B'$, whence $B' \in \overline{\mathfrak{A}}$, i.e., $\overline{\mathfrak{A}} \leqslant \overline{\mathfrak{B}}$. ∎

As a corollary we have:

3.4 *For principal filters,* $\overline{M} \leqslant \overline{N}$ *is equivalent to* $M \subseteq N$. *If* $\mathfrak{a}$ *is a filter, then* $\mathfrak{a} \leqslant \overline{M}$ *is equivalent to* $M \in \mathfrak{a}$.

Since $F(R)$ is an ordered set, the concepts of "inf" and "sup," defined as in Section 2 by the symbols $\vee$ and $\wedge$ are meaningful in $F(R)$ and we have

3.5 *$F(R)$ is a complete lattice.*

Proof: Let M^* be an arbitrary subset of $F(R)$. We shall show that M^* has a sup in $F(R)$. By (2.7) this guarantees that $F(R)$ is a complete lattice. Let $\mathfrak{v}$ be the system of all subsets of R, which at the same time belong to all filters $\mathfrak{a} \in M^*$ (If M^* is empty, $\mathfrak{v}$ is formed of all the subsets of R. The system $\mathfrak{v}$ is the set-theoretic intersection of all filters of M^* considered as systems of sets.) If $V_1, V_2 \in \mathfrak{v}$ then $V_1, V_2 \in \mathfrak{a}$ and $V_1 \cap V_2 \in \mathfrak{a}$ for all $\mathfrak{a} \in M^*$, hence $V_1 \cap V_2 \in \mathfrak{v}$. Likewise, if $V \in \mathfrak{v}$ and $V \subseteq W$, then $V \in \mathfrak{a}$ and consequently $W \in \mathfrak{a}$ for all $\mathfrak{a} \in M^*$, i.e., $W \in \mathfrak{v}$. Hence $\mathfrak{v}$ is a filter and $\mathfrak{a} \leqslant \mathfrak{v}$ for all $a \in M^*$. Moreover, let $\mathfrak{w}$ be a second filter with $\mathfrak{a} \leqslant \mathfrak{w}$ for all $\mathfrak{a} \in M^*$. Then $W \in \mathfrak{w}$ implies $W \in \mathfrak{a}$ for all $\mathfrak{a} \in M^*$; hence $W \in \mathfrak{v}$. Therefore $\mathfrak{v} \leqslant \mathfrak{w}$ and thus $\mathfrak{v} = \bigvee \{\mathfrak{a} : \mathfrak{a} \in M^*\}$. ∎

The lattice $F(R)$ is called the filter lattice of the fundamental set R. Its zero element is evidently the filter $\mathfrak{o}$ and its unit element the principal filter $\overline{R}$, composed of the one set R itself. The above proof shows us that we can form the sup of the subset of filters M^* of $F(R)$ by taking the set-theoretic intersection of all the filters in M^*. However, it is not possible to find the inf of the filters in M^* by proceeding analogously and taking their set-theoretic union. The set-theoretic union will in general not obey (F_1) and consequently not be a filter. The relations between filter bases and the inf and sup of filters are shown in the following theorem.

3.6 *Let* $\{\mathfrak{a}_\iota : \iota \in I\}$ *be a subset of* $F(R)$ *and for each* $\iota \in I$ *let* $\mathfrak{A}_\iota$ *be a basis of* $\mathfrak{a}_\iota$. *The system* $\mathfrak{B}$ *of all sets* V *expressible in the form*

$$V = \bigcup_{\iota \in I} A_\iota \quad \textit{with} \quad A_\iota \in \mathfrak{A}_\iota$$

is a basis for the filter $\mathfrak{v}$, *sup of the given* $\mathfrak{a}_\iota$, i.e., $\mathfrak{v} = \bigvee \{\mathfrak{a}_\iota : \iota \in I\}$.

The system $\mathfrak{D}$ *of all sets expressible in the form*

$$D = A_{\iota_1} \cap \dots \cap A_{\iota_n}$$

with $A_{\iota_\nu} \in \mathfrak{A}_{\iota_\nu}$ *and* $\iota_\nu \in I$ *for* $\nu = 1, \dots, n$ $(n \in N)$, *is likewise a basis for* $\mathfrak{d}$, *the inf of the given* $\mathfrak{a}_\iota$, *i.e.*, $\mathfrak{d} = \bigwedge \{\mathfrak{a}_\iota : \iota \in I\}$.

Proof: Let

$$V_1 = \bigcup_{\iota \in I} A^1_\iota \quad \text{and} \quad V_2 = \bigcup_{\iota \in I} A^2_\iota$$

be two sets belonging to $\mathfrak{B}$. Since the $\mathfrak{A}_\iota$ are bases, there exist sets $A_\iota \in \mathfrak{A}_\iota$ such that $A_\iota \leqslant A^1_\iota \cap A^2_\iota$. Let $V = \bigcup \{A_\iota : \iota \in I\}$. Then $V \in \mathfrak{B}$ and

$$V \subseteq \bigcup_{\iota \in I} (A^1_\iota \cap A^2_\iota) \subseteq \bigcup_{\iota \in I} A^1_\iota \cap \bigcup_{\iota \in I} A^2_\iota = V_1 \cap V_2$$

Hence $\mathfrak{B}$ is a basis. From $V \in \mathfrak{B}$ we see that $V \in \mathfrak{a}_\iota$, hence $\overline{\mathfrak{B}} \geqslant \mathfrak{a}_\iota$ for all $\iota \in I$ and thus $\mathfrak{v} \leqslant \overline{\mathfrak{B}}$. Conversely, $V \in \mathfrak{v}$ implies $V \in \mathfrak{a}_\iota$ for all $\iota \in I$. Then there exist sets $A_\iota \in \mathfrak{A}_\iota$ with $A_\iota \subseteq V$. $V' = \bigcup \{A_\iota : \iota \in I\}$ is a set belonging to the family $\mathfrak{B}$ and $V' \subseteq V$. Hence $V \in \overline{\mathfrak{B}}$ and $\mathfrak{v} \geqslant \overline{\mathfrak{B}}$, i.e., finally $\mathfrak{v} = \overline{\mathfrak{B}}$.

Since the system $\mathfrak{D}$ is formed only by taking finite intersections of sets, $\mathfrak{D}$ possesses property (**B**) and is a basis. Any set of the basis $\mathfrak{A}_\iota$ also belongs to $\mathfrak{D}$. By the comparison test, $\overline{\mathfrak{D}} \leqslant \mathfrak{a}_\iota$ for all $\iota \in I$, hence $\overline{\mathfrak{D}} \leqslant \mathfrak{a} = \bigwedge \{\mathfrak{a}_\iota : \iota \in I\}$. Conversely, $\mathfrak{a}$ contains all sets A_ι belonging to the bases A_ι. By (F_1), all intersections of the form $A_{\iota_1} \cap \dots \cap A_{\iota_n}$ with $A_{\iota_\nu} \in A_{\iota_\nu}$, lie in $\mathfrak{a}$ and hence $\mathfrak{a} \leqslant \overline{\mathfrak{D}}$. Thus $\mathfrak{a} = \overline{\mathfrak{D}}$. ∎

Since every filter is a basis of itself, a small calculation will show that the above theorem implies:

We can form an arbitrary set belonging to the filter $\bigvee \{\mathfrak{a}_\iota : \iota \in I\}$ by taking a set belonging to the filter $\mathfrak{a}_\iota$, doing this for each and every ι, and then forming the union of the chosen sets. We need not, as above, invoke the axiom of choice, in order to form an arbitrary set belonging to the filter $\bigwedge \{\mathfrak{a}_\iota : \iota \in I\}$—we simply choose a set from each of only a finite number of the $\mathfrak{a}_\iota$ and then take the intersection of the chosen sets. The very fact that we employed the inverse of set-theoretic inclusion to order $F(R)$ has led to the above symmetric results, namely, that sups and infs of filters correspond to unions and intersections of the sets belonging to filters. This symmetry of lattice operations is even more striking in the case of principal filters.

3.7 *The inf of a finite number of principal filters, and the sup of an arbitrary number of principal filters, is again a principal filter. We have*

$$\overline{\cup M_\iota} = \bigvee \overline{M_\iota}$$

$$\overline{M_1 \cap \ldots \cap M_n} = \overline{M_1} \wedge \ldots \wedge \overline{M_n}$$

Note that the inf of an arbitrary family of principal filters is not necessarily a principal filter. All we can say in this case is $\overline{\cap M_\iota} \leqslant \bigwedge \overline{M_\iota}$.

Proof: $\mathfrak{A} = \{M_\iota\}$ is a basis of $\overline{M_\iota}$. By 3.6, $\{\cup M_\iota\}$ is therefore a basis of $\bigvee \overline{M_\iota}$. Hence $\overline{\cup M_\iota} = \bigvee \overline{M_\iota}$ and $\bigvee \overline{M_\iota}$ is a principal filter. Now $\cap M_\iota \subseteq M_{\iota_0}$, and by 3.4 $\overline{\cap M_\iota} \leqslant \overline{M_{\iota_0}}$ for all indices ι_0; hence $\overline{\cap M_\iota} \leqslant \bigwedge \overline{M_\iota}$. If the indexing set I is finite, then by 3.6 $\{M_1 \cap \ldots \cap M_n\}$ is a basis of $\overline{M_1} \wedge \ldots \wedge \overline{M_n}$, whence $\overline{M_1 \cap \ldots \cap M_n} = \overline{M_1} \wedge \ldots \wedge \overline{M_n}$. ∎

The following example shows that the inf of an infinity of principal filters need not be a principal filter: Let A_n be the set of natural numbers $\geqslant n$. Then $\cap \{A_n : n \in N\} = \emptyset$, hence $\overline{\cap A_n} = \overline{\emptyset} = \mathfrak{o}$. However, the filter $\bigwedge \overline{A_n}$ is the Fréchet-filter associated with the natural numbers (cf. 3.8). This filter contains the empty set and is distinct from the nullfilter.

3.8 *The filter* $\overline{\mathfrak{A}}$ *generated by the basis* $\mathfrak{A}$ *is describable by*

$$\overline{\mathfrak{A}} = \bigwedge \{\overline{A} : A \in \mathfrak{A}\}$$

Proof: Let $\mathfrak{d} = \bigwedge \{\overline{A} : A \in \mathfrak{A}\}$. Now $A \in \mathfrak{A}$ implies $A \in \mathfrak{a}$ and hence $\mathfrak{d} \leqslant \overline{\mathfrak{A}}$. On the other hand, by 3.6, the sets of the form $D = A_1 \cap \ldots \cap A_n$, where $A_1, \ldots, A_n \in \mathfrak{A}$, constitute a basis of $\mathfrak{d}$. Moreover, (F_1) implies $D \in \overline{\mathfrak{A}}$, whence $\overline{\mathfrak{A}} \leqslant \mathfrak{d}$. ∎

3.9 *The lattice of filters, $F(R)$, is distributive. In fact, we even have the sharper equality*

$$\mathfrak{a} \vee \bigwedge \mathfrak{b}_\iota = \bigwedge (\mathfrak{a} \vee \mathfrak{b}_\iota)$$

Proof: For all indices ι_0, we have $\mathfrak{a} \vee \bigwedge \mathfrak{b}_\iota \leqslant \mathfrak{a} \vee \mathfrak{b}_{\iota_0}$. This implies

$$\mathfrak{a} \vee \bigwedge \mathfrak{b}_\iota \leqslant \bigwedge (\mathfrak{a} \vee \mathfrak{b}_\iota) \qquad (*)$$

In order to prove equality in (*), we must show that every set C belonging to the left-hand filter of (*) also belongs to the right-hand filter. By 3.6 we have

$$C = A \cup (B_1 \cap \ldots \cap B_n) = (A \cup B_1) \cap \ldots \cap (A \cup B_n)$$

with $A \in \mathfrak{a}$ and $B_\nu \in \mathfrak{b}_{\iota_\nu}$ for $\nu = 1, \ldots, n$.

By 3.6 again, the expression to the right in the above equality is a set belonging to $\bigwedge (\mathfrak{a} \vee \mathfrak{b}_\iota)$; this proves equality in (*). The distributivity condition $(\mathbf{D}_2)$ is just a special case of this equality, while $(\mathbf{D}_1)$ follows by 2.10. Hence $F(R)$ is a distributive lattice. ∎

Let us emphasize that it is not legitimate to pass from the equality proved above to its dual. Dualization can not be validly applied because the intersections occurring are not finite, but infinite in number, cf. 2.10. The following theorem points up the subtleties of the general case.

3.10 *The filter* $\mathfrak{a}$ *satisfies the equation*

$$\mathfrak{a} \wedge \left(\bigvee \mathfrak{b}_\iota\right) = \bigvee (\mathfrak{a} \wedge \mathfrak{b}_\iota)$$

for families of subsets $\{\mathfrak{b}_\iota : \iota \in I\}$ *of* $F(R)$ *if and only if* $\mathfrak{a}$ *is a principal filter.*

Proof: In general we have (by dualizing result (*) in the proof of Theorem 3.9)

$$\mathfrak{a} \wedge \left(\bigvee \mathfrak{b}_\iota\right) \geqslant \vee (\mathfrak{a} \wedge \mathfrak{b}_\iota) \tag{**}$$

Any set C belonging to the right-hand filter of (**) must be of the form $C = \cup (A_\iota \cap B_\iota)$ with $A_\iota \in \mathfrak{a}$ and $B_\iota \in \mathfrak{b}$. If $\mathfrak{a}$ is a principal filter—say $\mathfrak{a} = \ulcorner A \urcorner$—then $A \subseteq A_\iota$, and hence

$$C \supseteq \cup (A \cap B_\iota) = A \cap \cup B_\iota$$

But $A \cap \cup B_\iota \in \mathfrak{a} \wedge \bigvee \mathfrak{b}_\iota$ and hence C belongs to $\mathfrak{a} \wedge \bigvee \mathfrak{b}_\iota$, which establishes equality in (**).

Conversely, let $\mathfrak{a}$ verify the equation of this theorem. Let $D = \cap \{A : A \in \mathfrak{a}\}$. Then $\ulcorner D \urcorner \leqslant \mathfrak{a}$. Moreover, by hypothesis and 3.7

$$\mathfrak{a} \wedge \ulcorner \complement D \urcorner = \mathfrak{a} \wedge \ulcorner \complement \bigcap_{A \in \mathfrak{a}} A \urcorner = \mathfrak{a} \wedge \ulcorner \bigcup_{A \in \mathfrak{a}} \complement A \urcorner = \bigvee_{A \in \mathfrak{a}} (\mathfrak{a} \wedge \ulcorner \complement A \urcorner) = \mathfrak{o}$$

Hence $\mathfrak{a} \leqslant \ulcorner D \urcorner$ and thus $\mathfrak{a} = \ulcorner D \urcorner$. This proves that $\mathfrak{a}$ is a principal filter. ∎

3.11 *The lattice of filters* $F(R)$ *is an atomic lattice.*

Proof: We must show, that for a filter $\mathfrak{a} \neq \mathfrak{o}$, the set $M^* = \{\mathfrak{b} : \mathfrak{o} < \mathfrak{b} \leqslant \mathfrak{a}\}$ possesses at least one minimal element. By Zorn's lemma it suffices to show that the inf of any chain K^* of M^* belongs to M^*. Let $\mathfrak{d} = \wedge \{\mathfrak{b} : \mathfrak{b} \in K^*\}$. We must show that $\mathfrak{d} > \mathfrak{o}$. Suppose $\mathfrak{d} = \mathfrak{o}$. Then by 3.6 there exist filters $\mathfrak{b}_1, \ldots, \mathfrak{b}_n \in K^*$ and sets $B_1 \in \mathfrak{b}_1, \ldots, B_n \in \mathfrak{b}_n$ such that $B_1 \cap \ldots \cap B_n = \varnothing$. Since K^* is a chain, we may assume that $\mathfrak{b}_1 \leqslant \mathfrak{b}_2 \leqslant \ldots \leqslant \mathfrak{b}_n$. Then $\varnothing = B_1 \cap \ldots \cap B_n$, i.e., $\mathfrak{b}_1 = \mathfrak{o}$, in contradiction to $b_1 \in M^*$. ∎

The atoms of $F(R)$ are called *ultrafilters*.

3.12 *Let $\mathfrak{v}$ be a filter. The following properties of $\mathfrak{v}$ are equivalent to each other.*

(1) *$\mathfrak{v}$ is an ultrafilter.*

(2) *$\mathfrak{v} \neq \mathfrak{o}$ and $\mathfrak{v} \wedge \mathfrak{a} \neq \mathfrak{o}$ imply $\mathfrak{v} \leqslant \mathfrak{a}$.*

(3) *$\mathfrak{v} \neq \mathfrak{o}$ and $M_1 \cup \ldots \cup M_n \in \mathfrak{v}$ imply $M_\nu \in \mathfrak{v}$ for at least one index ν.*

(4) *For any set M, either $M \in \mathfrak{v}$ or else $\mathbf{C}M \in \mathfrak{v}$, but not both.*

Proof: (1) → (2): This was shown in 2.8.

(2) → (3): $M_1 \cup \ldots \cup M_n \in \mathfrak{v}$ implies $\mathfrak{v} \leqslant \overline{M_1} \vee \ldots \vee \overline{M_n}$; hence

$$\mathfrak{o} < \mathfrak{v} = \mathfrak{v} \wedge (\overline{M_1} \vee \ldots \vee \overline{M_n}) = (\mathfrak{v} \wedge \overline{M_1}) \vee \ldots \vee (\mathfrak{v} \wedge \overline{M_n}).$$

Hence $\mathfrak{v} \wedge (\overline{M_\nu}) \neq \mathfrak{o}$ for at least one ν and by (2), $\mathfrak{v} \leqslant \overline{M_\nu}$, i.e., $M_\nu \in \mathfrak{v}$.

(3) → (4): Since $M \cup \mathbf{C}M \in \mathfrak{v}$, either $M \in \mathfrak{v}$ or $\mathbf{C}M \in \mathfrak{v}$. If both cases were to occur simultaneously, then $\emptyset = M \cap \mathbf{C}M \in \mathfrak{v}$ and $\mathfrak{v} = \mathfrak{o}$, contradiction.

(4) → (1): For the nullfilter $\mathfrak{o}$ we have $R \in \mathfrak{o}$ and $\mathbf{C}R \in \mathfrak{o}$, simultaneously. Hence $\mathfrak{v} \neq \mathfrak{o}$. Let $\mathfrak{o} < \mathfrak{a} \leqslant \mathfrak{v}$. Since $\mathbf{C}A \in \mathfrak{v}$ is false for every $A \in \mathfrak{a}$, we must have $A \in \mathfrak{v}$ for all $A \in \mathfrak{a}$ and hence $\mathfrak{v} \leqslant \mathfrak{a}$. ∎

If $\mathfrak{H}$ is a system of sets then the sets of form $\mathbf{C}(H_1 \cup \ldots \cup H_n)$ with $H_1, \ldots, H_n \in \mathfrak{H}$ form a basis $\mathfrak{C}_\mathfrak{H}$. If $\mathfrak{H}$ is finite, so is $\mathfrak{C}_\mathfrak{H}$. If however $|\mathfrak{H}| \geqslant \aleph_0$, the equation $|\mathfrak{C}_\mathfrak{H}| = |\mathfrak{H}|$ follows from 1.16. The filter $\mathfrak{c}_\mathfrak{H}$ generated by $\mathfrak{C}_\mathfrak{H}$ is called the complementary filter to the system of sets $\mathfrak{H}$. We have $\mathrm{Rk}\,(\mathfrak{c}_\mathfrak{H}) \leqslant |\mathfrak{H}|$.

A system $\mathfrak{H}$ of sets is called a *covering*, or *cover*, of the set M if

$$M \subseteq \cup (H : H \in \mathfrak{H}).$$

The equation $\mathfrak{c}_\mathfrak{H} \wedge \overline{M} = \mathfrak{o}$ is manifestly equivalent to the fact that $\mathfrak{H}$ possesses a finite subcover, i.e., there exists a finite subsystem of sets of $\mathfrak{H}$ which already cover M.

3.13 *Let $\mathfrak{H}$ be a covering of M. Then the following assertions are equivalent to each other:*

(1) *$\mathfrak{H}$ contains a finite subcover of M.*

(2) *For every ultrafilter $\mathfrak{v}$ such that $M \in \mathfrak{v}$ there exists an $H \in \mathfrak{H}$ with $H \in \mathfrak{v}$.*

(3) *For each filter $\mathfrak{a}$ with $\mathfrak{o} < \mathfrak{a} \leqslant \overline{M}$ and $\mathrm{Rk}(\mathfrak{a}) \leqslant |\mathfrak{H}|$ there exists an $H \in \mathfrak{H}$ with $\mathfrak{a} \wedge \overline{H} \neq \mathfrak{o}$.*

Proof:

(1) → (2): Let $\{H_1, \ldots, H_n\}$ be a cover of M. Then

$$\mathfrak{o} < \mathfrak{v} = \mathfrak{v} \wedge \overline{M} \leqslant \mathfrak{v} \wedge (\overline{H_1} \vee \ldots \vee \overline{H_n}) = (\mathfrak{v} \wedge \overline{H_1}) \vee \ldots \vee (\mathfrak{v} \wedge \overline{H_n}).$$

This implies that for at least one index ν, $\mathfrak{v} \wedge \overline{H_\nu} \neq \mathfrak{o}$. Hence, by 3.12 we have $\mathfrak{v} \leqslant \overline{H_\nu}$ or $H_\nu \in \mathfrak{v}$.

(2) → (3): It follows from $\mathfrak{o} < \mathfrak{a}$ that there exists an ultrafilter $\mathfrak{v}$ with $\mathfrak{v} \leqslant \mathfrak{a}$. Now $\mathfrak{a} \leqslant \overline{M}$, and using (2) above we see that there exists an $H \in \overline{\mathfrak{H}}$ with $\mathfrak{v} \leqslant H$. Therefore $\mathfrak{a} \wedge \overline{H} \geqslant \mathfrak{v} > \mathfrak{o}$.[13]

(3) → (1): The filter $\mathfrak{a} = \mathfrak{c}_{\mathfrak{H}} \wedge \overline{M}$ satisfies the equation $\mathfrak{a} \wedge \overline{H} = \mathfrak{o}$ for all $H \in \mathfrak{H}$. Moreover, $\mathrm{Rk}(\mathfrak{a}) \leqslant \mathrm{Rk}(\mathfrak{c}_{\mathfrak{H}}) \leqslant |\,\mathfrak{H}\,|$. By (3) we must have $\mathfrak{a} = \mathfrak{o}$, but $\mathfrak{a} = \mathfrak{o}$ is equivalent to (1). ∎

Let φ be a map of R into R'. If $\mathfrak{A}$ is a basis of R, then $\varphi\mathfrak{A} = \{\varphi A : A \in \mathfrak{A}\}$ is a basis in R'. Conversely, if $\mathfrak{A}'$ is a basis of R', then $\varphi^{-1}\mathfrak{A}' = \{\varphi^{-1}A' : A' \in \mathfrak{A}'\}$ is a basis of R. However, note that if $\mathfrak{a}$ and $\mathfrak{a}'$ are filters, then $\varphi\mathfrak{a}$ and $\varphi^{-1}\mathfrak{a}'$ are not in general filters, but only filter bases. Consider the map φ^* induced by φ and defined by $\varphi^*\mathfrak{a} = \overline{\varphi\mathfrak{a}}$.

The map φ^* sends $F(R) \to F(R')$ and is called a *filter map*. Likewise, $(\varphi^{-1})^*\mathfrak{a}' = \overline{\varphi^{-1}\mathfrak{a}'}$ defines a map $(\varphi^{-1})^* : F(R') \to F(R)$, called the filter map induced by φ^{-1}. We have $\varphi^*\mathfrak{o} = \mathfrak{o}$, $(\varphi^{-1})^*\mathfrak{o} = \mathfrak{o}$, and $(\varphi^{-1})^*\mathfrak{a}' = (\varphi^{-1})^*(\mathfrak{a}' \wedge \overline{\varphi R})$. Since no confusion will ever be possible, we shall subsequently not bother to use special typography to designate filter maps. Instead of $\varphi^*\mathfrak{a}$ we shall simply write $\varphi\mathfrak{a}$ and instead of $(\varphi^{-1})^*\mathfrak{a}'$ simply $\varphi^{-1}\mathfrak{a}'$. Whenever we apply φ and φ^{-1} to filters, they will always represent the induced filter maps, their images always being a filter. The following theorem is a direct consequence of definitions.

3.14 (1) If $\mathfrak{A}$ is a basis of the filter $\mathfrak{a}$, then $\varphi\mathfrak{A}$ is a basis of $\varphi\mathfrak{a}$. If $\mathfrak{A}'$ is a basis of $\mathfrak{a}'$, then $\varphi^{-1}\mathfrak{A}'$, is a basis of $\varphi^{-1}\mathfrak{a}'$. For principal filters, $\varphi\overline{M} = \overline{\varphi M}$ and $\varphi^{-1}\overline{M'} = \overline{\varphi^{-1}M'}$.

(2) $\mathfrak{a} \leqslant \mathfrak{b}$ *implies* $\varphi\mathfrak{a} \leqslant \varphi\mathfrak{b}$ *and* $\mathfrak{a}' \leqslant \mathfrak{b}'$ *implies* $\varphi^{-1}\mathfrak{a}' \leqslant \varphi^{-1}\mathfrak{b}'$.

(3) $\varphi \bigvee \mathfrak{a}_\iota = \bigvee \varphi\mathfrak{a}_\iota$ *and* $\varphi^{-1} \bigvee \mathfrak{a}'_\iota = \bigvee \varphi^{-1}\mathfrak{a}'_\iota$.

(4) $\varphi \bigwedge \mathfrak{a}_\iota \leqslant \bigwedge \varphi\mathfrak{a}_\iota$ *and* $\varphi^{-1} \bigwedge \mathfrak{a}'_\iota \doteq \bigwedge \varphi^{-1}\mathfrak{a}'_\iota$.

If φ *is a 1-1 map, then* $\varphi \bigwedge \mathfrak{a}_\iota = \bigwedge \varphi\mathfrak{a}_\iota$.

3.15 (1) $\varphi(\varphi^{-1}\mathfrak{a}' \wedge \mathfrak{b}) = \mathfrak{a}' \wedge \varphi\mathfrak{b}$.

(2) $\varphi^{-1}\varphi\mathfrak{a} \geqslant \mathfrak{a}$ and $\varphi\varphi^{-1}\mathfrak{a}' = \mathfrak{a}' \wedge \overline{\varphi R}$.

(3) *If* $\mathfrak{v}$ *is an ultrafilter, then so is* $\varphi\mathfrak{v}$.

Proof: (1) Let $A' \in \mathfrak{a}'$ and $B \in \mathfrak{b}$. Then we have

$$\varphi((\varphi^{-1}A') \cap B) \subseteq \varphi(\varphi^{-1}A') \cap \varphi B \subseteq A' \cap \varphi B$$

[13] We did not use the hypothesis $\mathrm{Rk}(\mathfrak{a}) \leqslant |\,\mathfrak{H}\,|$. Hence (3) holds without this limitative stipulation.

It follows from $x' \in A' \cap \varphi B$ that there exists a $b \in B$ such that $x' = \varphi b$, and since $\varphi b = x' \in A'$, we see that $b \in \varphi^{-1}A'$. Hence $x' \in \varphi((\varphi^{-1}A') \cap B)$ and thus $\varphi((\varphi^{-1}A') \cap B) = A' \cap \varphi B$. Since the sets appearing in these equations represent bases of the corresponding filters, assertion (1) has been proved.

(2) The first inequality follows from $\varphi^{-1}\varphi A \supseteq A$. Using (1), we have

$$\varphi\varphi^{-1}\mathfrak{a}' = \varphi(\varphi^{-1}\mathfrak{a}' \wedge \overline{R}) = \mathfrak{a}' \wedge \overline{\varphi R}.$$

(3) From $\mathfrak{v} \neq \mathfrak{o}$ it follows that $\varphi\mathfrak{v} \neq \mathfrak{o}$. From $\mathfrak{a}' \wedge \varphi\mathfrak{v} \neq \mathfrak{o}$, we see, by (1), that first $\varphi(\varphi^{-1}\mathfrak{a}' \wedge \mathfrak{v}) \neq \mathfrak{o}$ and then that $\varphi^{-1}\mathfrak{a}' \wedge \mathfrak{v} \neq \mathfrak{o}$. By (3.12) we have $\mathfrak{v} \leqslant \varphi^{-1}\mathfrak{a}'$; hence $\varphi\mathfrak{v} \leqslant \varphi\varphi^{-1}\mathfrak{a}' \leqslant \mathfrak{a}'$. Hence $\varphi\mathfrak{v}$ is an ultrafilter (by 3.12 again). ∎

Notes and Exercises

3A (On the interrelations between sequences and filters.)

Exercises

1. If M is a countable, nonempty subset of the fundamental set R, then the principal filter $\overline{M}$ is a Fréchet-filter.

2. The sup of a finite number of Fréchet filters is again a Fréchet filter.

3. A filter $\mathfrak{a}$ of R is a Fréchet filter if and only if $0 \leqslant Pz(\mathfrak{a}) \leqslant \aleph_0$ and if and only if there exists an $A \in \mathfrak{a}$ such that $| A \cap \complement A' | < \aleph_0$ for all $A' \in \mathfrak{a}$.

4. Every filter $\mathfrak{a}$ with $\text{Rk}(\mathfrak{a}) \leqslant \aleph_0$ is a sup of Fréchet filters.

3B (Moore-Smith convergence) Let G be a nonempty set. Let the relation $<$ be defined in $G \times G$, be transitive and possess the following properties:

For every two elements $x, y \in G$ there exists a $z \in G$ with $x < z$ and $y < z$. The couple $(G, <)$ is called a *directed set*. Suppose φ is a map of G into the fundamental set R, then the couple $(\varphi, G, <)$ is called a *Moore-Smith sequence* of R. The following relationships link Moore-Smith sequences and filters.

Let $(\varphi, G, <)$ be a Moore-Smith sequence. For every $x \in G$ let the subset A_x of R be defined by $A_x = \{\varphi y : x < y\}$.

Exercise: The system of sets $\mathfrak{A} = \{A_x : x \in G\}$ is a filter basis. The filter $\mathfrak{a}$ generated by $\mathfrak{A}$ is $\neq \mathfrak{o}$. Conversely, let $\mathfrak{a}$ be a filter of R, $\mathfrak{a} \neq \mathfrak{o}$. Let G be the set of all ordered couples (p, A) with $p \in A$ and $A \in \mathfrak{a}$. Let us define $(p, A) < (q, B)$ if $B \subseteq A$. Then $(G, <)$ is a directed set. Finally, let us define by $\varphi(p, A) = p$ a map $\varphi : G \to R$. Then $(\varphi, G, <)$ is a Moore-Smith sequence and if we form the filter associated with it by the construction given in the exercise, this filter is none other than $\mathfrak{a}$, the filter from which we started in order to construct $(\varphi, G, <)$.

3C Let φ be a map from R to R' and M^* a subset of $F(R)$. Let us suppose that for any two filters $\mathfrak{a}, \mathfrak{b} \in M^*$ we can find a third filter $\mathfrak{c} \in M^*$ such that $\mathfrak{c} \leqslant \mathfrak{a} \wedge \mathfrak{b}$.

Exercise: With the above suppositions, we have the equation

$$\varphi \bigwedge_{\mathfrak{a} \in M^*} \mathfrak{a} = \bigwedge_{\mathfrak{a} \in M^*} \varphi\mathfrak{a}.$$

3D (Cardinality of $F(R)$) Let R be a fundamental set, of infinite cardinality. Let T be the set of all ordered couples $t = (E_t, \varphi_t)$ where E_t is a finite subset of R and φ_t a map from $P(E_t)$ into the system $\mathfrak{E}$ of all finite subsets of R. Finally, let Ψ be the set of all maps $\psi : P(R) \to P(R)$ and for $\psi \in \Psi$, $E \in \mathfrak{E}$ and any finite family of subsets $M_1, \ldots, M_n$ of R let $A_\psi(M_1, \ldots, M_n; E)$ denote the set of all $t \in T$ such that $(\varphi_t(M_\nu \cap E_t)) \cap E = (\psi M) \cap E$ for $\nu = 1, \ldots, n$.

Exercises

1. We have $| T | = | R |$ and $| \psi | \geqslant 2^{(2^{|R|})}$.
2. The sets $A_\psi(M_1, \ldots, M_n; E)$ are not empty.
3. Let $\psi \in \Psi$ be fixed. Then the system $\mathfrak{A}_\psi$ of all sets of the form $A_\psi(M_1, \ldots, M_n; E)$ is a basis. Let us denote by $\mathfrak{a}_\psi$ the filter generated by this basis. Then (2) implies $\mathfrak{a}_\psi \neq \mathfrak{o}$.
4. $\psi_1, \psi_2 \in \Psi$ and $\psi_1 \neq \psi_2$ imply $\mathfrak{a}_{\psi_1} \wedge \mathfrak{a}_{\psi_2} \neq \mathfrak{o}$.
5. $| F(R) | \leqslant 2^{(2^{|R|})}$.

The mapping $\psi \to \mathfrak{a}_\psi$ is 1-1, from Ψ into $F(T)$. Hence (1) implies

$$2^{(2^{|R|})} \leqslant | \Psi | \leqslant | F(T) | = | F(R) | \leqslant 2^{(2^{|R|})}$$

and therefore $| F(R) | = 2^{(2^{|R|})}$.

Exercise: The set of all ultrafilters of R has the same cardinality as $F(R)$.

II

Topological Spaces

The problem of defining a notion of convergence appropriate to any set whatsoever, a notion which lends itself to easy rules of calculation (like the rules for calculating with limits in the classical theory), leads almost inescapably to the concept of topological space. In Section 4 of this chapter we shall introduce and study topological spaces from the viewpoint of the problem proposed above. In the later sections we shall treat allied questions and concepts, all stemming directly from the definition of topological space.

4. Topology and Limits

As we all know, the definition of limit used in real analysis reads as follows: The sequence of numbers $(a_\nu)_{\nu \in \mathbf{N}}$ converges to the limit p if and only if, given an arbitrary $\epsilon > 0$, we can find an integer n such that $\nu \geqslant n$ implies $| a_\nu - p | < \epsilon$. The set $U_\epsilon(p) = \{x : | x - p | < \epsilon\}$ is called an ϵ-neighborhood of p. The convergence of the sequence $\{a_\nu\}_{\nu \in \mathbf{N}}$ to the limit p is equivalent to the following:

Given any $\epsilon > 0$ there exists a terminating section A_n of the given sequence such that $A_n \subseteq U_\epsilon(p)$. The system $\mathfrak{U}_p = \{U_\epsilon(p) : \epsilon > 0\}$ is manifestly a filter-basis. We shall call the filter $\mathfrak{u}_p$ generated by this basis the *neighborhoods-filter* of p. Moreover, if we designate by $\mathfrak{a}$ the Fréchet filter of the sequence $(\mathfrak{a}_\nu)_{\nu \in N}$, then the convergence of our original sequence to p is also equivalent to the condition $\mathfrak{a} \leqslant \mathfrak{u}_p$. It is this final formulation of convergent sequence which can be carried over to arbitrary sets; i.e., if $\mathfrak{a} \leqslant \mathfrak{u}_p$ we define the filter $\mathfrak{a}$ to be *convergent* to the point p.

To each point p of a fundamental set R let there be uniquely associated a filter $\mathfrak{u}_p$, called the neighborhoodsfilter of p. This correspondence allows us to define convergence in R, as follows: A filter $\mathfrak{a}$ of R is said to converge to p if and only if $\mathfrak{a} \leqslant \mathfrak{u}_p$. In order to make sure that the above definition of convergence is reasonable and fruitful we must impose certain further conditions upon the neighborhoodsfilter associated with each point p.

First, let us note that, in the classical case of the real line, given any ϵ-neighborhood of the point p, $\epsilon > 0$, it is always true that $p \in U_\epsilon(p)$, i.e., $\overline{p} \leqslant \mathfrak{u}_p$. The first condition that we shall therefore impose in the general case is

$$\overline{p} \leqslant \mathfrak{u}_p, \quad \textit{for all} \quad p \in R. \tag{1}$$

Again, let U be any set belonging to $\mathfrak{u}_p$. Then there exists an $\epsilon > 0$ such that $U_\epsilon(p) \subseteq U$. For any $q \in U_\epsilon(p)$ we have $\eta = \epsilon - |q - p| > 0$ and moreover, $U_\eta(q) \subseteq U_\epsilon(p) \subseteq U$; hence $U \in \mathfrak{u}_q$. Let us take this property as the second condition that we impose in the general case.

$$\begin{gathered}\textit{For any } p \in R \textit{ and for each } U \in \mathfrak{u}_p, \textit{ there exists a}\\ V \in \mathfrak{u}_p \textit{ such that } U \in \mathfrak{u}_q \textit{ for each } q \in V.\end{gathered} \tag{2}$$

This formulation of condition (2) mentions both filters and the sets belonging to them. We can re-express this condition purely in terms of filters of neighborhoods:

$U \in \mathfrak{u}_q$ implies $\mathfrak{u}_q \leqslant \overline{U}$ and since this holds for all $q \in V$ we have

$$\bigvee\{\mathfrak{u}_q : q \in V\} \leqslant \overline{U}$$

and, all the more so,

$$\bigwedge_{V \in \mathfrak{u}_q} \bigvee_{q \in V} \mathfrak{u}_q \leqslant \overline{U}.$$

This last relation holds for all $U \in \mathfrak{u}_p$, so that finally we have the equivalent condition:

$$\bigwedge_{V \in \mathfrak{u}_q} \bigvee_{q \in V} \mathfrak{u}_q \leqslant \mathfrak{u}_p. \tag{2'}$$

If we reverse the steps of the above argument we get a proof of the converse, i.e., (2′) implies (2).

The correspondence $p \to \mathfrak{u}_p$ is a map τ from R into the lattice of filters $F(R)$. This map associates with $p \in R$ an image $\tau_p = \mathfrak{u}_p$, where $\mathfrak{u}_p$ is a neighborhoods-filter of p. Such a map, provided that in addition to (1), verifies (2) or (2′), is called a *topology* (defined on the fundamental set R).

Definition 4a: *A* topology *of the fundamental set R is a map τ: $R \to F(R)$, satisfying, for each $p \in R$, the following two conditions:*

$$\overline{p} \leqslant \tau p$$

$$\bigwedge_{U \in \tau_p} \bigvee_{q \in U} \tau q \leqslant \tau p.$$

The filter τp *is called the* neighborhoodsfilter *of* p. *The sets belonging to* τp *are called* neighborhoods[1] *of the point* p. *A filter* $\mathfrak{a}$ *of* R *is said to be* τ-convergent *to* p *if* $\mathfrak{a} \leqslant \tau p$.

Whenever τ *is a topology on* R, *the couple* (R, τ) *is called a* topological space.

If the topology τ defined on the set R has previously been mentioned and no confusion can occur, we shall often write R rather than (R, τ) and call R itself the topological space. Since in this book we shall deal almost exclusively with topological spaces, the adjective "topological" will often also be dropped and the word space used alone. Note also that we shall always consider the empty set to be a topological space—this will allow us to state certain results without special provisos to cover exceptional cases.

Examples

4.I If we associate with each point p on the real line R the filter generated by the set of all ϵ-neighborhoods of p, we define a topology on R, called the *usual* or *standard topology*. The theory of convergence associated with this topology is none other than the usual "topology of real variables." Whenever we speak of the real line as a space without specifically mentioning the underlying topology we shall always mean the standard topology.

4.II For an arbitrary set R the map defined by $\tau p = \ulcorner p \urcorner$ is a topology known as the *discrete* topology on R. Under this topology the only filters that converge to p are the nullfilter and the principal filter $\ulcorner p \urcorner$.

For all $p \in R$, the map $\tau p = \ulcorner R \urcorner$ also defines a topology on R, called the *trival topology*. Under this topology all filters converge to each point of R.

4.III As in 3.I, let R be an infinite set and let k be a cardinal number with $\aleph_0 \leqslant k \leqslant |R|$. The map $\tau^k p = \{M : p \in M, |\complement M| < k\}$ defines a topology on R.

The above procedures, based as they were on a simple and appropriate transcription of specific properties of classical real analysis, enabled us to define the general concept of topological space. However, it cannot be said that the properties chosen for transcription pertain to convergence per se; rather, they pertain to the special example of convergence of sequences on the real line. There is therefore an element of the arbitrary in our procedure. Nevertheless, we shall show that the concept of topological spaces is derivable from a few extra conditions, imposable in a natural way on any notion of convergence.

In any case, each type of convergence in the set R can be completely described by associating, uniquely, to each point $p \in R$ a subset K_p^* of $F(R)$, composed of exactly those filters of R which converge to p. If now (a_ν) and (b_ν) are two sequences of real numbers both convergent to the same limit p, then we know

[1] In less modern literature the word *neighborhood* often means what we shall call "open neighborhood" in Section 5.

that any mixed sequence of a_ν's and b_ν's (i.e., any sequence decomposable into the two subsequences (a_ν) and (b_ν)) also converges to p. Let $\mathfrak{a}$ and $\mathfrak{b}$ be the Fréchet filters of these two sequences; then every possible mixed sequence of a_ν's and b_ν's has associated with it the Fréchet filter $\mathfrak{a} \vee \mathfrak{b}$. Note that the lattice-theoretic sup of any infinity of Fréchet filters need not be a Fréchet filter, and hence, in real analysis only finite sups of Fréchet filters play an important role. But when we come to define convergence of filters in full generality, we can drop the stipulation limiting sups to a finite number of terms and impose instead the following natural condition:

(α) *If the set K_p^* contains an arbitrary number of filters, then it also contains their sup.*

If the sequence of numbers $(a_\nu)_{\nu \in \mathbf{N}}$ converges to p, then any of its subsequences $(b_\nu)_{\nu \in N}$ also converges to p. If $\mathfrak{a}$ and $\mathfrak{b}$ denote respectively the Fréchet filters associated with the above sequences, then $\mathfrak{b} \leqslant \mathfrak{a}$. The selection of a subsequence leads to a smaller, or finer, associated filter. We therefore impose the following second natural condition:

(β) *If* $\mathfrak{a} \in K_p^*$ *and* $\mathfrak{b} \leqslant \mathfrak{a}$, *then* $\mathfrak{b} \in K_p^*$.

The third and last condition is suggested by the fact that the constant sequence of numbers $(p)_{\nu \in N}$ always converges to p. The associated Fréchet filter is the principal filter $\ulcorner p \urcorner$. Hence the condition

(γ) $\ulcorner p \urcorner \in K_p^*$.

Condition (α) implies that $\mathfrak{u}_p = \bigvee \{\mathfrak{a} : \mathfrak{a} \in K_p^*\}$ is again a filter in K_p^*, while (β) tells us that K_p^* consists of exactly those filters $\mathfrak{a}$ for which $\mathfrak{a} \leqslant \mathfrak{u}_p$. Condition (γ) therefore implies $\ulcorner p \urcorner \leqslant \mathfrak{u}_p$. Hence, if we are given a concept of convergence verifying (α)-(γ) and we define the map τ: $R \to F(R)$ by $\tau p = \mathfrak{u}_p$, then τ will verify the first of the two conditions needed to stipulate a topology, and the resulting convergence in this topology would necessarily have to be identical with the convergence from which we started. If therefore behooves us to mirror the second condition defining a topology by an appropriate property of convergence.

In the case of convergence of real numbers the following *diagonal property* holds true: Given the double sequence $(a_{\mu,\nu})_{\mu,\nu \in \mathbf{N}}$ and suppose $\lim_{\nu\to\infty} a_{\mu,\nu} = b_\mu$ and $\lim_{\mu\to\infty} b_\mu = p$. Then we can be certain that there exists a diagonal subsequence $(a_{\mu,\eta_\mu})_{\mu \in \mathbf{N}}$ which also convergences to p. If we impose a single limitative condition, then the diagonal property can be shown to be equivalent to the second defining condition for a topology. More precisely, let R be a set wherein a convergence concept verifying (α)-(γ) has been given, and hence wherein we can define filters $\mathfrak{u}_p$. The necessary limitative condition to be imposed is that all filters $\mathfrak{u}_p$ possess a denumerable basis. This guarantees that at least one sequence converges to any given point. When these conditions are verified we have:

4.1 *The map τ from $R \to F(R)$ defined by $\tau p = \mathfrak{u}_p$ is a topology if and only if the diagonal property for double sequences holds for the underlying concept of convergence.*

Proof: For each $p \in R$ let $\{U_s(p) : s \in \mathbf{N}\}$ be a basis of $\mathfrak{u}_p$. Without loss of generality we can assume that $U_{s+1}(p) \subseteq U_s(p)$, for otherwise, let

$$U'_s = \cap \{U_\sigma(p) : \sigma \leqslant s\};$$

then $\{U'_s(p) : s \in \mathbf{N}\}$ is also a basis of $\mathfrak{u}_p$ and $U'_{s+1}(p) \subseteq U'_s(p)$ certainly holds.

Let τ denote a topology. Given any $s \in \mathbf{N}$ there exists an $x(s) \in \mathbf{N}$ such that, for each $q \in U_{x(s)}(p)$ we can find a $y(s) \in \mathbf{N}$ with $U_{y(s)}(q) \subseteq U_s(p)$. (Cf. Eq. (2), p. 34). Consider the double sequence $(a_{\mu,\nu})_{\mu,\nu \in \mathbf{N}}$ and let us assume that for fixed μ the subsequence $(a_{\mu,\nu})_{\nu \in \mathbf{N}}$ converges to b_μ (i.e., its Fréchet filter is τ-convergent to b_μ). Let us further assume that the sequence $(b_\mu)_{\mu \in \mathbf{N}}$ converges to p. Then, given any $s \in \mathbf{N}$, we can find an $m(s)$ such that $b_\mu \in U_{x(s)}(p)$ for all $\mu \geqslant m(s)$. Moreover, given any $s \in \mathbf{N}$ and $\mu \geqslant m(s)$, we can find an $n(\mu, s)$ such that $a_{\mu,\nu} \in U_{y(s)}(b_\mu)$ for all $V \geqslant n(\mu, s)$. For fixed μ let s_μ denote the largest number such that $\mu \geqslant m(s_\mu)$. Define $n_\mu = \max\{n(\mu, s) : s \leqslant s_\mu\}$. It follows from $\mu \geqslant m(s)$ that $b_\mu \in U_{x(s)}(p)$ and further that $a_{\mu,n_\mu} \in U_{y(s)}(b_\mu) \subseteq U_s(p)$. Hence the diagonal sequence $(a_{\mu,n_\mu})_{\mu \in \mathbf{N}}$ converges to p.

To prove the converse, let us suppose that τ is not a topology; i.e., τ does not possess the second property characterizing a topology.

Then there exist $p \in R$ and $s \in \mathbf{N}$ such that the following property holds (negation of Eq. (2), p. 34): For each $\mu \in \mathbf{N}$ there exists a point b_μ belonging to $U_\mu(p)$, such that given any $\nu \in \mathbf{N}$ we can find a point $a_{\mu,\nu} \in U_\nu(b_\mu)$ such that $a_{\mu,\nu} \notin U_s(p)$. Since $a_{\mu,\nu} \in U_\nu(b_\mu) \subseteq U_n(b_\mu)$ for all $\nu \geqslant n$, it follows that for a fixed μ the sequence $(a_{\mu,\nu})_{\mu \in \mathbf{N}}$ converges to b_μ. And since $b_\mu \in U_\mu(p) \subseteq U_m(p)$ for $\mu \geqslant m$ the sequence $(b_\mu)_{\mu \in \mathbf{N}}$ converges to p. But, for all μ and ν, $a_{\mu,\nu} \notin U_s(p)$ and hence no diagonal subsequence of $(a_{\mu,\nu})_{\mu,\nu \in \mathbf{N}}$ can converge to p. ∎

In the general case where $\mathfrak{u}_p$ does not possess a denumerable basis we can not hope to characterize a topology by the use of convergent sequences. The above proof tells us, however, that the second defining property of a topology is an appropriate generalization of the diagonal property of real sequences. Any concept of convergence which verifies conditions (α)-(γ), and in which the generalized diagonal property holds, can be thought of as having been generated by a topology.

In our discussion of convergence we have as yet made no mention of uniqueness of limits: In a topological space a filter can, in general, converge to several points. Uniqueness of limits can only be guaranteed by imposing additional conditions, a question which we shall examine later in Section 8.

Notes and Exercises

4A *Exercise:* In the previously defined topological space (R, τ^k), cf. 4.III, all ultrafilters $\mathfrak{v}$ such that $Pz(\mathfrak{v}) \geqslant k$ are convergent; they even converge to every point of R. An ultrafilter $\mathfrak{v}$ with $Pz(\mathfrak{v}) < k$ can still be convergent, but only when it is a principal filter.

4B Let τ be the standard topology of the real line $\mathbf{R}$.

Exercise: The map τ', from $\mathbf{R} \to F(\mathbf{R})$ defined by

$$\tau' p = \begin{cases} \tau p & \text{for} \quad p \neq 0 \\ \tau p \wedge \overline{\mathbf{Q}} & \text{for} \quad p = 0 \end{cases}$$

is not a topology. ($\mathbf{Q}$ = set of all rational numbers.)

4C Let A be the set of all maps $\varphi : \mathbf{R} \to \mathbf{R}$. ($A$ is the set of all real functions.) For each $x \in R$, we shall consider the map $\epsilon_x : A \to \mathbf{R}$ defined by $\epsilon_x \varphi = \varphi x$. Finally, let τ designate the standard topology on $\mathbf{R}$.

Exercise

1. The map $\tau' \varphi = \bigwedge \{\epsilon_x^{-1} \tau \epsilon_x \varphi : x \in \mathbf{R}\}$ defines a topology on A.

2. For any finite set of points $x_1, \ldots, x_n \in \mathbf{R}$, for every $\varphi \in A$, and for every positive real number α let

$$U_{x_1, \ldots, x_n; \alpha}(\varphi) = \{\psi : \psi \in A, \mid \psi x_\nu - \varphi x_\nu \mid < \alpha \ \text{ for } \ \nu = 1, \ldots, n\}$$

For fixed φ the sets of this form constitute a basis of the neighborhoods filter $\tau' \varphi$.
The convergence associated with this topology τ' on A, is none other than the usual pointwise convergence of functions.

5. Fundamental Concepts and Their Reformulations

In what follows we shall be concerned exclusively with a single, fixed topological space (R, τ). In the previous section we showed how a topology can always be introduced once an appropriate type of convergence concept has been defined. In this section we study other ways of characterizing topological spaces and, in so doing, we show the connections between our definition of a topology and the several other definitions often employed in the literature. These characterizations will, of course, depend upon new concepts, the most important of which are those of *open* and *closed* sets.

Definition 5a: *Given the space (R, τ). A subset M of R is called an* open *set if $p \in M$ implies $M \in \tau p$ (or equivalently, $\tau p \leqslant \ulcorner M \urcorner$).*[2] *A filter of R is called open if it possesses a basis all of whose members are open sets.*

A set M is called closed *if its complement $\complement M$ is an open set. A filter of R is called closed if it possesses a basis all of whose members are closed sets.*[3]

As an immediate consequence of these definitions we have

5.1 *A set M is open if and only if it is a neighborhood of each of its points.*

Definition 5b: *In (R, τ) we say that the point p is* adherent *to the set M if $\tau p \wedge \ulcorner M \urcorner \neq \mathfrak{o}$. If p is an adherent point to the set $M \cap \complement p$, then we say that p is a* point of accumulation *of the set M.*

Evidently p is an adherent point of M if and only if each neighborhood of τp has a nonempty intersection with M, e.g., p is adherent to the one-point set $\{p\}$ but it is not a point of accumulation of this set.

5.2 *A set M is closed if and only if M contains all its adherent points.*

Proof: Let M be closed. Then $\complement M$ is open and $p \notin M$ implies $\tau p \leqslant \ulcorner \complement M \urcorner$. Hence a point not belonging to M cannot be adherent to M. Conversely, suppose M contains all its adherent points. Then $p \in \complement M$ implies $\tau p \wedge \ulcorner M \urcorner = \mathfrak{o}$, hence $\complement M \in \tau p$; i.e., $\complement M$ is open. ∎

The concepts "open" and "closed" are complementary to each other, yet they are not negations of each other. In general a set need be neither open nor closed. On the other hand, there exists sets which are both open and closed, as the following theorem shows.

5.3 *The empty set and the whole space are two subsets of any space R which are simultaneously both open and closed.*

The intersection of a finite number of open sets is open, while the union of any number of open sets is open.

The union of a finite number of closed sets is closed, while the intersection of any number of closed sets is closed.

Proof: For all $p \in R$ we have $\tau p \wedge \ulcorner \emptyset \urcorner = \mathfrak{o}$. Hence the empty set can have no adherent points and must therefore be closed. Since R contains all points, it also is closed. But $\complement \emptyset = R$ and $\complement R = \emptyset$ and both these sets must therefore also be open.

Let $M_1, \ldots, M_n$ be open sets and $p \in M_1 \cap \ldots \cap M_n$. Then $p \in M_\nu$ and $M_\nu \in \tau p$, both for $\nu = 1, \ldots, n$. Therefore $M_1 \cap \ldots \cap M_n \in \tau p$ (Definition 3a, $\mathbf{F}_1$).

[2] A point p such that $M \in \tau p$ is also called *inner point* of M. A set is open if and only if all its points are inner points.

[3] When we wish to emphasize the underlying topology τ without which, of course, we could not define our new concepts, we speak of "τ-open" and "τ-closed" sets, etc.

Let $\{M_\iota : \iota \in I\}$ be a family of open sets and $p \in \bigcup M_\iota$. Then $p \in M_{\iota_0}$ for some index ι_0 and hence $M_{\iota_0} \in \tau p$. We therefore certainly have $\bigcup M_\iota \in \tau p$ (Definition 3a, $\mathbf{F}_2$).

The third of the above assertions then follows by taking complements. ∎

5.4 *The sup of a finite number of closed filters is closed. The intersection of any number of closed filters is closed.*

But for open filters we have: The sup and the inf of any number of open filters is again open.

Proof: These assertions are direct consequences of 3.6 and 5.3. In the case of open filters we can even take the inf of an arbitrary family of open filters and still be assured that it is open. But in constructing a basis for the inf of an arbitrary family of filters we are only allowed to take intersections of a finite number of sets. ∎

5.5 *Every neighborhoodsfilter τp is open: The family of all open sets M such that $p \in M$ is a basis of τp.*

Proof: $\mathfrak{U}_p$ is a basis, by 5.3. $M \in \mathfrak{U}_p$ implies $M \in \tau p$. Hence $\tau p \leqslant \overline{\mathfrak{U}_p}$. Conversely, let U be a neighborhood belonging to τp and $M = \{q : \tau q \leqslant \overline{U}\}$. Then $p \in M$, $M \subseteq U$, and from $q \in M$ it follows that

$$\bigwedge_{V \in \tau q} \bigvee_{x \in V} \tau x \leqslant \tau q \leqslant \overline{U}$$

Hence there exists a $V \in \tau q$ such that $\tau x \leqslant \overline{U}$ for all $x \in V$. Thus $V \subseteq M$ and moreover $M \in \tau q$. This means that M is open and belongs to $\mathfrak{U}_p$. Hence every $U \in \tau p$ contains an $M \in \mathfrak{U}_p$, i.e., $\overline{\mathfrak{U}_p} \leqslant \tau p$. ∎

This was the first occasion where we used the second characteristic property of a topology. Conversely, this property is a consequence of the neighborhoodsfilter being open. The open basis of a neighborhoodsfilter can also be used to characterize topological spaces.

5.6 *Let R be a fundamental set and p a point of R. To each point p let there be associated a system of sets $\mathfrak{U}_p$ in such a way that the following* Hausdorf Neighborhood Axioms *hold:*

($\mathbf{U}_1$) *Every system $\mathfrak{U}_p$ is a filter basis.*

($\mathbf{U}_2$) *$U \in \mathfrak{U}_p$ implies $p \in U$.*

($\mathbf{U}_3$) *Given $U \in \mathfrak{U}_p$ and $q \in U$. Then there exists a $V \in \mathfrak{U}_q$ such that $V \subseteq U$.*

Then the map $\tau p = \overline{\mathfrak{U}_p}$ defines a topology τ of R. Conversely, every topology on R can be obtained in this way.

Proof: Since $\mathfrak{U}_p$ is a filter basis by ($\mathbf{U}_1$), $\tau p = \overline{\mathfrak{U}_p}$ defines a filter, for which $\overline{p} \leqslant \tau p$, by ($\mathbf{U}_2$). If $U \in \mathfrak{U}_p$ is fixed and $q \in U$ then ($\mathbf{U}_3$) implies $\tau q \leqslant \overline{U}$. Hence $\bigvee \{\tau q : q \in U\} \leqslant \overline{U}$ and, by (3.8),

$$\bigwedge_{U \in \tau p} \bigvee_{q \in U} \tau q \leqslant \bigwedge_{U \in \tau p} \overline{U} = \tau p$$

Thus τ is a topology on R. To prove the converse, assume τ to be a topology on R. Then, by 5.5, each neighborhood filter τp possesses a basis $\mathfrak{U}_p$ of open sets. These systems of sets verify $(\mathbf{U}_1)$ and $(\mathbf{U}_2)$, and by 5.1, $(\mathbf{U}_3)$. They therefore define, in their own right, a topology on R, which must however be identical with the topology from which we started, because $\tau p = \overline{\mathfrak{U}_p}$. ∎

A second way of characterizing topological spaces follows from the results of 5.3.

5.7 *Let R be a given fundamental set and $\mathfrak{S}$ a system of subsets of R such that*

(**a**) $\emptyset \in \mathfrak{S}$ *and* $R \in \mathfrak{S}$.

(**b**) *If $\mathfrak{S}$ contains a finite family then it also contains their intersection (union).*

(**c**) *If $\mathfrak{S}$ contains an arbitrary family then it also contains their union (intersection).*

There then exists exactly one topology τ of R, having $\mathfrak{S}$ as the family of its τ-open (τ-closed) sets. Conversely, any topology on R can be obtained in this way.[4]

Proof: Note that $\mathfrak{S}$ verifies the conditions given above in parentheses if and only if the complementary system $\mathfrak{S}' = \{\complement S : S \in \mathfrak{S}\}$ obeys the conditions not in parentheses. Since open and closed sets also form complementary systems, we need only prove the assertions not in parentheses.

Hence let $\mathfrak{S}$ possess the properties not in parentheses. Let us associate with each $p \in R$ the system $\mathfrak{U}_p = \{S : S \in \mathfrak{S}, p \in S\}$. The system $\mathfrak{U}_p$ certainly verifies the Hausdorf neighborhood axiom $(\mathbf{U}_2)$. By (**a**), $R \in \mathfrak{U}_p$ and so $\mathfrak{U}_p$ is nonempty. Moreover, $\mathfrak{U}_p$ is a basis, by (**b**), and hence (U_1) is also verified. From $S \in \mathfrak{U}_p$ and $q \in \mathfrak{S}$, it follows that $S \in \mathfrak{U}_q$ and hence $(\mathbf{U}_3)$ is verified with $U = V = S$. By 5.6 we see that $\tau p = \overline{\mathfrak{U}_p}$ defines a topology τ on R. From $S \in \mathfrak{S}$ and $p \in S$ we have $S \in \mathfrak{U}_p$, hence $S \in \tau p$. Thus $\mathfrak{S}$ contains only τ-open sets. On the other hand, let M be an arbitrary τ-open set. If $M = \emptyset$, then $M \in \mathfrak{S}$ by (**a**); while if $M \neq \emptyset$, then for each $p \in M$ we can find, by 5.1, an $S_p \in \mathfrak{S}$ such that $p \in S_p$ and $S_p \subseteq M$. By (**c**) we likewise have $M = \cup \{S_p : p \in M\} \in \mathfrak{S}$ This proves that $\mathfrak{S}$ contains all possible τ-open sets.

To prove the converse, let τ be a topology on R. Then by 5.3 the system $\mathfrak{S}$ of all τ-open sets possesses properties (**a**)–(**c**) and thus defines, in the manner previously shown, a topology τ' on R. By 5.5 we must have $\tau p = \overline{\mathfrak{U}_p} = \tau' p$ and thus the topologies τ and τ' are identical and the last contention of this theorem has been established. ∎

Definition 5c: *A system of sets $\mathfrak{B}$ is called a* basis *of* (R, τ)[5] *if it is composed*

[4] Conditions (**b**) and (**c**) do not exclude from consideration the empty family of sets. The intersection of the empty family of sets is the whole space, while the union of the empty family is the empty set. This shows that (**b**) and (**c**) imply (**a**); cf. Section 1.

[5] Or also simply a basis of τ.

only of τ-open sets and if every nonempty τ-open set of R is expressible as a union of sets of $\mathfrak{B}$.[6]

Every space possesses at least one basis, namely the system of all open sets. The concept of basis will enable us to give yet another characterization of topological space.

5.8 *Let the system $\mathfrak{B}$ of subsets of R possess the following property:*

(*) *For each $p \in R$ the system $\mathfrak{U}_p = \{B : B \in \mathfrak{B}, p \in B\}$ is a filter basis.*

Then there exists one and only one topology τ on R, having $\mathfrak{B}$ as basis. For this topology, $\tau p = \overline{\mathfrak{U}}_p$. Conversely, every topology on R can be obtained in this way.

Proof: By its very definition $\mathfrak{U}_p$ satisfies the Hausdorff-neighborhood axiom ($\mathbf{U}_2$), and by (*), it also satisfies ($\mathbf{U}_1$). From $B \in \mathfrak{U}_p$ and $q \in B$ we have $B \in \mathfrak{U}_q$; i.e. ($\mathbf{U}_3$) is also satisfied. Hence by 5.6, $\tau p = \overline{\mathfrak{U}}_p$ defines a topology τ on R. From $B \in \mathfrak{B}$ and $p \in B$ we have $B \in \tau p$. Thus $\mathfrak{B}$ is composed only of τ-open sets. On the other hand, if M is a nonempty τ-open set, for each $p \in M$ we can find a $B_p \in \mathfrak{B}$ such that $p \in B_p$ and $B_p \subseteq M$. Hence $M = \cup \{B_p : p \in M\}$ and $\mathfrak{B}$ is a basis of (R, τ).

Conversely, let τ be a topology on R and $\mathfrak{B}$ a basis of (R, τ). By 5.3 R itself is an open set and hence $R = \cup \{B : B \in \mathfrak{B}\}$. For all $p \in R$, $\mathfrak{U}_p$ is therefore nonempty. If $B_1, B_2 \in \mathfrak{U}_p$ then $D = B_1 \cap B_2$ is an open set and $p \in D$. Hence D is a union of sets of $\mathfrak{B}$ and there exists a $B_3 \in \mathfrak{B}$ such that $p \in B_3$ and $B_3 \leqslant D$, i.e., $\mathfrak{U}_p$ is a filter basis. By what we have proved above $\mathfrak{B}$ defines a topology τ' on R, but $\tau' p = \overline{\mathfrak{U}}_p = \tau p$, and hence τ and τ' are identical. ∎

Definition 5d: *Let $\mathfrak{C}$ be a system of subsets of R. Consider the new system obtainable by taking all possible finite intersections of sets in $\mathfrak{C}$. If this new system is a basis of (R, τ) we say that $\mathfrak{C}$ itself is a* subbasis *of (R, τ).*

Every basis is also a subbasis; every subbasis is a covering of R. Conversely, let $\mathfrak{C}$ be a covering of R. The system formed by taking all possible finite intersections of sets in $\mathfrak{C}$ verifies property (*) of 5.8 and is therefore a basis for a unique topology on R. We have thus proved the following theorem.

5.9 *A system $\mathfrak{C}$ of subsets of R is a subbasis of a topology τ on R if and only if $\mathfrak{C}$ is a covering of R. The topology τ is uniquely determined by $\mathfrak{C}$. Conversely, any topology can be obtained in this way.*

Examples

5.I The system of *all* subsets of R satisfies the two conditions of 5.7 (those in parentheses and those not in parentheses). This system therefore defines a

[6] Instead of "basis" some authors often speak of "open basis," in contrast to possible "closed bases" (cf. 5C).

topology on R for which all sets are simultaneously open and closed. This is the discrete topology.

The system $\{\emptyset, R\}$ also satisfies the above-mentioned conditions. It defines the trival topology, whose only open (and closed) sets are $\emptyset$ and R.

5.II Let R be an infinite set. The system $\{\complement p : p \in R\}$ is a cover of R and hence a subbasis of a topology τ on R. We must have $\tau = \tau^{\aleph_0}$ (cf. 4.III). For this topology, the only closed sets, besides $\emptyset$ and R, are the finite sets.

5.III For every $n \in \mathbf{N}$ and $p \in \mathbf{R}$ let $U(n, p) = \{q : |p - q| < 1/n\}$. The systems of sets $\mathfrak{U}_p = \{U(n, p) : n \in \mathbf{N}\}$ satisfy the Hausdorff neighborhood axioms and define none other than the standard topology on $\mathbf{R}$.

Let M be any subset of the space (R, τ). M always contains at least one open set: namely, the empty set. The union of all open sets contained in M is, by 5.3, an open set and evidently the largest open set contained in M. It is called the *interior* of M and denoted by $\mathring{M}$. The interior is sometimes also called the *Kernel*, *open kernel* or *inner kernel* of M. If we wish to emphasize the underlying topology we write $k_\tau(M)$ instead of $\mathring{M}$.

5.10 $\mathring{M} = \{p : p \in \tau p\}$ and $\tau p = \{M : p \in \mathring{M}\}$. M is open if and only if $M = \mathring{M}$.

Proof: Since $\mathring{M}$ is an open set contained in M, $p \in \mathring{M}$ implies $\tau p \leqslant \overline{\mathring{M}} \leqslant \overline{M}$. If, on the other hand, $M \in \tau p$, then by 5.5 there exists an open neighborhood $U \in \tau p$ such that $U \leqslant M$. Since $\mathring{M}$ is the largest open set contained in M, it follows that $U \leqslant \mathring{M}$ and hence $p \in \mathring{M}$.

If $M \in \tau p$, the first equation above, just proved, implies $p \in \mathring{M}$. Conversely, $p \in \mathring{M}$ and $\mathring{M}$ open imply $\mathring{M} \in \tau p$, and hence, a fortiori, $M \in \tau p$.

The third assertion is an immediate consequence of the definition of interior of a set. ∎

5.11 *The operation of passing from a set to its interior possesses the following properties*[7]:

$$(\mathbf{K}_1) \quad \mathring{M} \subseteq M, \qquad k_\tau(M) = M$$

$$(\mathbf{K}_2) \quad \overset{\circ}{\widehat{M \cap N}} = \mathring{M} \cap \mathring{N}, \qquad k_\tau(M \cap N) = k_\tau(M) \cap k_\tau(N)$$

$$(\mathbf{K}_3) \quad \mathring{R} = R, \qquad k_\tau(R) = R$$

$$(\mathbf{K}_3) \quad \overset{\circ}{\widehat{\mathring{M}}} - \mathring{M}, \qquad k_\tau(k_\tau(M)) = k_\tau(M)$$

[7] We have given the properties of this operation in two columns, using different but equivalent notations. The second column will later serve to define the notion of "kernel operator."

Proof: $(\mathbf{K}_1)$ follows from the very definition of kernel. Since R is open, $(\mathbf{K}_2)$ also holds; while $(\mathbf{K}_4)$ follows from $\mathring{M}$ being open. $\mathring{M} \cap \mathring{N}$ is open, being the intersection of two open sets, and it is contained in $M \cap N$, by $(\mathbf{K}_1)$. Consequently, $\mathring{M} \cap \mathring{N} \subseteq \overset{\circ}{\widehat{M \cap N}}$. Conversely, $\overset{\circ}{\widehat{M \cap N}}$ is open and contained in both M and N hence

$$\overset{\circ}{\widehat{M \cap N}} \subseteq \mathring{M} \quad \text{and} \quad \overset{\circ}{\widehat{M \cap N}} \subseteq \mathring{N}, \quad \text{whence} \quad \overset{\circ}{\widehat{M \cap N}} \subseteq \mathring{M} \cap \mathring{N}. \blacksquare$$

Apart from any considerations of topology, we shall always call a map $k : P(R) \to P(R)$, verifying conditions $(\mathbf{K}_1)$-$(\mathbf{K}_4)$, a *Kernel operator* of R.

5.12 *For any kernel-operator k, we have:*

$$M \subseteq N \qquad \textit{implies} \qquad k(M) \subseteq k(N).$$

$$k(M) \cup k(N) \subseteq k(M \cup N).$$

Proof: $M \subseteq N$ is equivalent to $M \cap N = M$. By $(\mathbf{K}_2)$, $k(M) \cap k(N) = k(M \cap N) = k(M)$; hence $k(M) \subseteq k(N)$. $\blacksquare$

5.13 *Given any kernel operator k on R there exists one and only one topology τ such that $k(M) = k_\tau(M)$ for all subsets M of R.*

Proof: By $(\mathbf{K}_3)$ the system $\mathfrak{U}_p = \{M : p \in k(M)\}$ is nonempty, by $(\mathbf{K}_2)$ it is a filter basis, and by 5.12 it is even a filter. From $(\mathbf{K}_1)$, we see that $p \in M$ for all $M \in \mathfrak{U}_p$. Let us now assume $M \in \mathfrak{U}_p$. By $(\mathbf{K}_4)$, we have $p \in k(M) = k(k(M))$; hence $k(M) \in \mathfrak{U}_p$. For each $q \in k(M)$ we have $k(M) \in \mathfrak{U}_q$. Thus the systems $\mathfrak{U}_p$ satisfy the Hausdorff neighborhood axioms and consequently, by 5.6, $\tau p = \ulcorner\!\urcorner\mathfrak{U}_p = \mathfrak{U}_p$ defines a topology on R. By 5.10, $p \in k_\tau(M)$ is equivalent to $M \in \tau p = \mathfrak{U}_p$ and hence also equivalent to $p \in k(M)$. Thus $k_\tau(M) = k(M)$. Conversely, this last equation, together with 5.10, implies $\tau p = \{M : p \in k(M)\}$. Hence τ is the only topology for which this last equation holds.

From $M \subseteq M \cup N$ and $N \subseteq M \cup N$ we deduce, using the validity of the first assertion above, that $k(M) \leqslant k(M \cup N)$ and $k(N) \subseteq k(M \cup N)$, thus proving the second assertion. $\blacksquare$

Let M be an arbitrary subset of the space (R, τ). By a *neighborhood of M* we shall mean any set U such that $M \subseteq \mathring{U}$. By $(\mathbf{K}_3)$ R is always a neighborhood of any M, and by $(\mathbf{K}_2)$ and 5.12 we see that the neighborhoods of M constitute a filter, the *neighborhoods filter of the set M*. By $(\mathbf{K}_4)$, this filter must be open.

Let M be an arbitrary subset of the space (R, τ). R itself is a closed set containing M. The intersection of all closed sets containing M is again closed, by 5.3; it is the smallest closed set containing M and is called the *closure* of M. Some authors use *hull* or *closed hull*, as a synonym for closure. The closure of

M is designated by $\bar{M}$ or $h_\tau(M)$. Interior of a set and closure of a set (kernel and hull, respectively) are complementary notions; we have as an immediate consequence of definitions

$$\textbf{5.14} \qquad \bar{M} = \complement\,\overset{\circ}{\widehat{\complement M}} \quad \textit{and} \quad \mathring{M} = \complement\,\widehat{\complement M}$$

M is closed if and only if $M = \bar{M}$.

5.15 $\bar{M}$ *is the set of all points adherent to* M:

$$\bar{M} = \{p : \tau p \wedge \overset{\sqcap}{M} \neq \mathfrak{o}\}$$

Proof: $\tau p \wedge \overset{\sqcap}{M} = \mathfrak{o}$ is equivalent to $\complement M \in \tau p$ and hence equivalent to

$$p \in \overset{\circ}{\widehat{\complement M}}$$

Consequently, $\tau p \wedge \overset{\sqcap}{M} \neq \mathfrak{o}$ is equivalent to

$$p \in \complement\,\overset{\circ}{\widehat{\complement M}} = \mathring{M}. \blacksquare$$

5.16 *The operation of passing from a set to its closure possesses the following properties*[8]:

$(\mathbf{H}_1)$	$M \subseteq \bar{M}$,	$M \subseteq h_\tau(M)$.
$(\mathbf{H}_2)$	$\overline{M \cup N} = \bar{M} \cup \bar{N}$,	$h_\tau(M \cup N) = h_\tau(M) \cup h_\tau(N)$.
$(\mathbf{H}_3)$	$\bar{\varnothing} = \varnothing$,	$h_\tau(\varnothing) = \varnothing$.
$(\mathbf{H}_4)$	$\bar{\bar{M}} = \bar{M}$,	$h_\tau(h_\tau(M)) = h_\tau(M)$.

Proof: The above results follow immediately from 5.11, by taking complements. $\blacksquare$

Independently of any topology, any map $h : \mathfrak{P}(R) \to \mathfrak{P}(R)$ verifying conditions $(\mathbf{H}_1)$-$(\mathbf{H}_4)$ is called a *closure operator* or hull operator). If h is a hull operator, the map k defined by $k(M) = \complement h(\complement M)$ is a kernel operator, and conversely if k is a kernel operator, the map h defined by $h(M) = \complement k(\complement M)$ is a hull operator. The following theorem follows directly from 5.12 and 5.13 on taking complements.

[8] We give the properties of this operation in two columns, using different but equivalent notations. The second column will then serve to define the notion of "closure or hull operator."

5.17 *For any closure (hull) operator h we have:*

$$M \subseteq N \quad \textit{implies} \quad h(M) \subseteq h(N).$$

$$h(M \cap N) \subseteq h(M) \cap h(N).$$

If h is a closure operator on R, then there exists one and only one topology τ *on R such that* $h(M) = h_\tau(M)$ *for all subsets M of R.*

We can, therefore, take the closure properties as axioms characterizing a topological space. Considered from this viewpoint, they are called the Kuratowski Closure Axioms. Using 5.10, the topology τ defined by the closure operator h can be described variously, e.g.,

$$\tau p = \{M : p \in k_\tau(M)\} = \{M : p \notin h_\tau(\complement M)\} = \{M : p \notin h(\complement M)\}.$$

Let $\mathfrak{a}$ be an arbitrary filter in the space (R, τ). By $(\mathbf{K}_2)$, we see that $\mathring{\mathfrak{A}} = \{\mathring{A} : A \in \mathfrak{a}\}$ is a filter basis. By 5.17 $\bar{\mathfrak{A}} = \{\bar{A} : A \in \mathfrak{a}\}$ is also a filter basis. For the filters $\mathring{\mathfrak{a}} = \ulcorner\mathring{\mathfrak{A}}\urcorner$ and $\bar{\mathfrak{a}} = \ulcorner\bar{\mathfrak{A}}\urcorner$ we manifestly have the following results.

5.18 $\mathring{\mathfrak{a}}$ *is an open filter and the sup of all open filters* $\mathfrak{b}$ *where* $\mathfrak{b} \leqslant \mathfrak{a}$.
$\bar{\mathfrak{a}}$ *is a closed filter and the inf of all closed filters* $\mathfrak{b}$ *where* $\mathfrak{b} \geqslant \mathfrak{a}$.
A filter $\mathfrak{a}$ *is open (closed) if and only if* $\mathfrak{a} = \mathring{\mathfrak{a}} (\mathfrak{a} = \bar{\mathfrak{a}})$.

Complements and Exercises

5A Let R be a set of mapping s from set X into Y. Let M be any subset of R and let $h(M)$ be the set of all maps φ verifying the following condition: For any finite set of points $x_1, \ldots, x_n \in X$ there exists a $\psi \in M$ such that $\varphi x_1 = \psi x_1, \ldots, \varphi x_n = \psi x_n$.

Exercise

1. h is a closure operator on R.

2. For every $x \in X$ and $y \in Y$ let $C_{x,y} = \{\varphi : \varphi \in R, \varphi x = y\}$. The system of sets $\mathfrak{C} = \{C_{x,y} : x \in X, y \in Y\}$ is a sub-basis for the topology on R defined by h.

5B Let M be a subset of the topological space R.

Exercise: Starting from M and taking complements and closures we can form at most 13 new sets. There exists a subset M of the real line for which the maximum number 13 is actually attained [24, p. 57].

5C A system of sets $\mathfrak{B}$ is called a *closed basis* of the topological space R if its members are all closed sets and if each closed set $\neq R$ can be represented as an intersection of members of $\mathfrak{B}$.

Exercise: A system of sets $\mathfrak{B}$ is a closed basis for a topology on R if and only if the following two conditions hold:

(a) $\cap \{B : B \in \mathfrak{B}\} = \emptyset$.

(b) Given any two sets $B_1, B_2 \in \mathfrak{B}$ and any point $p \in R$ where $p \notin B_1 \cup B_2$, we can find a third set $B_3 \in \mathfrak{B}$ such that $B_1 \cup B_2 \subseteq B_3$ and $p \notin B_3$.

5D Let $\mathfrak{a}$ be an arbitrary filter in the space (R, τ).

Exercise: There exists a smallest (finest) open filter $\mathfrak{b}$ such that $\mathfrak{a} \leqslant \mathfrak{b}$.

Moreover,

$$\mathfrak{b} = \bigwedge_{A \in \mathfrak{a}} \bigvee_{p \in A} \tau p.$$

If $\mathfrak{a} = \overset{\frown}{M}$, then $\mathfrak{b}$ is identical with the neighborhoods filter of M.

5E Let M be a subset of the topological space R. The set $r(M) = M \cap \overline{\complement M}$ is called the *rim*[9] of M while the set $b(M) = \bar{M} \cap \overline{\complement M}$ is called the *boundary*[9] of M.

Exercise

1. $b(M) = r(M) \cup r(\complement M)$.

2. $r(M)$ is closed if and only if M is expressible as the union of a closed set and an open set.

3. $b(b(M)) \subseteq b(M) \cup b(\overline{\complement M})$; in general we do not have equality.

4. The passage from M to $r(M)$ possesses the following properties:

(a) $r(M) \cap N) = (M \cap r(N)) \cup (N \cap r(M))$.

(b) $r(R) = \emptyset$.

(c) $r(\complement r(\complement M)) \subseteq M$.

5. Conversely, let r be any map from $\mathfrak{P}(R) \to \mathfrak{P}(R)$ verifying conditions (*a*)–(*c*). Then $\bar{M} = M \cup r(\complement M)$ defines a closure operator on R, and the rim associated with this closure operator coincides with r itself [42, p. 58].

5F Let p be a point of the space R and M a subject of R. If $p \in \overline{M \cap \complement p}$ then p is a point of accumulation of M. The set $d(M)$ of all the points of accumulation of M is called the *derived set* of M.

[9] *Translator's note:* The concepts of *rim* and *boundary* of a set are so akin to each other as to cause confusion as to terminology. They are the translator's choice for *Rang* and *Begrenzung* of a set, respectively. Some authors, notably Vaidyanathaswamy, cited in reference [42], use "boundary" for "Rang" and "frontier" for "Begrenzung." Nevertheless we have used *boundary* as a translation for "Begrenzung," for modern usage clearly dictates this—while for "Rang" we have chosen *rim*. No one else uses the word and hence no confusion should result.

Exercise

1. The passage from set to derived set possesses the following properties:

(a) $d(M \cup N) = d(M) \cup d(N)$.

(b) $d(dM) \subseteq M \cup d(M)$.

(c) $d(\emptyset) = \emptyset$.

(d) For all $p \in R$, we have $p \notin d(\{p\})$. We can not in general sharpen (b) to $d(dM) \subseteq d(M)$.

2. Conversely, let *a* mapping $d: \mathfrak{P}(R) \to \mathfrak{P}(R)$ verify conditions (*a*)–(*d*). Then $\bar{M} = M \cup d(M)$ defines a closure operator on R, and the derived set of M defined by this closure operator ($\{p : p \in \overline{M \cap \complement p}\}$) coincides with $d(M)$.

3. $r(M) = M \cap d(\complement M)$. From this equality one can deduce the properties of the rim operator as consequences of the properties of the derivation operator [34, p. 66].

5G A set M is said to be *dense-in-itself* if $M \subseteq d(M)$ (cf. 5F); i.e., whenever each point $p \in M$ is also a point of accumulation of M. A set which is closed and dense-in-itself is called *perfect*. On the real line the subset of rationals and the subset of reals are each dense-in-itself, but neither one is perfect. Every closed interval on the line is however perfect.

Exercise

1. The closure of a set which is dense-in-itself is a perfect set.

2. The union of an arbitrary family of sets, each of which is dense-in-itself, is also dense-in-itself.

A set is called *scattered* or *dispersed* if it contains no nonempty dense-in-itself subset.

Exercise

1. Every subset M of a space can be expressed in the form $M = A \cup B$, where A is dense-in-itself, B is scattered, and $A \cap B = \emptyset$. (Take A to be the union of all sets each of which is dense-in-itself.)

2. Every closed set M can be uniquely expressed in the form $M = A \cup B$ where A is perfect, B is scattered, and $A \cap B = \emptyset$ [34, p. 68].

6. The Order Topology and Metric Spaces

In the first part of this section we assume that the fundamental underlying set R contains at least two elements and is totally ordered by the order $\leqslant$. Then the system $\mathfrak{C}$ of intervals,

$$\mathfrak{C} = \{[\leftarrow, x[\; : x \in R\} \cup \{]x, \rightarrow] : x \in R\}$$

is a covering of R, and hence by 5.9 there exists exactly one topology τ on R having $\mathfrak{C}$ as subbasis. This topology is called the *order topology* on R. If $a < p < b$, then $]a, \rightarrow[$, $]\leftarrow, b[$ and hence $]a, b[\,=\,]a, \rightarrow] \cap [\leftarrow, b[$ are neighborhoods of τp. If p is neither the initial nor terminal element of R, then τp is composed of all sets which contain an interval of form $]a, b[$ with $p \in \,]a, b[$. If p is the initial point of R, the system $\{[\leftarrow, x[\,:\, p < x\}$ forms a basis of τp, while if p is the terminal point of R, the system $\{]x, \rightarrow] : x < p\}$ constitutes a basis of τp. All open intervals $]a, b[$, $]a, \rightarrow]$ and $[\leftarrow, a]$ are also τ-open sets. Likewise all closed intervals $[a, b]$, $[\leftarrow, a]$ and $[a, \rightarrow]$ are also τ-closed sets.

Definition 6a: *A filter* $\mathfrak{a}$ *of* R *is said to be* bounded *if there exist points* $p, q \in R$ *such that* $[p, q] \in \mathfrak{a}$.

If R has a first and last element, then all filters of R are necessarily bounded. Every filter which is convergent under the order topology is bounded.

6.1 *R is conditionally complete if and only if all bounded ultrafilters are convergent under the order topology.*

Proof: Let R be conditionally complete and let $\mathfrak{v}$ be a bounded ultrafilter of R. Then there exist points $p, q \in R$ such that $[p, q] \in \mathfrak{v}$. The set $M = \{x : [x, \rightarrow] \in \mathfrak{v}\}$ is not empty because $p \in M$. Since $[\leftarrow, q] \in \mathfrak{v}$ the point q is an upper bound of M. This implies that M has a sup, s. The inequality $x < s < y$ implies $[x, \rightarrow] \in \mathfrak{v}$, $[y, \rightarrow] \notin \mathfrak{v}$, hence $[\leftarrow, y[\,\in \mathfrak{v}$ and finally $]x, y[\,\in \mathfrak{v}$. Thus $\mathfrak{v} \leqslant \tau s$ and $\mathfrak{v}$ converges to s.

Conversely, let every bounded ultrafilter be convergent in R and let M be a nonempty subset of R possessing an upper bound. Then $\mathfrak{A} = \{[x, \rightarrow] \cap M : x \in M\}$ is a filter basis, which generates a bounded filter $\mathfrak{a} \neq \mathfrak{o}$. Hence there exists a bounded ultrafilter $\mathfrak{v}$ with $\mathfrak{v} \leqslant \mathfrak{a}$ and we may assume that $\mathfrak{v}$ converges to some point s. If $x \in M$ were such that $x > s$ then, since $[x, \rightarrow] \in \mathfrak{v}$ it would follow that $\mathfrak{v} \wedge \tau s \leqslant \overline{[x, \rightarrow]} \wedge \overline{[\leftarrow, x[} = \mathfrak{o}$, but this contradicts $\mathfrak{v} \leqslant \tau s$. Hence s is an upper bound of M. If s' is another upper bound of M and if $s' < s$ then, as above, it would follow that $\mathfrak{v} \wedge \tau s \leqslant \overline{[\leftarrow, s']} \wedge \overline{[s', \rightarrow]} = \mathfrak{o}$, a contradiction. Hence s is the sup of M. The existence of the inf of any set bounded below follows by similar reasoning. ∎

6.2 *The following assertions are equivalent:*

(1) *R is gap-free and conditionally complete.*

(2) *If under the order topology on R, a subset M is both open and closed, then $M = \emptyset$ or else $M = R$.*[10]

Proof: (1)→(2): Let M be a set different from $\emptyset$ and from R. Then there exist points $p \in M$ and $q \in \complement M$ with $p < q$ (if $q < p$, the proof would be analogous

[10] Condition (1) will serve later to characterize the notion of connectedness (Section 14).

to what follows). Let N be the set of all $x \in M$ such that $p \leqslant x < q$ and such that if $p \leqslant y \leqslant x$, then $y \in M$. Since $p \in N$, $N \neq \emptyset$. And since q is an upper bound of N, N possesses a sup s. Every neighborhood belonging to τs contains a point of N and also a point of $\complement M$. Hence $s \in \bar{N} \subseteq \bar{M}$ and $s \in \overline{\complement M}$. If M were both open and closed, we would have $s \in \bar{M} \cap \overline{\complement M} = M \cap \complement M$, a contradiction.

(2)→(1): Suppose $a < b$. If there exists no $p \in R$ such that $a < p < b$, then $[\leftarrow, a]$ equals $[\leftarrow, b[$. Hence the interval $[\leftarrow, a]$ is both open and closed; since $a \in [\leftarrow, a]$ this interval is nonempty and since $a < b$, $[\leftarrow, a] \neq R$. But this is contrary to hypothesis and thus R must be gap-free. Moreover, let M be a nonempty subset, bounded above, and suppose M possesses no sup. Then the set S of all the upper bounds of M is nonempty and not equal to R. Since $p \in \complement S$ there must exist an $m \in M$ such that $p < m$. The interval $[\leftarrow, m[$ is therefore a neighborhood of p and disjoint to S. Hence $\complement S$ is open and S is closed. On the other hand, because M possesses no sup in R, we can associate to each $s \in S$ an $s' \in S$ such that $s' < s$, and $]s', \rightarrow]$ is therefore a neighborhood of s contained in S.

Hence S is also open, contradicting (2). The proof of the existence of the inf of a set bounded below follows by similar reasoning. ∎

Examples

6.I The real line **R** in its usual order, is gap-free and conditionally complete. Its order topology is identical with the standard topology. All bounded ultra filters of **R** are convergent. The empty set and **R** itself are the only subsets of **R** which are both open and closed.

6.II Each closed interval $[a, b]$ of **R** is likewise gap-free and conditionally complete. Since $[a, b]$ possesses a first and last element, each of its ultrafilters is bounded and therefore convergent.

6.III The set of all rational numbers, in its usual order, is gap-free but not conditionally complete. The rational subset $\{x : x < \sqrt{2}\}$ is both open and closed under the order topology.

Along with totally order sets the most important examples of topological spaces are furnished by sets which possess a *metric*.

A metric is an appropriate generalization of the intuitive notion of distance between two points.

Definition 6b: *A* metric *defined over the fundamental set R is a mapping $\delta : R \times R \rightarrow \mathbf{R}$ possessing the following properties:*

($\mathbf{M}_1$) $\delta(p, q) \geqslant 0$ *and* $\delta(p, p) = 0$

($\mathbf{M}_2$) $\delta(p, q) = \delta(q, p)$ (*symmetry*).

($\mathbf{M}_3$) $\delta(p, r) \leqslant \delta(p, q) + \delta(p, r)$ (*triangle inequality*).

($\mathbf{M}_4$) $\delta(p, q) = 0$ *implies* $p = q$.

If δ *only satisfies properties* ($\mathbf{M}_1$)-($\mathbf{M}_3$) *it is called a* pseudometric. *The couple* (R, δ) *is called a* metric (pseudo-metric) *space.*

Thus a pseudometric δ of R associates with each pair of points $p, q \in R$ a nonnegative real number $\delta(p, q)$, called the *distance* between these two points. We can also define a distance between two subsets M and N of R by

$$\delta(M, N) = \inf \{\delta(p, q) : p \in M, q \in N\}$$

Here $\delta(M, N)$ must be considered an element of $\mathbf{R}^+ = \{x : x \in \mathbf{R}, x \geqslant 0\}$, cf. 2.IV. If either M or N is empty, then $\delta(M, N) = +\infty$. Instead of writing $\delta(\{p\}, M)$ we shall simply write $\delta(p, M)$. Accordingly, $\delta(\{p\}, \{q\}) = \delta(p, q)$, i.e. given two points, we can consider the distance between them as if they were one-point sets and this distance between sets is nothing but the old distance between the two given points. If $M \cap N \neq \emptyset$ then $\delta(M, N) = 0$. We can also define the *diameter* of a given set M, as follows:

$$\delta(M) = \sup \{\delta(p, q) : p, q \in M\}$$

Again we must consider $\delta(M)$ as an element of $\mathbf{R}^+$. A set M such that $\delta(M) < +\infty$ is said to be *bounded*. The diameter of the empty set is 0, $\delta(\emptyset) = 0$.

Every pseudometric (and all the more so, every metric) space (R, δ) is a topological space, as the following considerations show:

Let ϵ be a positive real number. Then the set

$$U_\epsilon(p) = \{q : \delta(p, q) < \epsilon\}$$

is called the ϵ-neighborhood of the point p. Whenever we wish to emphasize the pseudometric δ in question, we can write $U_\epsilon^\delta(p)$ rather than $U_\epsilon(p)$. Consider the system $\mathfrak{U}_p$ of all the ϵ-neighborhoods of p. We shall show that the systems $\mathfrak{U}_p$ verify the Hausdorff neighborhood axioms of 5.6 : Let $\epsilon = \min \{\epsilon_1, \epsilon_2\}$. Then

$$U_\epsilon(p) \subseteq U_{\epsilon_1}(p) \cap U_{\epsilon_2}(p)$$

which proves axiom ($\mathbf{U}_1$).($\mathbf{U}_2$) follows from the fact that $\delta(p, p) = 0$. If $q \in U_\epsilon(p)$, then $\epsilon_1 > 0$, where $\epsilon_1 = \epsilon - \delta(p, q)$ by definition. Hence, if $r \in U_{\epsilon_1}(q)$ we can deduce, using ($\mathbf{M}_3$), that

$$\delta(p, r) \leqslant \delta(p, q) + \delta(q, r) < \delta(p, q) + \epsilon_1 = \epsilon,$$

i.e., $\tau \in U_\epsilon(p)$. Thus $U_{\epsilon_1}(q) \subseteq U_\epsilon(p)$ and ($\mathbf{U}_3$) holds. By 5.6, $\tau_\delta p = \overline{\mathfrak{U}_p}$ therefore defines a topology τ_δ on R; called the topology on R induced by the pseudometric δ. In this way we can associate with each pseudometric space (R, δ) a unique topological space (R, τ_δ). We shall often simply write (R, δ) rather than (R, τ_δ); likewise, rather than speak of τ_δ-convergence, τ_δ-closure, etc; we shall simply speak of δ-convergence, δ-closure, etc.

6.3 *The following assertions are valid for any subset M of the pseudometric space (R, δ):*

(1) *The ϵ-neighborhoods of M are δ-open sets.*

(2) *The set $\{p : \delta(p, M) \leqslant \epsilon\}$ is δ-closed and contains $\overline{U_\epsilon(M)}$ as a subset.*

(3) $\bar{M} = \{p : \delta(p, M) = 0\} = \cap \{U_\epsilon(M) : \epsilon > 0\}$ *and* $\mathring{M} = \{p : \delta(p, \complement M) > 0\}$.

Proof: (1) If $p \in U_\epsilon(M)$, then $\epsilon_1 > 0$, where $\epsilon_1 = \epsilon - \delta(p, M)$. Using the triangle inequality, we see that $U_{\epsilon_1}(p) \subseteq U_\epsilon(M)$.

(2) Suppose $\delta(q, M) > \epsilon$ and let $\epsilon_1 = \delta(q, M) - \epsilon$. Then $p \in U_{\epsilon_1}(q)$ and $\delta(p, M) \leqslant \epsilon$ imply the contradiction

$$\delta(q, M) \leqslant \delta(p, M) + \delta(p, q) < \delta(q, M).$$

Hence q cannot lie in the closure of the set $\{p : \delta(p, M) \leqslant \epsilon\}$. This last set is therefore closed and since it contains $U_\epsilon(M)$ it must contain $\overline{U_\epsilon(M)}$.

(3) $p \in \bar{M}$ implies $U_\epsilon(p) \cap M \neq \emptyset$ and hence $\delta(p, M) < \epsilon$ for every $\epsilon > 0$. Thus $\delta(p, M) = 0$. Conversely, if $\delta(p, M) = 0$, each ϵ-neighborhood of p contains a point of M and so $p \in \bar{M}$. Lastly, $\delta(p, M) = 0$ is equivalent to $p \in U_\epsilon(M)$ for all $\epsilon > 0$, hence equivalent to $p \in \cap \{U_\epsilon(M) : \epsilon > 0\}$. These remarks show that $\delta(p, \complement M) > 0$ is equivalent to $p \notin \overline{\complement M}$, i.e., $p \in \complement\overline{\complement M} = \mathring{M}$. ∎

Examples

6.IV If we define $\delta(p, q) = | p - q |$ we obtain a metric on the real line, the *absolute-value metric*. The topology it induces on $\mathbf{R}$ is the standard topology. Whenever we speak of $\mathbf{R}$ as a metric space we mean $\mathbf{R}$ under the absolute-value metric.

6.V The metric space $\mathbf{R}$ is a special case of the more general concept of *normed space:* Let R be a vector space over the field of real or complex numbers. A map which associates to each $x \in R$ a real number $\| x \|$ is called a *norm* on R if the following axioms are verified:

(N_1) $\| x \| \geqslant 0.$

(N_2) $\| \lambda x \| = | \lambda | \, \| x \|$ (λ *real or complex*).

(N_3) $\| x + y \| \leqslant \| x \| + \| y \|$

(N_4) $\| x \| = 0$ *implies* $x = 0.$

A map like that for a norm, but which only verifies $(\mathbf{N}_1)$-$(\mathbf{N}_3)$ is called a *seminorm*. Any seminorm on R defines a pseudo metric on R, by $\delta(x, y) = \| x - y \|$ If this seminorm is really a norm, then $\delta(x, y) = \| x - y \|$ is a metric.

Notes and Exercises

6A In the totally ordered set $(R, \leqslant)$ the following family of sets

$$\mathfrak{C} = \{[\leftarrow, x[\, : x \in R\} \cup \{[x, \rightarrow] : x \in R\}$$

is a subbasis of a topology on R, called the *right-sided order topology*. We can similarly define the *left-sided order topology*.

Exercise

1. The half-open intervals $]a, b[$ are both open and closed in the right-sided order topology.

2. Theorem 6.1 does not hold for one-sided order topologies.

6B *Exercise:* Among the subgroups of the additive group of real numbers, $\mathbf{R}$ is the only one which is an open set in the standard topology. $\mathbf{R}$ is also the only uncountable closed subgroup of itself [24, p. 58].

6C Let $\mathbf{C}$ = set of complex numbers. Then the usual absolute value of complex numbers is a norm of $\mathbf{C}$ (viewed as a real or complex vector space). In n-dimensional (real or complex) space $\mathbf{R}^n$, resp. $\mathbf{C}^n$ the definition

$$\| x \| = \sqrt{| x_1 |^2 + \dots + | x_n |^2} \qquad (x = (x_1, \dots, x_n))$$

also stipulates a norm. Whenever we speak of $\mathbf{R}^n$ and $\mathbf{C}^n$ as topological spaces we shall always mean the space induced by the above so-called Euclidean norm and its associated metric.

6D Let H be the set of all sequences $x = (\xi_n)_{n \in N}$ of real (complex) numbers for which $\sum | \xi_n |^2 < \infty$ (convergence in the usual sense). Let $\bar{a}$ be the conjugate of the complex number a.

Exercise: The following Cauchy-Schwarz inequality always holds.

$$\left| \sum \xi_n \bar{\eta}_n \right|^2 \leqslant \left(\sum \left| \xi_n \right|^2 \right) \left(\sum \left| \eta_n \right|^2 \right)$$

If we define $\| x \| = \sqrt{\sum | \xi_n |^2}$, then $\| x \|$ is effectively a norm and hence $\delta(x, y) = \| x - y \|$ is a metric on H.

The metric space (H, δ) is called Hilbert Space (real or complex, as the case may be).[11] In the real case, the set Q of all sequences $(\xi_n) \in H$ where $0 \leqslant \xi_n \leqslant 1/n$ is called the *unit cube*.

Exercise

The unit cube Q is a closed subset of (H, δ).

6E Consider the vector space of all bounded, real-valued functions φ defined on the interval $[a, b]$ of $\mathbf{R}$. Then

$$\| \varphi \| = \sup \{ | \varphi(t) | : a \leqslant t \leqslant b \}$$

effectively defines a norm over this space and hence $\delta(\varphi, \psi) = \| \varphi - \psi \|$ is a metric over this space. Convergence with respect to this metric is the familiar *uniform convergence of functions*.

The above space contains as vector subspace the set of all bounded Riemann-integrable functions on $[a, b]$. We can define a seminorm and associated pseudometric for these functions by

$$\| \varphi \| = \left[\int_a^b \varphi(t)^2 \, dt \right]^{1/2}, \qquad \delta(\varphi, \psi) = \left[\int_a^b (\varphi t - \psi t)^2 \, dt \right]^{1/2}$$

This seminorm is not a norm because $\delta(\varphi, \psi) = 0$ holds whenever φ and ψ differ on a set of Lebesgue measure zero. Convergence in the sense of the pseudometric δ is none other than *convergence in the mean*, a type of convergence well-known for its use in the theory of Fourier series. Note that the pseudometric δ becomes a metric on the set of all continuous functions over $[a, b]$.

6F *Exercise:* Given a pseudometric δ of R and suppose there exists a total order on R whose order topology coincides with τ_δ. Then δ must be a metric.

Given a metric δ of R, then there exists a total order on R whose induced order topology coincides with τ_δ if and only if at least one of the following three equations holds:

$$\delta(x, y) = \delta(x, z) + \delta(z, y),$$
$$\delta(y, z) = \delta(y, x) + \delta(x, z),$$
$$\delta(z, x) = \delta(z, y) + \delta(y, x).$$

6G We can give a definition of *dimension* for topological spaces by the following inductive procedure [22, 29]:

(a) The empty set, and only this set, shall possess dimension -1.

(b) $\dim_p (R, \tau) \leqslant n$ $(n \in \mathbf{N}, n \geqslant 0)$ if and only if τp has a filter basis $\mathfrak{U}$ such that $\dim b(U) \leqslant n - 1$ for all $U \in \mathfrak{U}$ (cf. 5E).

[11] More generally, a Hilbert Space is defined to be any normed (real or complex) vector space which is complete in the norm-induced topology and whose norm is obtainable from a scalar product.

(c) $\dim(R, \tau) \leqslant n$ if and only if $\dim_p(R, \tau) \leqslant n$ for all $p \in R$.

(d) $\dim_p(R, \tau) = n$ is equivalent to the double condition: $\dim_p(R, \tau) \leqslant n$ but $\dim_p(R, \tau)$ is not less than or equal to $(n - 1)$.

(e) $\dim(R, \tau) = n$ is equivalent to the double condition: $\dim(R, \tau) \leqslant n$ but $\dim(R, \tau)$ not less than or equal to $(n - 1)$.

(f) If $\dim(R, \tau) \leqslant n$ is false for all $n \in \mathbf{N}$, we say $\dim(R, \tau) = +\infty$.

For the real vector spaces $\mathbf{R}^n$ (6C) the topological dimension coincides with the usual algebric dimension.

7. The Lattice of Topologies

All the topologies we mention in this section are to be considered topologies over the same fundamental set R. Let $\mathfrak{T}$ represent the totality of topologies that can be put on R. Then $\mathfrak{T}$ can be ordered, as the following considerations will show.

Definition 7a: *Consider the topologies $\tau_1, \tau_2 \in \mathfrak{T}$. Define $\tau_1 \leqslant \tau_2$ if and only if $\tau_1 p \leqslant \tau_2 p$ for all $p \in R$.*

The fact that this relation is really an order is a direct consequence of its having been defined by means of the very ordering of filters. The relation $\tau_1 \leqslant \tau_2$ is read "τ_1 is finer than τ_2" or "τ_2 is coarser (less fine) than τ_1." For example, on a totally ordered set the order topology is coarser than either the right-order topology or the left-order topology (6A).

7.1 *The following statements are pairwise equivalent:*

(1) $\tau_1 \leqslant \tau_2$.

(2) *Every τ_2-open (τ_2-closed) set is also τ_1-open (τ_1-closed).*

(3) *For interiors (or kernels) we have $k_{\tau_1}(M) \supseteq k_{\tau_2}(M)$.*

(4) *For closures (or hulls) we have $h_{\tau_1}(M) \subseteq h_{\tau_2}(M)$.*

Proof: (1) → (2) If $\tau_1 p \leqslant \tau_2 p$, then $\tau_2 p \leqslant \overbracket{M}$ certainly implies $\tau_1 p \leqslant \overbracket{M}$. Hence every τ_2-open set is also τ_1-open. On taking complements we obtain the proof of the assertion pertaining to closed sets.

(2) → (3): The set $k_{\tau_i}(M)$ is the union of all τ_i-open sets contained in M. Since every τ_2-open set is also τ_1-open, assertion (3) follows immediately.

(3) → (1) : Let $\mathfrak{U}_p = \{M : p \in k_\tau(M)\}$ where $i = 1, 2$. Then $\mathfrak{U}_p^1 \geqslant \mathfrak{U}_p^2$ and by 5.10, $\tau_1 p = \overbracket{\mathfrak{U}_p^1} \leqslant \overbracket{\mathfrak{U}_p^2} = \tau_2 p$. Assertions (3) and (4) are equivalent, by taking complements. $\blacksquare$

7.2 *The ordered set $\mathfrak{T}$ is a complete lattice. Its zero element is the discrete topology, its unit element the trivial topology. In this* lattice of topologies *the operation of taking infs obeys*

$$(\bigwedge \tau_\iota)\, p = \bigwedge (\tau_\iota p).$$

Proof: Let $\{\tau_\iota : \iota \in I\}$ be a nonempty subset of $\mathfrak{T}$. The mapping $\tau : R \to F(R)$ defined by $\tau p = \bigwedge \{\tau_\iota p : \iota \in I\}$ is then a topology: Since $\ulcorner p \urcorner \leqslant \tau_\iota p$ for all $\iota \in I$ we must have $\ulcorner p \urcorner \leqslant \tau p$. Moreover we have (cf. 4a)

$$\bigwedge_{U \in \tau p} \bigvee_{q \in U} \leqslant \bigwedge_{U \in \tau_\iota p} \bigvee_{q \in U} \tau_\iota q \leqslant \tau_\iota p$$

for all $\iota \in I$, and therefore

$$\bigwedge_{U \in \tau p} \bigvee_{q \in U} \tau q \leqslant \bigwedge_{\iota \in I} \tau_\iota p = \tau p$$

From the definition of τ we have $\tau \leqslant \tau_\iota$ for all $\iota \in I$. If furthermore there exists a topology τ' such that $\tau' \leqslant \tau_\iota$ for all ι, then $\tau' p \leqslant \bigwedge \tau_\iota p = \tau p$ for all $p \in R$ and $\tau' \leqslant \tau$. This proves the existence of the inf of any set of topologies and also proves $(\bigwedge \tau_\iota)\, p = \bigwedge (\tau_\iota p)$.

For the discrete topology τ_d, the trivial topology τ_t, the arbitrary topology τ, and each $p \in R$, we have $\tau_d p = \ulcorner p \urcorner \leqslant \tau p \leqslant \ulcorner R \urcorner = \tau_t p$, hence $\tau_d \leqslant \tau \leqslant \tau_t$. Thus τ_d is the zero element and τ_t the unit element of $\mathfrak{T}$. Since the inf of the empty family of topologies must be the unit element we can assert, by 2.7, that $\mathfrak{T}$ is a complete lattice. ∎

The representation given above of the inf of a set of topologies immediately leads to

7.3 *A filter converges to p in the inf topology $\bigwedge \tau_\iota$ if and only if it converges to p in each of the topologies τ_ι.*

We cannot describe the formation of the sup of a set of topologies in terms of the concept of neighborhoodsfilter as simply as we could for the inf of topologies. Nevertheless, we can characterize the sup of topologies using open sets.

7.4 *A subset M of R is open (closed) in the sup topology $\tau = \bigvee \{\tau_\iota : \iota \in I\}$ if and only if it is τ_ι-open (τ_ι-closed) for all $\iota \in I$.*

Proof: Suppose M is τ-open. Since $\tau \geqslant \tau_\iota$, 7.1 implies M is τ_ι-open for all $\iota \in I$. Conversely, let $\mathfrak{S}$ be the family of all subsets S of R which are τ_ι-open in all the topologies τ_ι. Then $\mathfrak{S}$ manifestly verifies conditions (a)-(c) of 5.7. Hence there exists a unique topology τ' on R, having $\mathfrak{S}$ as the system of its open sets. By 7.1, we have $\tau' \geqslant \tau_\iota$ for all $\iota \in I$ and therefore $\tau' \geqslant \tau$. On the other hand, every τ-open set belongs to $\mathfrak{S}$ and is thereby τ'-open, i.e., $\tau' \leqslant \tau$ also. If in this proof we systematically replace the word "open" by the word "closed" we obtain a proof of the assertion in parentheses. ∎

Notes and Exercises

7A *Exercise:* For the topology τ^k defined in 4.III we have:

$$\aleph_0 \leqslant k_1 < k_2 \leqslant |R| \qquad \text{implies} \qquad \tau^{k_2} < \tau^{k_1}$$

7B *Exercise:* The order topology on a totally ordered set is the sup of the right-sided and left-sided order topologies (cf. 6A).

7C *Exercise:* The lattice of all topologies of a fundamental set R is atomic whenever R contains at least two points. There exist maximal elements in the subset of all nontrivial topologies. This subset also contains chains, having the trivial topology as their sup.

7D *Given the following families of subsets of* **R**:

$$\mathfrak{S}_1 = \{\emptyset, [-1, 0], \mathbf{R}\}, \qquad \mathfrak{S}_2 = \{\emptyset, [0, +1], \mathbf{R}\}$$

$$\mathfrak{S}_3 = \{\emptyset, \{-2\}, \{0\}, \{+2\}, \mathbf{R}\}$$

Exercise: Each of these families can be taken as the open sets defining a topology on **R**. Using these topologies, one can show that the lattice of all topologies is not necessarily distributive.

8. Axioms of Separation

As we already have seen in Section 4, in a topological space a filter can certainly converge to several different points as limits. In order to guarantee uniqueness of limits it is therefore necessary to add further axiomatic conditions, called *axioms of separation*. We have already embodied in our definition of topological space such a condition—namely, the principal filter $\ulcorner p \urcorner$ generated by the single point p must converge to a unique limit, i.e., $\ulcorner p \urcorner$ has been postulated convergent to p. Proceeding further, we have

Definition 8a: *A topology* τ *is said to be a* T_1-topology *and the space* (R, τ) *a* T_1-space *if the following* Fréchet Separation Axiom *holds:*

$$(\mathbf{T}_1) \quad \ulcorner p \urcorner \leqslant \tau q \quad \textit{implies} \quad p = q$$

T_1-spaces are also called *Fréchet Spaces.* If the fundamental set R contains at least two points, then the trivial topology on R is not a T_1-topology. On the contrary, the discrete topology on any set is always a T_1-topology. If R in finite, then the discrete topology is the only τ_1-topology possible, because in such a case all filters are principal.

8.1 *The following assertions are pairwise equivalent:*

(1) *(R, τ) is a τ_1-space.*

(2) *Given two distinct points p and q, there exists a (open) neighborhood $U \in \tau p$ such that $q \notin U$.*

(3) $\cap \{U : U \in \tau(p)\} = \{p\}$.

(4) *Every subset of R consisting of a single point is closed.*

Proof: (1) → (2): Suppose $q \in U$ for all $U \in \tau p$. Then $\ulcorner q \urcorner \leqslant \tau p$ and by (1) we must have $q = p$. Hence there must exist a $U \in \tau p$ such that $q \notin U$. By 5.5, we can choose U open.

(2) → (3): Since $\ulcorner p \urcorner \leqslant \tau p$ we have $p \in \cap \{U : U \in \tau p\}$. By (2) we have $q \notin \cap \{U : U \in \tau p\}$ for all $q \neq p$.

(3) → (4): If $p \neq q$, then $p \notin \cap \{U : U \in \tau q\}$ and hence $\tau q \wedge \ulcorner p \urcorner = \mathfrak{o}$. Thus $q \notin \overline{\{p\}}$ and $\overline{\{p\}} \neq p$.

(4) → (1): $\ulcorner p \urcorner \leqslant \tau q$ implies $\tau q \wedge \ulcorner p \urcorner \neq \mathfrak{o}$, hence $q \in \overline{\{p\}} = \{p\}$ and therefore $q = p$. ∎

8.2 *If $\tau' \leqslant \tau$ and τ is a T_1-topology, then so is τ'. The sup of any number of T_1-topologies is again a T_1-topology.*

Proof: $\tau' \leqslant \tau q$ and $\ulcorner p \urcorner \leqslant \tau' q$ imply $\ulcorner p \urcorner \leqslant \tau q$. If τ is a T_1-topology, then $p = q$ and hence τ' is also a T_1-topology. By 8.1, one-point sets are closed in any T_1-topology. They are also closed in the sup topology of any family of T_1-topologies, using 7.4, and hence the sup topology of T_1-topologies is itself a T_1-topology. ∎

Definition 8b: *A topology τ is called a* T_2-topology *and the space (R, τ) a* T_2-space *if the following* Hausdorff Separation Axiom *is verified:*

($\mathbf{T}_2$) *If $p \neq q$, then $\tau p \wedge \tau q = \mathfrak{o}$.*

T_2-spaces are also called *Hausdorff spaces* or *separated spaces.* Note that in any topology the nullfilter converges to all points. Ignoring this trival phenomenon we can say that axiom ($\mathbf{T}_2$) holds if and only if all filters $\neq \mathfrak{o}$ converge to at most one point. The Hausdorff axioms guarantees uniqueness of the limit (if the limit exists). The standard topology on the real line is a T_2-topology. The discrete topology on any space is a T_2-topology.

8.3 *Every T_2-topology is a T_1-topology.*

Proof: From $\ulcorner p \urcorner \leqslant \tau q$ we deduce $\ulcorner p \urcorner \leqslant \tau p \wedge \tau q$, hence $\tau p \wedge \tau q \neq \mathfrak{o}$. If τ is a T_2-topology, we therefore have $p = q$. ∎

The converse is not true. There exists T_1-topologies which are not T_2-topologies: The topologies τ^k defined in 4.III are examples of this phenomenon.

8.4 *The following statements are pairwise equivalent:*

(1) *(R, τ) is a T_2-space.*

(2) *Any two distinct points of R possess disjoint (open) neighborhoods.*

(3) *If $p \neq q$, there exists a (open) neighborhood $U \in \tau p$ such that $q \notin \bar{U}$.*

Proof: $\tau p \wedge \tau q = \mathfrak{o}$ is equivalent to the existence of neighborhoods $U \in \tau p$ and $V \in \tau q$ such that $U \cap V = \emptyset$. This last condition is itself equivalent to $\tau q \wedge \lceil U \rceil = \mathfrak{o}$, and hence to $q \notin \bar{U}$. By 5.5, all neighborhoods can be taken to be open. ∎

The only T_2-topology we can put on a finite set is the discrete topology, because this result already holds for τ_1-topologies. Note that if $\tau' \leqslant \tau$ and τ is a T_2-topology, then so is τ': for $p \neq q$ implies $\tau' p \wedge \tau' q \leqslant \tau p \wedge \tau q = \mathfrak{o}$. Contrary to the case for T_1-topologies, the sup of a set of T_2-topologies need not be a T_2-topology, as the following example shows.

8.I Let p_1 and p_2 be two distinct points belonging to the infinite set R. For $i = 1, 2$ let the system $\mathfrak{S}_i$ consist of all those subsets S of R for which $p_i \in \complement S$ or $|\complement S| < \aleph_0$. Each of these systems satisfies the hypotheses of 5.7. Hence there exist topologies τ_i of R, having the $\mathfrak{S}_i$ as their system of all open sets. For $q \neq p_i$ we have $\tau_i q = \lceil q \rceil$. The neighborhoods filter $\tau_i p_i$ is composed of all sets containing p_i and having finite complement. This immediately implies that τ_1 and τ_2 are both T_2-topologies. Their sup $\tau = \tau_1 \vee \tau_2$ is, however, not a T_2-topology: for if it were, then there would exist neighborhoods $U \in \tau p_1$ and $V \in \tau p_2$ with $U \cap V = \emptyset$. Since $\tau_1 \leqslant \tau$, we have $U \in \tau_1 p_1$, and therefore $|\complement U| < \aleph_0$. Likewise, $|\complement V| < \aleph_0$. This implies

$$|R| = |\complement U \cup \complement V| \leqslant |\complement U| + |\complement V| < \aleph_0$$

which contradicts the earlier assumption that R was infinite.

Besides the above axioms, so important for the theory of convergence of filters, there exist many situations in which other conditions of separation are important enough to be axiomatized.

Definition 8c: *A topology τ is said to be a T_3-topology and (R, τ) a T_3-space whenever the following separation axiom holds:*

(T_3) $\overline{\tau p} = \tau p$; *i.e., each neighborhoods filter is closed and therefore possesses a closed basis.*

A space which is both a T_3-space and a T_1-space is said to be *regular*; a similar definition holds for regular topologies.[12]

[12] Some authors occasionally call T_3-spaces regular, not insisting upon the (T_1)-condition.

8.5 *The following assertions are pairwise equivalent:*

(1) *(R, τ) is a T_3-space.*

(2) *Given any $U \in \tau p$, there exists a (open) neighborhood $V \in \tau p$ such that $\bar{V} \subseteq U$.*

(3) *Given any closed set A and any point $p \notin A$ then there exist open sets U and V such that $p \in U$, $A \subseteq V$ and $U \cap V = \emptyset$.*

Proof: The equivalent of (1) and (2) follows directly from the definitions.

(2) $\rightarrow$ (3): $\complement A$ is an open set containing p and therefore $\complement A \in \tau p$. There then exists an open neighborhood $U \in \tau p$ such that $\bar{U} \subseteq \complement A$. $V = \complement \bar{U}$ is open, $A \subseteq V$, and $U \cap V = \emptyset$.

(3) $\rightarrow$ (2): $U \in \tau p$ implies $\overset{\circ}{U} \in \tau p$. $\complement(\overset{\circ}{U})$ is a closed set not containing p and hence there exist open sets V and W such that $p \in V$, $\complement(\overset{\circ}{U}) \subseteq W$, and $V \cap W = \emptyset$. Since $p \in V$ we have $V \in \tau p$. In addition, $\bar{V} \subseteq \complement W \subseteq \overset{\circ}{U} \subseteq U$. ∎

The third assertion of this theorem can be reformulated with the aid of the notion of neighborhood of a set, as follows: Every closed set and every point not in this closed set possess disjoint neighborhoods. In this reformulation $(\mathbf{T}_3)$ truly appears as a separation condition. The standard topology of the real line is an example of a regular topology. There are, however, T_3-spaces which are not regular, i.e., not T_1-spaces; e.g., the trivial topology is always a T_3-topology but if the fundamental set R contains more than two points it cannot be a T_1-topology. Conversely, there exist T_2-spaces which are not T_3-spaces (cf. 8C). This means that axioms $(\mathbf{T}_2)$ and $(\mathbf{T}_3)$ are independent of each other. However, the following theorem shows that $(\mathbf{T}_3)$ and $(\mathbf{T}_1)$, taken together, imply $(\mathbf{T}_2)$.

8.6 *In a T_3-space the condition $\tau p \wedge \tau q \neq \mathfrak{o}$ implies $\tau p = \tau q$. Every regular space is a T_2-space.*

Proof: $\tau p \wedge \tau q \neq \mathfrak{o}$ implies $q \in \bar{U}$ for every $U \in \tau p$, hence $\ulcorner q \urcorner \leqslant \ulcorner \tau p \urcorner = \tau p$. Since τp is open we must have $\tau q \leqslant \tau p$. But our hypotheses are symetrical in p and q and we therefore also have $\tau p \leqslant \tau q$, and hence $\tau p = \tau q$. If $(\mathbf{T}_1)$ holds, then p and q must be identical, i.e., $(\mathbf{T}_2)$ holds. ∎

8.7 *The inf τ of any family of T_3-*topologies *τ_ι is itself a T_3-topology. If at least one of the τ_ι is regular, then so is τ.*

Proof: The filters $\tau_\iota p$ are τ_ι-closed and since $\tau \leqslant \tau_\iota$ they are also τ-closed (7.1). By 5.4, $\tau p = \bigwedge \tau_\iota p$ is therefore a τ-closed filter also. Moreover, $\tau \leqslant \tau_\iota$ implies that if the τ_ι are T_1-topologies, then so is τ. ∎

Axiom $(\mathbf{T}_2)$ separates two points while axiom $(\mathbf{T}_3)$ separates a point and a closed set. The following axioms will serve to separate sets themselves.

Definition 8d: *A topology τ is said to be a T_4-topology and the space (R, τ) a T_4-space if the following axiom is verified:*

$(\mathbf{T}_4)$ *Any two disjoint closed sets possess disjoint neighborhoods.*
A space is said to be normal *if it is both a T_4- and a T_1-space.*

In other words, axiom $(\mathbf{T}_4)$ says the following: If A and B are closed sets and $A \cap B = \emptyset$, then there exist sets U and V such that $A \subseteq \mathring{U}$, $B \subseteq \mathring{V}$, and $U \cap V = \emptyset$. The following example serves to show that $(\mathbf{T}_4)$ does not imply any of the axioms $(\mathbf{T}_1)$–$(\mathbf{T}_3)$.

8.II Let the fundamental set R consist of two points: $R = \{p, q\}$ and let $\tau p = \ulcorner p \urcorner$ and $\tau q = \ulcorner R \urcorner$ define a topology τ on R. This topology is evidently not a T_1-topology and, *a fortiori*, not a T_2-topology. Since $\overline{\tau p} = \ulcorner R \urcorner$, τ is also not a T_3-topology. But τ is, however, a T_4-topology: Since every nonempty closed set contains the point p, if we are given two disjoint closed sets at least one of them must surely be empty. But then $\emptyset$ and R can be taken as disjoint neighborhoods of these two closed sets.

Conversely, there exist regular spaces which are not T_4-spaces (cf. 8D). Hence $(\mathbf{T}_4)$ cannot be a consequence of axioms $(\mathbf{T}_1)$-$(\mathbf{T}_3)$. However, $(\mathbf{T}_4)$ and $(\mathbf{T}_1)$ taken together do imply axioms $(\mathbf{T}_2)$ and $(\mathbf{T}_3)$.

8.8 *Every normal space is regular.*

Proof: In a normal space every one-point set is closed (8.1); axiom $(\mathbf{T}_3)$ then follows immediately from $(\mathbf{T}_4)$. ∎

8.9 *A space is a T_4-space if, and only if, given any closed set A and any neighborhood U of A, we can find a (open) neighborhood V of A such that $\bar{V} \subseteq U$.*[13]

Proof: Let us first assume that $(\mathbf{T}_4)$ holds. It follows from $A \subseteq \mathring{U}$, that A and $\complement \mathring{U}$ are disjoint, closed sets. Hence there exist (open) neighborhoods V of A and W of $\complement \mathring{U}$ such that $V \cap W = \emptyset$. This implies $\bar{V} \subseteq \complement \mathring{W} \subseteq U \subseteq \mathring{U}$.

To prove the converse, let us suppose the conditions of the theorem to hold. If then A and B denote two disjoint closed sets, $\complement B$ must be a neighborhood of A, and, by hypothesis, there exists a neighborhood U of A such that $\bar{U} \subseteq \complement B$. The set $V = \complement \bar{U}$ is a neighborhood of B and we have $U \cap V = \emptyset$. ∎

Two sets A and B are said to be *separated* whenever $A \cap \bar{B} = \bar{A} \cap B = \emptyset$. With the aid of this concept, we can sharpen $(\mathbf{T}_4)$ as follows:

Definition 8e: *A topology τ is called a T_5-topology and the space (R, τ) a T_5-space if the following separation condition is verified:*

$(\mathbf{T}_5)$ *Any two separated sets possess disjoint neighborhoods. A space is said to be* completely normal *if it is both a T_5-space and a T_1-space; its topology is likewise said to be completely normal.*

[13] For a characterization of T_4-spaces using the theory of coverings, cf. 13.16.

There exist normal spaces which are not T_5-spaces (cf. 8E). Hence $(\mathbf{T}_5)$ cannot be a consequence of axioms $(\mathbf{T}_1)$-$(\mathbf{T}_4)$. Conversely $(\mathbf{T}_5)$ does not imply the axioms $(\mathbf{T}_1)$-$(\mathbf{T}_3)$; the space defined in 8.II is even a T_5-space. Since, however, any two closed sets are always separated, we have

8.10 *Every T_5-space is a T_4-space; every completely normal space is normal.*

We can summarize the relations between the diverse separations axioms as follows: In full generality only two implications are valid

$$(\mathbf{T}_5) \to (\mathbf{T}_4) \quad \text{and} \quad (\mathbf{T}_2) \to (\mathbf{T}_1).$$

If, however, we assume that $(\mathbf{T}_1)$ holds, then the following chain of implications also holds

$$(\mathbf{T}_5) \to (\mathbf{T}_4) \to (\mathbf{T}_3) \to (\mathbf{T}_2) \to (\mathbf{T}_1).$$

We shall finish this section with an examination of the separation properties of the two special cases—order topology and metric space.

8.11 *The order topology on any totally ordered set is a completely normal topology.*

Proof: Let $(R, \leqslant)$ be a totally ordered set with order topology τ. If $p < q$, then $[\leftarrow, q[$ must be a neighborhood belonging to τp and not containing q. Likewise, $]p, \to]$ is a member of τq not containing p. Hence τ is at least a T_1-topology.

Let now A and B be two separated sets and for each $a \in A$ define

$$S_a = \{x : [x, a] \cup [a, x] \subseteq \complement \bar{B}\}.$$

Then $a \in S_a$ and $S_{a_1} \cap S_{a_2} \neq \emptyset$ implies $S_{a_1} = S_{a_2}$. In order to complete the proof of our theorem we need two lemmas whose proof we shall give subsequently.

Lemma 1. *Given any $b \in B$ we can find a neighborhood $W \in \tau b$ such that W has a nonempty intersection with at most two of the sets S_a.*

Lemma 2. *Given any set S_a we can find an open set T_a such that*

$$A \cap S_a \subseteq T_a \subseteq S_a \quad \text{and} \quad B \subseteq \complement \bar{T}_a.$$

Then the following two sets are manifestly open:

$$U = \cup \{T_a : a \in A\} \quad \text{and} \quad V = \complement \overline{\cup \{T_a : a \in A\}}$$

We also have

$$A = \cup \{A \cap S_a : a \in A\} \subseteq \cup \{T_a : a \in A\} = U.$$

For each $b \in B$ we can find, by Lemma 1, a neighborhood $W \in \tau b$ which has a nonempty intersection with at most two of the sets S_a. W therefore also has a nonempty intersection with at most two of the T_a. But $B \subseteq \complement \bar{T}_a$ and therefore we can find a neighborhood of b which is disjoint from all the sets T_a. This means that $B \subseteq \complement \bar{U} = V$, and shows that τ is a T_5-topology.

Proof of Lemma 1: Since $B \cap \bar{A} = \emptyset$, for each $b \notin B$ we can find an $x \in R$ such that $b < x$ and $[b, x[\cap A = \emptyset$ (save when b is an endpoint of R). Let $a_1, a_2 \in A$ and suppose $a_1 < a_2$. Suppose further that $s_1 \in S_{a_1} \cap [b, x[$ and that $s_2 \in S_{a_2} \cap [b, x[$. If $a_1 < b$, then $b \in [a_1, s_1] \subseteq S_{a_1}$, but this contradicts $S_{a_1} \subseteq \complement \bar{B}$. Hence we must have $s_2 < a_1 < a_2$, and consequently $a_1 \in S_{a_1} \cap S_{a_2}$. Therefore $S_{a_1} = S_{a_2}$; i.e., $[b, x[$ can have a nonempty intersection with at most one of the sets S_a. Likewise (if b is not an initial point) there exists a $y < b$ such that $]y, b]$ has a nonempty intersection with at most one of the S_a. Then $]y, x]$ is a neighborhood in the filter τb possessing the property being sought. If b is an initial or end point, instead of the neighborhood $]y, x[$ we can use $[b, x[$ or $]y, b]$, respectively.

Proof of Lemma 2: If $x \in S_a$, then $x \in \complement \bar{B}$, and hence there exists an open interval $J \in \tau x$ with $J \subseteq \complement \bar{B}$. For each $p \in J$ we have

$$[a, p] \cup [p, a] \subseteq [a, x] \cup [x, a] \cup [x, p] \cup [p, x] \subseteq \complement \bar{B},$$

which implies $J \subseteq S_a$. Hence S_a is an open set. If there exists an $x \in S_a$ such that $A \cap S_a = [\leftarrow, x[$, define $M_a = S_a \cap [\leftarrow, x[$; otherwise let $M_a = S_a$. If likewise there exists a $y \in S_a$ such that $A \cap S_a \subseteq]y, \rightarrow]$, let $N_a = S_a \cap]y, \rightarrow]$; otherwise take $N_a = S_a$. In all possible cases, the set $T_a = M_a \cap N_a$ is open and $A \cap S_a \subseteq T_a \subseteq S_a$.

Let $b \in B$ and assume $a < b$. Since $b \notin \bar{A}$ there exists a $z \in R$ such that $]z, b] \cap A = \emptyset$. If in addition $b \in \bar{S}_a$, then $z \in S_a$ and $A \cap S_a \subseteq [\leftarrow, z]$. Consequently, $M_a = S_a \cap [\leftarrow, x[$ for some $x \in S_a$. If $b \leqslant x$ we have the contradiction $b \in [a, x] \subseteq S_a \subseteq \complement \bar{B}$. Hence $]x, \rightarrow]$ is a neighborhood of b and disjoint from T_a. In any case we always have $b \in \complement \bar{T}_a$. We get similar conclusions if we suppose $b < a$. ∎

8.12 *A pseudometric space (R, δ) is a T_1-space if and only if δ is a metric. Every pseudometric space is a T_3-, T_4- and T_5-space. Every metric space is completely normal.*

Proof: $\delta(p, q) = 0$ is equivalent to $\overline{p} \leqslant \tau_\delta q$. Hence if one of these relations implies $p = q$, so does the other one.

Since $\overline{U_\epsilon(p)} \leqslant U_{2\epsilon}(p)$ we see that $\{\overline{U_\epsilon(p)} : \epsilon > 0\}$ is a basis of $\tau_\epsilon p$. Hence $\overline{\tau_\delta p} = \tau_\delta p$ and τ_δ is a T_3-topology. Let us moreover assume A and B to be separated sets in (R, δ). Since $A \cap \bar{B} = \emptyset$, given any $a \in A$ there exists $\epsilon_a > 0$

such that $U_{2\epsilon_a}(a) \cap B = \emptyset$. Similarly, since $\bar{A} \cap B = \emptyset$, given any $b \in B$ we can find an $\epsilon_b > 0$ such that $U_{2\epsilon_b}(b) \cap A = \emptyset$. Let

$$V = \bigcup_{a \in A} U_{\epsilon_a}(a) \qquad \text{and} \qquad W = \bigcup_{b \in B} U_{\epsilon_b}(b).$$

The sets V and W are open and $A \subseteq V$, $B \supseteq W$. Moreover $V \cap W = \emptyset$: for $p \in V \cap W$ implies $p \in U_{\epsilon_a}(a) \cap U_{\epsilon_b}(b)$ for suitable $a \in A$ and $b \in B$, and this in turn implies

$$\delta(a, b) \leqslant \delta(a, p) + \delta(p, b) < \epsilon_a + \epsilon_a \leqslant 2 \max \{\epsilon_a, \epsilon_b\},$$

hence $a \in U_{2\epsilon_b}(b)$ or $b \in U_{2\epsilon_a}(a)$, contradicting the very way in which ϵ_a and ϵ_b were chosen. Hence τ_δ is also a T_5-topology and, even more so, a T_4-topology. If δ is a metric, then $(\mathbf{T}_1)$ is also verified; i.e., every metric space is completely normal. ∎

Notes and Exercises

8A A weaker axiom than the Fréchet Separation Axiom is the following *Kolmogorov Separation Axiom.*

$(\mathbf{T}_0)$ *Given any two distinct points, then at least one of them possesses a neighborhood not containing the other point.*

Exercise: $(\mathbf{T}_0)$ and $(\mathbf{T}_3)$, taken together, imply $(\mathbf{T}_2)$.

8B *Exercise:* Describe the sup of all T_1-topologies on a given fundamental set R. By 8.2, we know that this sup must itself be a T_1-topology.

8C (Example of a T_2-topology which is not a T_3-topology) Let τ be the standard topology on the real line and let $\mathfrak{B}$ be the family of all subsets B of $\mathbf{R}$ for which there exists a open set U with $| U \cap \complement B | \leqslant \aleph_0$.

Exercise

1. $\mathfrak{B}$ is the basis of a topology τ' of $\mathbf{R}$, and $\tau' < \tau$. If τ is a T_2-topology then so is τ' (in the sense of 4.III we have $\tau' = \tau \wedge \tau^{\aleph_1}$).

2. The set $A = \{1/n; n \in \mathbf{N}\}$ is τ'-closed and does not contain the number 0.

3. In the τ' topology the set A and the set $\{0\}$ do not possess any disjoint neighborhoods.

Hence τ' cannot be a T_3-topology nor, *a fortiori*, a (T_4)- or (T_5)-topology, because $(\mathbf{T}_3)$ follows from $(\mathbf{T}_2)$ and $(\mathbf{T}_4)$ or $(\mathbf{T}_2)$ and $(\mathbf{T}_5)$.

8D (Example of a regular topology which does not satisfy (T_4)). Let τ again represent the standard topology of $\mathbf{R}$. For each $x \in [0, \frac{1}{2}]$ let

$$U_n(x) = \{x\} \cup \bigcup_{k=n}^{\infty} \left] x + k - \frac{1}{k}, x + k + \frac{1}{k} \right[$$

and $\mathfrak{U}(x) = \{U_n(x) : n \in N\}$ be the filter basis formed by these sets. Then

$$\tau' x = \begin{cases} \tau x \\ \overline{\mathfrak{U}_x} \end{cases} \quad \text{for} \quad \begin{matrix} x \notin [0, \frac{1}{2}] \\ x \in [0, \frac{1}{2}] \end{matrix}$$

defines a new topology τ' on **R**.

Exercise: τ' is a regular topology.

Let A be the set of all rationals on $[0, \frac{1}{2}]$ and B the set of all irrationals on $[0, \frac{1}{2}]$.

Exercise: A and B are τ'-closed.

However the sets A and B do not possess any disjoint τ'-neighborhoods: For suppose U and V were disjoint neighborhoods of A and B. Then it is possible to inductively define a sequence $(x_n)_{n\in\mathbf{N}}$ of real numbers and a sequence $(r_n)_{n\in N}$ of natural numbers such that the following conditions are verified:

(a) $x_1 = \frac{1}{4}$ and $U_{r_1}(x_1) \subseteq U$.

(b) $x_n \in A$ $(x_n \in B)$ implies $x_{n+1} \in B$ $(x_{n+1} \in A)$.

(c) $U_{r_n}(x_n) \subseteq U$ if $x_n \in A$; $U_{r_n}(x_n) \subseteq V$ if $x_n \in B$.

(d) $| x_{n+1} - x_n | < 1/r_n$.

Exercise

1. Since $U \cap V = \emptyset$, we must have $r_{n+1} \geqslant 2r_n$.

2. The sequence $(x_n)_{n\in\mathbf{N}}$ converges in the standard topology τ to a limit $x \in [0, \frac{1}{2}]$.

3. Contrary to hypothesis, this implies $x \in U \cap V$.

Hence τ' cannot be a T_4-, nor a T_5-topology. (For another formulation of the above example and the following one, we refer the reader to [34], p. 80).

8E (Example of a normal space which is not a T_5-space) Let the topology τ' on **R** be the one defined in the previous exercise. Let the system $\mathfrak{A}$ be composed of all τ'-open subsets of **R** which contain all points of $[0, \frac{1}{2}]$, save for possibly a finite number of said points. Then $\mathfrak{A}$ is a filter basis. Then

$$\tau'' x = \begin{cases} \tau' x \\ \overline{-1} \wedge \overline{\mathfrak{A}} \end{cases} \quad \text{for} \quad \begin{matrix} x \neq -1 \\ x = -1 \end{matrix}$$

defines a topology τ'' on **R**.

Exercise

1. In the topology τ'' the number -1 is a point of accumulation of every infinite subset of $[0, \frac{1}{2}]$.

2. τ'' is a normal topology.

3. The sets A and B of the previous exercise are separated in the τ''-topology, but nevertheless do not possess any disjoint neighborhoods.

8F *Exercise:* In a T_1-space (R, τ) the point p is an accumulation point of the set M if and only if every neighborhood of τp contains infinitely many points of M.

8G A system of sets $\mathfrak{B}$ is said to be a ***regular basis*** of the space R if every nonempty open subset M of R can be expressed as a union of sets $B \in \mathfrak{B}$ such that $\bar{B} \subseteq M$.

Exercise

1. Every basis of a T_3-space is regular.

2. Conversely, every space which has a regular basis must be a T_3-space [34, p. 82].

8H *Exercise:* In a T_1-space the derived set of any set is closed (cf. 5F). We have

$$d(d(M)) \subseteq d(M) \qquad \text{and} \qquad d(\overline{M}) = d(M).$$

Show by a counter example that $(\mathbf{T}_1)$ is a necessary hypothesis for this result [34, p. 66].

9. Subspaces

Let M be a subset of R and $\mathfrak{a}$ a filter of R. Then the system of sets

$$\mathrm{Tr}_M(\mathfrak{a}) = \{A \cap M : A \in \mathfrak{a}\}$$

is a filter of M; it is called the *trace* of the filter $\mathfrak{a}$ in M. $\mathrm{Tr}_M(\mathfrak{a})$ is composed of the sets of the filter $\mathfrak{a} \wedge \ulcorner M \urcorner$, less all those sets not contained in M. We therefore see immediately from 3.10 that

$$\mathrm{Tr}_M(\bigvee \mathfrak{a}_i) = \bigvee \mathrm{Tr}_M(\mathfrak{a}_i) \qquad \text{and} \qquad \mathrm{Tr}_M(\bigwedge \mathfrak{a}_i) = \bigwedge \mathrm{Tr}_M(\mathfrak{a}_i)$$
$$\mathrm{Rk}\,(\mathrm{Tr}_M(\mathfrak{a})) \leqslant \mathrm{Rk}\,(\mathfrak{a}).$$

Whenever the subset M, with respect to which the trace is to be taken, has been clearly stipulated and no confusion can arise, we shall simply write $\mathrm{Tr}\,(\mathfrak{a})$ rather than $\mathrm{Tr}_M(\mathfrak{a})$. In order to prove a result about traces of filters, say $\mathrm{Tr}_M(\mathfrak{a})$ or $\mathrm{Tr}_M(\mathfrak{b})$, etc., it is in general sufficient to prove the corresponding result for the filters $\mathfrak{a} \wedge \ulcorner M \urcorner$, $\mathfrak{b} \wedge \ulcorner M \urcorner$, etc.

Let τ again represent a topology of R. For every point $p \in M$ we then have $\ulcorner p \urcorner \leqslant \tau p \wedge \ulcorner M \urcorner$, hence $\ulcorner p \urcorner \leqslant \mathrm{Tr}(\tau p)$. We also have

$$\bigwedge_{U \in \tau p \wedge \ulcorner \overline{M} \urcorner} \bigvee_{q \in U \cap M} (\tau q \wedge \ulcorner M \urcorner) \leqslant \bigwedge_{U \in \tau p} \bigvee_{q \in U} (\tau q \wedge \ulcorner M \urcorner) = \ulcorner M \urcorner \wedge \bigwedge_{U \in \tau p} \bigvee_{q \in U} \tau q \leqslant \tau p \wedge \ulcorner M \urcorner$$

and therefore

$$\bigwedge_{U \in \mathrm{Tr}(\tau p)} \bigvee_{q \in U} \mathrm{Tr}(\tau q) \leqslant \mathrm{Tr}(\tau p).$$

Thus $\tau_M p = \mathrm{Tr}_M(\tau p)$ defines a topology τ_M on M, called the *trace topology* of τ, the (relative) induced *topology of τ on M*, or the *topology induced by τ on M*.

Definition 9a: *Let M be a subset of the topological space (R, τ) and τ_M the topology induced by τ on M. Then the space (M, τ_M) is said to be a* subspace *of (R, τ).*

Rather than speak of the subspace (M, τ_M) of (R, τ) we usually simply speak of the subspace M of R. A subset A of M is said to be *open* (*closed*) in M whenever it is a τ_M-open (τ_M-closed) set. A is open in M if and only if $p \in A$ implies $\tau p \wedge \overset{\sqcap}{M} \leqslant \overset{\sqcap}{A}$. Likewise, A is closed in M if and only if $p \in M$ and $\tau p \wedge \overset{\sqcap}{A} \neq \mathfrak{o}$ imply $p \in A$.

9.1 *Let M be a subset of the space (R, τ). A subset A of M is open (closed) in M if and only if there exists an open (closed) subset S of (R, τ) such that $A = S \cap M$.*

Proof: Let $A = S \cap M$. If S is open, $p \in A$ implies $\tau p \leqslant \overset{\sqcap}{S}$ and then $\tau p \wedge \overset{\sqcap}{M} \leqslant \overset{\sqcap}{S} \wedge \overset{\sqcap}{M} = \overset{\sqcap}{A}$. Conversely, let A be open in M. For any $p \in A$ we can find an open neighborhood $U_p \in \tau p$ such that $U_p \cap M \subseteq A$. Hence

$$S = \cup \{U_p : p \in A\}$$

is an open subset of R, and we have $S \cap M = \cup \{U_p \cap M : p \in A\} = A$. The theorem on closed subsets follows by complementation, using

$$\complement_M(S \cap M) = (\complement_R S) \cap M. \blacksquare$$

9.2 *Let A be a subset of the subspace M of R. The following relationships link the hull and kernel operators of (R, τ) and (M, τ_M):*

$$h_{\tau_M}(A) = h_\tau(A) \cap M \qquad \text{and} \qquad k_\tau(A) = k_{\tau_M}(A) \cap k_\tau(M).$$

Proof: $p \in h_{\tau_M}(A)$ is equivalent to the two conditions, $p \in M$ and

$$(\tau p \wedge \overset{\sqcap}{M}) \wedge \overset{\sqcap}{A} = \tau p \wedge \overset{\sqcap}{A} \neq \mathfrak{o},$$

and hence also equivalent to $p \in h_\tau(A) \cap M$.

If $p \in k_\tau(A)$, it follows that $\tau p \leqslant \overset{\sqcap}{A} \leqslant \overset{\sqcap}{M}$ and from $\tau p \leqslant \overset{\sqcap}{M}$ we have $p \in k_\tau(M)$. Moreover, $\tau p \wedge \overset{\sqcap}{M} \leqslant \overset{\sqcap}{A} \wedge \overset{\sqcap}{M} = \overset{\sqcap}{A}$ and hence $p \in k_{\tau_M}(A)$. To prove the reverse inequality, let $p \in k_{\tau_M}(A) \cap k_\tau(M)$. Then $\tau p \leqslant \overset{\sqcap}{M}$ and consequently $\tau p = \tau p \wedge \overset{\sqcap}{M} \leqslant \overset{\sqcap}{A}$, whence $p \in k_\tau(A)$. $\blacksquare$

Every set open (closed) in R is also open (closed) in M, by 9.1. The converse is, in general, false: for example, M itself is open and closed in M. Hence M cannot be an arbitrary set of R, for such sets need not be open nor closed. Nevertheless we have

9.3 *If M is open (closed) in R, then every set which is open (closed) in M is also open (closed) in R.*

Proof: Let A be open in M. By 9.1 there exists an S, open in R, such that $A = S \cap M$. If now M is also open in R, then the intersection $S \cap M$ is likewise open in R. Similar reasoning will prove the contention concerning closed sets. ∎

9.4 *Every subspace of a T_i-space, where $i = 1, 2, 3$ is itself a T_i-space. In particular, every subspace of a regular space is regular.*

Proof: Let M be a subspace of R. If $p, q \in M$ and $\ulcorner p \urcorner \leqslant \tau_M q$, then $\ulcorner p \urcorner \leqslant \tau q$. Hence if τ is a T_1-topology, we must have $p = q$; i.e., τ_M is also a T_1-topology. If T is a T_2-topology, $p, q \in M$ and $p \neq q$ implies

$$\mathfrak{o} = \tau p \wedge \tau q \geqslant (\tau p \wedge \ulcorner M \urcorner) \wedge (\tau q \wedge \ulcorner M \urcorner),$$

hence $\tau_M p \wedge \tau_M q = \mathfrak{o}$, which in turn implies that τ_M is also a T_2-topology. Finally let τ be a T_3-topology. Then every neighborhoods-filter τp possesses a basis $\mathfrak{U}_p$ made up of τ-closed sets. If $p \in M$, then $\{U \cap M : U \in \mathfrak{U}_p\}$ is a basis of $\tau_M p$, whose member sets are closed in M, by 9.1. Hence $\tau_M p$ is a closed filter in M and τ_M is therefore a T_3-topology. ∎

Separation axiom ($\mathbf{T}_4$) is not preserved on passage to subspaces: Even a normal space can have nonnormal subspaces. In 9.7 we shall show that if all subspaces of a space are T_4-spaces, then the whole space must be a T_5-space. Since there do exist, however, T_4-spaces which are not T_5-spaces (cf. 8E), these spaces must have subspaces which are not T_4-spaces (in 8E, for example, the complement of -1 is such a subspace). But by adding a further hypothesis, we have:

9.5 *Every closed subspace of a T_4-space is again a T_4-space.*

Proof: Let M be a closed subset of the T_4-space (R, τ). If now A and B are two disjoint sets, closed in M, then they must be closed in R (9.3) and hence possess in R the disjoint neighborhoods U and V. The sets $U' = U \cap M$ and $V' = V \cap M$ are then disjoint neighborhoods of A and B in M. ∎

For T_5-spaces, however, we have a more general preservation theorem.

9.6 *Every subspace of a T_5-space is again a T_5-space. The subspaces of a completely normal space are themselves completely normal.*

Proof: Let M be a subspace of the T_5-space R and let the subsets A and B of M be separated, i.e., $A \cap h_{\tau_M}(B) = h_{\tau_M}(A) \cap B = \emptyset$. Since $A \subseteq M$, we have

$$A \cap h_\tau(B) = A \cap M \cap h_\tau(B) = A \cap h_{\tau_M}(B) = \emptyset$$

and similarly $h_\tau(A) \cap B = \emptyset$. Hence there exist open subsets U and V of R such that $A \subseteq U$, $B \subseteq V$, and $U \cap V = \emptyset$. The sets $U' = U \cap M$ and $V' = V \cap M$ are then open in M and possess the same properties as U and V above; i.e., τ_M is also a T_5-topology. ∎

This result tells us that every subspace of a T_5-space is at least a T_4-space. The following converse is true.

9.7 *If every subspace of (R, τ) is a T_4-space, then (R, τ) itself is a T_5-space. If all subspaces are normal, the whole space is completely normal.*

Proof: Let A and B be separated subsets of R. The set $M = \complement(\bar{A} \cap \bar{B})$ is open, while the sets $\bar{A} \cap M$, $\bar{B} \cap M$ are disjoint and, by 9.1, closed in M. By hypothesis the subspace M is a T_4-space and hence there exist in M open sets U and V such that $\bar{A} \cap M \subseteq U$, $\bar{B} \cap M \subseteq V$ and $U \cap V = \emptyset$. But M is open and hence, by 9.3, U and V must even be open in R. From $A \cap \bar{B} = \emptyset$ we see that $A \subseteq \complement\bar{B} \subseteq \complement\bar{B} \cup \complement\bar{A} = \complement(\bar{A} \cap \bar{B}) = M$, and thus $A \subseteq \bar{A} \cap M \subseteq U$. Likewise, $B \subseteq V$. ∎

9.8 *Let $\{\tau_\iota : \iota \in I\}$ be a system of topologies on the fundamental set R, and let M be a subset of R. Then we have*

$$\bigvee (\tau_\iota)_M \leqslant (\bigvee \tau_\iota)_M$$

and

$$\bigwedge (\tau_\iota)_M = (\bigwedge \tau_\iota)_M .$$

Proof: From $\tau \leqslant \tau'$ we see that $\tau_M \leqslant \tau'_M$. Hence $(\tau_{\iota_0})_M \leqslant (\bigvee \tau_\iota)_M$ for all $i_0 \in I$ and thus $\bigvee (\tau_\iota)_M \leqslant (\bigvee \tau_\iota)_M$. For each $p \in M$ we have

$$\bigwedge (\tau_\iota p \wedge \overline{\underline{M}}) = \overline{\underline{M}} \wedge \bigwedge (\tau_\iota p) = \overline{\underline{M}} \wedge (\bigwedge \tau_\iota)\, p.$$

This implies

$$(\bigwedge (\tau_\iota)_M)\, p = \bigwedge ((\tau_\iota)_M)\, p = (\bigwedge \tau_\iota)_M\, p. \ \blacksquare$$

Notes and Exercises

9A Let $(R, \leqslant)$ be a totally ordered set with order topology. Then every subset M of R is also totally ordered by the order relation of R and hence M possesses its own order topology τ'. Nevertheless, this last topology is in general not the same as the topology τ_M induced by τ on M. For instance, let τ be the standard topology on the real line and let $M = \{0\} \cup \{x : |x| > 1\}$. Every τ'-neighborhood of 0 then also contains numbers x with $|x| > 1$, but $\tau_M 0 = \overline{\underline{0}}$.

9B Let δ be a pseudometric of R and M a subset of R. Then the restriction δ' of δ to $M \times M$ is a pseudometric of M.

Exercise: We have $(\tau_\delta)_M = \tau_{\delta'}$.

9C We generally do not have equality in the first formula of 9.8, as the following example shows: Let M, S_1, and S_2 be pairwise disjoint, nonempty subsets of R such that $R = M \cup S_1 \cup S_2$ and let A be a nonempty proper subset of M.

Exercise

1. For $i = 1, 2$ let $\mathfrak{S}_i$ consist of R, ø, and all sets U such that

$$A \cup S_i \subseteq U \subseteq M \cup S_i .$$

Then there exist topologies τ_1 and τ_2 on R whose open sets are $\mathfrak{S}_1$ and $\mathfrak{S}_2$, respectively.

2. A is a $(\tau_1)_M$-open and $(\tau_2)_M$-open set. By 7.4, A is then also $((\tau_1)_M \vee (\tau_2)_M)$-open.

3. Let $\tau = \tau_1 \vee \tau_2$. Then A is not τ_M-open. The only τ_M-open sets are ø and M.

10. Cardinality Conditions

The set of the ranks of all the filters of R is bounded above, for 3.2 tells us $\text{Rk}(\mathfrak{a}) \leqslant 2^{P_z(\mathfrak{a})} \leqslant 2^{|R|}$. For every topology on R there therefore exists a cardinal number k such that $\text{Rk}(\tau p) \leqslant k$ for all $p \in R$. This cardinal number is called the *rank* of the space (R, τ) with respect to the topology τ.

Definition 10a:

$$\text{Rk}(R, \tau) = \text{Rk}(\tau) = \sup \{\text{Rk}(\tau p) : p \in R\}.$$

The stipulation $\text{Rk}(R, \tau) \leqslant \aleph_0$ *is known as the* first countability condition.

Examples

10.I The discrete and the trivial topologies on any fundamental set are of rank 1.

10.II Every pseudometric space (R, δ) satisfies the first countability condition, $\{U^{\delta}_{1/n}(p) : n \in \mathbf{N}\}$ being a countable basis for $\tau_{\delta} p$.

10.III If the fundamental set R is uncountable, the topology $\tau^{\aleph_0}$ (cf. 4.III) cannot have a countable rank: for if $\{U_n : n \in \mathbf{N}\}$ were a basis of the neighborhoods filter of p, and because $|\complement U_n| < \aleph_0$ and $\cap \{U_n : n \in \mathbf{N}\} = p$, we would have

$$|R| = |\{p\} \cup \cup \complement U_n| \leqslant 1 + \sup \{|\complement U_n| : n \in \mathbf{N}\} = \aleph_0 ,$$

a contradiction.

In a space (R, τ), $p \in \overline{M}$ is equivalent to the existence of a filter $\mathfrak{a}$ such that $\mathfrak{o} < \mathfrak{a} \leqslant \overline{M}$ with $\mathfrak{a}$ converging to p (i.e., $\mathfrak{a} = \tau p \wedge \overline{M}$). In general, there does not exist in M any sequence of points converging to p (i.e., Fréchet filter convergent to p). If, however, given any subset M of R and any point $p \in \overline{M}$, we can always find a sequence of points converging to p, then we say that τ is a *sequence topology*.

10.1 *Every topology τ satisfying the first countability condition is a sequence topology.*

Proof: Let $\{U_n : n \in \mathbf{N}\}$ be a basis of τp. From $p \in \bar{M}$ it follows that $U_n \cap M \neq \emptyset$ for all $n \in \mathbf{N}$. Hence there exist points $q_n \in U_n \cap M$, and the Fréchet filter of the sequence $(q_n)_{n \in \mathbf{N}}$ converges to p. ∎

However, there do exist sequence topologies having uncountable rank.

10.2 *τ is a sequence topology if and only if each neighborhoods filter τp is a sup of Fréchet filters.*

Proof: Let us first suppose $\tau p = \bigvee \mathfrak{f}_\iota$, where the $\mathfrak{f}_\iota$ are Frechet filters. From $p \in \overline{M}$ it follows that

$$\mathfrak{o} \neq \overline{M} \wedge \tau p = \overline{M} \wedge \bigvee \mathfrak{f}_\iota = \bigvee (\overline{M} \wedge \mathfrak{f}_\iota).$$

Hence $\mathfrak{f} = \overline{M} \wedge \mathfrak{f}_{\iota_0} \neq \mathfrak{o}$ for at least one index ι_0. This means that $\mathfrak{f}$ is also a Fréchet filter, associated with a sequence of points of M and converging to p.

Conversely, let τ be a sequence topology. Let F_p^* denote the set of all Fréchet filters $\mathfrak{f}$ such that $\mathfrak{f} \leqslant \tau p$. Then $\mathfrak{v} = \bigvee \{\mathfrak{f} : \mathfrak{f} \in F_p^*\} \leqslant \tau p$. If $V \in \mathfrak{v}$ and we suppose that $p \in \overline{\complement V}$, then there must exist an $\mathfrak{f} \in F_p^*$ such that $\mathfrak{f} \leqslant \overline{\complement V}$, which contradicts $\mathfrak{f} \leqslant \overline{V}$. Hence $p \in \mathring{V}$, $V \in \tau p$, and finally $\mathfrak{v} = \tau p$. ∎

Among all the bases (cf. 5C) of a space (R, τ) there are those of minimal cardinality. We call their cardinality the *Basis degree* of (R, τ) or τ.

Definition 10b:

$$\mathrm{Bd}\,(R, \tau) = \mathrm{Bd}\,(\tau) = \min \{|\, \mathfrak{B} \,| : \mathfrak{B}, \text{ a basis of } (R, \tau)\}$$

Any basis $\mathfrak{B}$ for which $|\, \mathfrak{B} \,| = \mathrm{Bd}\,(R, \tau)$ is called a minimal basis. *The stipulation* $\mathrm{Bd}\,(R, \tau) \leqslant \aleph_0$ *is known as the* second countability condition. *Spaces which satisfy the second countability condition are also called* spaces with countable basis.

The basis degree of a space is evidently equal to the minimal cardinality of all the subbases of the space, provided that this cardinality is infinite.

Examples

10.IV The basis degree of the trivial topology is always 1, while that of the discrete topology is the same as the cardinality of the underlying space.

10.V The standard topology of the real line satisfies the second countability condition: The intervals $]a, b[$, where both a and b are rational, constitute a countable basis.

10.VI The space of 10.III does not possess a countable basis. This is a direct consequence of the following theorem.

10.3 $\text{Rk}\,(R, \tau) \leqslant \text{Bd}\,(R, \tau)$. *The second countability condition implies the first.*

Proof: Let $\mathfrak{B}$ be a minimal basis of (R, τ). Since $\mathfrak{B}_p = \{B : B \in \mathfrak{B}, p \in B\}$ is a basis of τp, we have $\text{Rk}\,(\tau p) \leqslant |\mathfrak{B}_p| \leqslant |\mathfrak{B}| = \text{Bd}\,(R, \tau)$ for all $p \in R$, and hence $\text{Rk}\,(R, \tau) \leqslant \text{Bd}\,(R, \tau)$. ∎

10.4 *For every subspace* (M, τ_M) *of* (R, τ) *we have*

$$\text{Rk}\,(M, \tau_M) \leqslant \text{Rk}\,(R, \tau) \qquad \textit{and} \qquad \text{Bd}\,(M, \tau_M) \leqslant \text{Bd}\,(R, \tau).$$

Proof: The first contention is a direct consequence of $\text{Rk}\,(Sd_M(\tau p)) \leqslant \text{Rk}\,(\tau p)$.

Let $\mathfrak{B}$ be a minimal basis of (R, τ). If A is an open set in M then, by 9.1, there exists a set S, open in R, such that $A = S \cap M$. We have $S = \cup\{B : B \in \mathfrak{B}, B \subseteq S\}$ and thus $A = S \cap M = \cup\{B \cap M : B \in \mathfrak{B}, B \subseteq S\}$. Hence $\mathfrak{B}' = \{B \cap M : B \in \mathfrak{B}\}$ is a basis of the subspace (M, τ_M) and we have

$$\text{Bd}\,(M, \tau_M) \leqslant |\mathfrak{B}'| \leqslant |\mathfrak{B}| = \text{Bd}\,(R, \tau). \blacksquare$$

10.5 *Every* T_3*-space (regular space) having a countable basis, is a* T_5*-space (completely normal).*

Proof: Let (R, τ) be a T_3-space with countable basis $\mathfrak{B}$ and let A and B be separated subsets of R. From $p \in \complement \bar{A}$ we deduce, by 8.5, the existence of an open neighborhood $V \in \tau p$ with $\bar{V} \cap \bar{A} = \emptyset$. Since $\mathfrak{B}$ is a basis, we can assume that $V \in \mathfrak{B}$. Because $|\mathfrak{B}| \leqslant \aleph_0$ we have $\complement \bar{A} = \cup \{V_n : n \in \mathbf{N}\}$ with $V_n \in \mathfrak{B}$ and $\bar{V}_n \cap \bar{A} = \emptyset$. Similarly we have $\complement \bar{B} = \cup \{U_n : n \in \mathbf{N}\}$ with $U_n \in \mathfrak{B}$ and $\bar{U}_n \cap \bar{B} = \emptyset$. The sets

$$U = \bigcup_{n \in \mathbf{N}} (U_n \cap \complement \bar{V}_1 \cap \ldots \cap \complement \bar{V}_n) \quad \text{and} \quad V = \bigcup_{n \in \mathbf{N}} (V_n \cap \complement \bar{U}_1 \cap \ldots \cap \complement \bar{U}_n)$$

are open. Since $A \cap \bar{B} = \emptyset$, we have $A \subseteq \complement \bar{B} = \cup U_n$, and since $\bar{V}_n \cap A = \emptyset$ we have $A \subseteq \complement \bar{V}_n$ for all $n \in \mathbf{N}$. This implies $A \subseteq U$. Similarly, we see that $B \subseteq V$. We finally also see that $U \cap V = \emptyset$: for $p \in U \cap V$ implies

$$p \in U_n \cap \complement \bar{V}_1 \cap \ldots \cap \complement \bar{V}_n \qquad \text{and} \qquad p \in V_m \cap \complement \bar{U}_1 \cap \ldots \cap \complement \bar{U}_m$$

for appropriate n and m. Let us say that $n \leqslant m$. Then we must have $p \in U_n$ and $p \in \complement \bar{U}_n$, which is a contradiction. ∎

Definition 10C: *Let A and B be subsets of the space R. Then A is said to be* dense over B *if $B \subseteq \bar{A}$. If A is a subset of B and A is dense over B, then we say A is* dense in B. *If A is dense in R we simply say that A is* dense (*i.e.*, $R = \bar{A}$).[14]

In any space (R, τ) we can consider all dense sets and their cardinalities. The smallest possible cardinality of such sets is called the *separability degree* of (R, τ), or simply of τ.

Definition 10d:

$$\text{Sd}\,(R, \tau) = \text{Sd}\,(\tau) = \min\{| A | : A \textit{ dense in } R\}$$

If $\text{Sd}\,(R, \tau) \leqslant \aleph_0$, we say that the space and topology are *separable.*

Examples

10.VII Under the trivial topology every nonempty set is dense. The separability degree of this topology is therefore always 1. Under the discrete topology, however, the only dense set is the whole space R. Hence the separability degree of the discrete topology in R is equal to $| R |$.

10.VIII The set of all rational numbers is dense in the real line. The real line is therefore separable.

10.IX In the space (R, τ^k) of 4.III all sets whose cardinality is $\geqslant k$ are dense, and only such sets are dense. Hence $\text{Sd}\,(R, \tau^k) = k$.

10.6 *If $\tau \leqslant \tau'$, then* $\text{Sd}\,(\tau) \geqslant \text{Sd}\,(\tau')$. *If τ is separable, so is τ'.*

Proof: By 7.1 (4) every τ-dense set is also τ'-dense. ∎

10.7
$$\text{Sd}\,(R, \tau) \leqslant \text{Bd}\,(R, \tau)$$

Every space with a countable basis is separable.

Proof: Let $\mathfrak{B}$ be a minimal basis of (R, τ). With each nonempty set $B \in \mathfrak{B}$ let us associate a unique point $p_B \in \mathfrak{B}$ and let $A = \{p_B : B \in \mathfrak{B}\}$. Given any $q \in R$ and any $U \in \tau q$ we can find a $B \in \mathfrak{B}$ such that $q \in B$ and $B \subseteq U$. This implies $p_B \in U$, hence $\tau q \wedge \ulcorner A \urcorner \neq o$ and therefore $q \in \bar{A}$. Thus A is dense in R, and we have $\text{Sd}(R, \tau) \leqslant | A | \leqslant | \mathfrak{B} | = \text{Bd}(R, \tau)$. ∎

10.8 *In (R, τ) let M be an open set and A a dense set. Then*

$$\overline{M} = \overline{M \cap A},$$
$$\text{Sd}\,(\tau_M) \leqslant \text{Sd}\,(\tau).$$

[14] This concept is quite different from that of a set being *dense-in-itself* (cf. 5g). Any set A is dense in A, but not, in general, dense-in-itself.

Proof: Since A is dense, every nonempty open set must contain a point of A, and for every filter $\mathfrak{b} \neq \mathfrak{o}$, we have $\mathfrak{b} \wedge \ulcorner A \urcorner \neq \mathfrak{o}$. From $p \in \bar{M}$ it follows that $\tau p \wedge \ulcorner M \urcorner \neq \mathfrak{o}$. And since M is open, $\tau p \wedge \ulcorner M \urcorner$ must be an open filter. Consequently, $\tau p \wedge \ulcorner M \cap A \urcorner = (\tau p \wedge \ulcorner M \urcorner) \wedge \ulcorner A \urcorner \neq \mathfrak{o}$; hence $p \in \overline{M \cap A}$. This proves the desired equality and moreover

$$h_{\tau_M}(M \cap A) = h_\tau(M \cap A) \cap M = h_\tau(M) \cap M = M.$$

Hence $M \cap A$ is dense in the subspace M, and we have

$$\operatorname{Sd}(M, \tau_M) \leqslant |M \cap A| \leqslant A.$$

Since this holds for all sets A dense in R, the second assertion above is also proved. ∎

Definition 10e: *A family $\mathfrak{H}$ of sets is called an open (closed) covering of the subset M of (R, τ) if it is a cover of M and if all its members are open (closed) subsets of (R, τ).*

For example, every basis is an open covering of the whole space.

10.9 *Let M be a subset of (R, τ). Then every open covering $\mathfrak{H}$ of M contains a subcovering $\mathfrak{H}'$ of M such that* $|\mathfrak{H}'| \leqslant \operatorname{Bd}(\tau_M)$.

Proof: Let $\mathfrak{B}$ be a minimal basis of (M, τ_M), and let $\mathfrak{B}'$ be the family of all $B \in \mathfrak{B}$ such that B is contained in at least one set belonging to $\mathfrak{H}$. To each $B \in \mathfrak{B}'$ let us associate a unique $H_B \in \mathfrak{H}$ such that $B \subseteq H_B$. The system of sets $\mathfrak{H}' = \{H_B : B \in \mathfrak{B}'\}$ is such that $\mathfrak{H}' \subseteq \mathfrak{H}$ and $|\mathfrak{H}'| \leqslant |\mathfrak{B}'| \leqslant |\mathfrak{B}| = \operatorname{Bd}(M, \tau_M)$. But $\mathfrak{H}'$ is also a covering of M: given any $p \in M$ there exists an $H \in \mathfrak{H}$ such that $p \in H$. Since $H \cap M$ is open in M, there also exists a $B \in \mathfrak{B}$ such that $p \in B$ and $B \subseteq H$. Hence $B \in \mathfrak{B}'$ and $p \in H_B$. ∎

A space R, such that every one of its open coverings contains a countable subcover, is said to be a *Lindelöf space*. The above theorem implies:

10.10 *Every space with a countable basis is a Lindelof space.*

10.11 *For every pseudometric δ we have* $\operatorname{Sd}(\tau_\delta) = \operatorname{Bd}(\tau_\delta)$. *A pseudometric space is separable if and only if it possesses a countable basis.*

Proof: Let A denote a dense subset of (R, δ) such that $|A| = \operatorname{Sd}(\tau_\delta)$ and let us abbreviate $\operatorname{Sd}(\tau_\delta)$ by s, i.e., put $s = \operatorname{Sd}(\tau, \delta)$. We shall examine the two cases, $s < \aleph_0$ and $s \geqslant \aleph_0$.

Case 1: $s < \aleph_0$

We must have $A = \{a_1, \ldots, a_s\}$ and $R = \bar{A} = \overline{\{a_1\}} \cup \ldots \cup \overline{\{a_s\}}$. The closed sets $A_\sigma = \overline{\{a_\sigma\}}$ are composed of all points $p \in R$ such that $\delta(a_\sigma, p) = 0$. Therefore $A_{\sigma'} \cap A_{\sigma''} \neq \emptyset$ implies $A_{\sigma'} = A_{\sigma''}$. But this cannot occur if $\sigma' \neq \sigma''$, for in that case $s - 1$ of the points a_σ would already be dense in R. Thus the sets A are pairwise disjoint and open, for they are the complements of closed sets. If M is open, and $a_\sigma \in M$, then, since $\delta(A_\sigma) = 0$, we must have $A_\sigma \subseteq M$, and similarly, if $a_\sigma \notin M$, we must have $A_\sigma \cap M = \emptyset$. Thus $\mathfrak{B} = \{A_1, ..., A_s\}$ is a basis of (R, δ) and we have $\text{Bd}(\tau_\delta) \leqslant |\mathfrak{B}| = s = \text{Sd}(\tau_\delta)$, which, using 10.7, implies the equation being sought.

Case 2: $s \geqslant \aleph_0$

The system of sets $\mathfrak{B} = \{U^\delta_{1/n}(a) : a \in A, n \in \mathbf{N}\}$ is composed of open sets. If M is a nonempty open set, for each $p \in M$ we can find an $n \in \mathbf{N}$ such that $U^\delta_{2/n}(p) \subseteq M$. Since A is dense, there exists an $a \in A$ such that $a \in U^\delta_{1/n}(p)$ and consequently $p \in U^\delta_{1/n}(a) \leqslant U^\delta_{2/n}(p) \subseteq M$. Hence $\mathfrak{B}$ is a basis of (R, δ). Using 10.7, and because $s \geqslant \aleph_0$, we have

$$s \leqslant \text{Bd}(\tau_\delta) \leqslant |\mathfrak{B}| \leqslant |A| \cdot \aleph_0 = s \cdot \aleph_0 = s,$$

and thus once again $\text{Bd}(\tau_\delta) = \text{Sd}(\tau_\delta)$. ∎

10.12 *A totally ordered infinite set* $(R, \leqslant)$ *is order-isomorphic to one of the intervals of the real line* [0, 1],]0, 1], [0, 1[, or]0, 1[*if and only if it is gap-free, conditionally complete, and separable in the order topology.*

Proof: Let $(R, \leqslant)$ possess the listed properties and let A be a countable, dense set in the order topology. Then certainly A is infinite. If $a_1, a_2 \in A$ and $a_1 < a_2$ then $]a_1, a_2[$ is nonempty because R has been assumed gap-free. Since A is dense, there exists an $a \in A$ such that $a_1 < a < a_2$, i.e., A is gap-free also. By 2.2, there therefore exists an order isomorphism φ of A onto the rational numbers of [0, 1], or onto one of the rational intervals obtained from [0, 1] by deleting one or both of its end points, as the case may be. Then

$$\psi x = \sup \{\varphi a : a \in A, a \leqslant x\}$$

defines a mapping from R into [0, 1]. If $x < y$ then there exists an $a \in A$ such that $x < a < y$ and in addition, $\psi x < \psi y$; i.e., ψ is strictly monotonic. Since R is conditionally complete, given any $t \in [0, 1]$ (save perhaps $t = 0$ or $t = 1$), we can find in R the point $x = \sup \{a : a \in A, \varphi a \leqslant t\}$. The point x satisfies $\psi x = t$, and hence ψ is a mapping of R onto one of the four intervals listed in the statement of this theorem. By 2.1, ψ is therefore also an order isomorphism. Since every interval of the real line possess the properties we have listed and since these properties are all preserved under order isomorphism, we see that, conversely, they must also represent necessary conditions. ∎

Notes and Exercises

10A For the topology $\tau^{\aleph_0}$ on an uncountable set R we have $\mathrm{Rk}(\tau^{\aleph_0}) > \aleph_0$ (cf. 10.III) yet this is not a sequence topology: if $p \in \bar{M}$, then $p \in M$ or $|M| \geqslant \aleph_0$. In the first case, $(p)_{n \in \mathbf{N}}$ is a sequence in M whose Fréchet filter $\overline{p}$ converges to p. In the second case M contains a sequence all of whose terms are distinct. But under the topology $\tau^{\aleph_0}$ the Fréchet filter of such a sequence converges to every point.

10B *Exercise*

1. If A is dense over B and C is open, then $A \cap C$ is dense over $B \cap C$.
2. If A is dense and B is dense over A, then B is dense.
3. The intersection of a dense open set and a dense set is dense.
4. If A_ι is dense over B for all $\iota \in I$, then $\bigcup A_\iota$ is dense over $\bigcup B_\iota$ [34, p. 109].

10C A set A is said to be *nowhere dense* over B if $\complement \bar{A} \cap B$ is dense in B.

Exercise

1. If the sets $A_1, \ldots, A_n$ are *nowhere dense over* B, then so is their union.
2. If A and B are open, then the boundary $b(A)$ (cf. 5e) is nowhere dense over B.
3. A is nowhere dense over R if and only if $(\mathring{\bar{W}}) = \emptyset$.

A set nowhere dense over R is also said to be a *rare* set. A set is called *meager*, or *of the first category*, if it is the union of a countable number of rare sets. For example, the set of all rational numbers is meager. A set which is not meager is said to be *of the second category*. A space whose only meager open set is the empty set is called a *Baire Space*.

Exercise: The following statements are pairwise equivalent.

1. R is a Baire space.
2. Every meager subset of R has a complement dense in R.
3. The intersection of any countable number of dense open subsets of R is dense.
4. If the union of a countable number of closed subsets of R posses an interior point, then at least one of these closed sets posses an interior point (cf. footnote 2, p. 39) [8].

10D *Exercise:* In a space with a countable basis every system of open (closed) sets which is well-ordered under set-theoretic inclusion must be a countable system [42, p. 77].

10E *Exercise:* In a separable space every system of pairwise disjoint open sets must be countable. The converse is false (topology $\tau^{\aleph_1}$ on an uncountable set; cf. 4.III) [24, p. 60].

10F *Exercise:* In a space with countable basis every uncountable set has at least one point of accumulation [24, p. 48].

10G *Exercise:* Under its right- or left-order topology (cf. 6A) the real line is separable, yet it does not possess a countable basis [24, p. 59].

III

Compactness and Connectedness

The topics treated in this chapter are especially important for applications. They belong to two essentially different categories.

It is often very important to know whether, in a given topological space, there exist a sufficient number of convergent filters. Certain conditions, known as compactness conditions, furnish answers to this question. These conditions can be formulated in two ways; as convergence conditions, or else as conditions pertaining to coverings of a space. Each of these formulations leads in turn to important generalizations; that of a locally compact space and that of a paracompact space.

The second category of questions stems from the fact that certain spaces can be decomposed into two subspaces, whose topological structures are completely independent of each other. The impossibility of such a decomposition leads us directly to the definitions of connectedness and local connectedness.

11. Compactness

The definition already given of a point adherent to a set can be immediately generalized to filters.

Definition 11a: *In a space* (R, τ) *we say that the point* p *is* adherent *to the filter* $\mathfrak{a}$ *if* $\mathfrak{a} \wedge \tau p \neq \mathfrak{o}$.

11.1 *The following assertions are pairwise equivalent:*

(1) p *is an adherent point to* $\mathfrak{a}$.

(2) *There exists a filter* $\mathfrak{b}$ *converging to* p *and satisfying* $\mathfrak{o} < \mathfrak{b} \leqslant \mathfrak{a}$.

(3) $\ulcorner p \urcorner < \bar{\mathfrak{a}}$ (*or, in other words,* $p \in \bigcap \{\bar{A} : A \in \mathfrak{a}\}$.

Proof: If p is adherent to $\mathfrak{a}$, then the filter $\mathfrak{b} = \mathfrak{a} \wedge \tau p$ satisfies condition (2) above. If $\mathfrak{o} < \mathfrak{b} \leqslant \mathfrak{a}$ and $\mathfrak{b} \leqslant \tau p$, then $\tau p \wedge \ulcorner \bar{A} \urcorner \geqslant \mathfrak{b} > \mathfrak{o}$, hence $p \in \bar{A}$ for all

$A \in \mathfrak{a}$. Finally, if $\ulcorner p \urcorner \leqslant \bar{\mathfrak{a}}$, then $p \in \bar{A}$, hence $\tau p \wedge \ulcorner A \urcorner \neq \mathfrak{o}$ for all $A \in \mathfrak{a}$ and thus $\tau p \wedge \mathfrak{a} \neq \mathfrak{o}$. ∎

The particular case where $\mathfrak{a}$ is a principal filter leads us back to the concept of a point adherent to a set: p is adherent to the principal filter $\ulcorner M \urcorner$ if and only if p is adherent to the set M.

In general a filter need possess no adherent points. For example, under the discrete topology every filter is closed, and hence only a filter, the intersection of all of whose member sets is nonempty, will possess adherent points. Compactness conditions guarantee the existence of adherent points—if not always, then at least for certain classes of filters. In the definition which follows k stands for a transfinite cardinal.

Definition 11b: *A space* (R, τ) *and its topology are said to be* completely compact (k-compact) *if every filter* $\mathfrak{a}$ *of* R *where* $\mathfrak{a} \neq \mathfrak{o}$ (*and* $\operatorname{Rk}(\mathfrak{a}) \leqslant k$) *possesses at least one adherent point*[1]).

A space is evidently completely compact if and only if it is k-compact for all k. If $k' \leqslant k$ and a space is k-compact then it is also k'-compact. The concept of k-compactness is only important for transfinite cardinals: From $\operatorname{Rk}(\mathfrak{a}) < \aleph_0$ we deduce that $\mathfrak{a}$ is a principal filter. But a principal filter $\neq \mathfrak{o}$ always possesses an adherent point.

11.2 *A space* (R, τ) *is completely compact if and only if every one of its ultrafilters is convergent.*

Proof: Let (R, τ) be completely compact. Then every ultrafilter $\mathfrak{v}$ possesses an adherent point p. Hence $\mathfrak{v} \wedge \tau p \neq \mathfrak{o}$ and, by 3.12, $\mathfrak{v} \leqslant \tau p$. Conversely, let each ultrafilter converge in (R, τ). From $\mathfrak{o} < \mathfrak{a}$, we deduce the existence of an ultrafilter $\mathfrak{v}$ with $\mathfrak{v} \leqslant \mathfrak{a}$. If $\mathfrak{v} \leqslant \tau p$, then, by 11.1, the point p is adherent to $\mathfrak{a}$. ∎

Examples

11.I Every finite topological space is completely compact because finite sets will only admit of principal filters.

11.II A set R is completely compact under the discrete topology if and only if R is finite: In an infinite set there always exists a filter $\mathfrak{f}$, the sets of which have a nonempty intersection (cf., e.g., 3.I) and hence $\mathfrak{f}$ cannot have any adherent points under the discrete topology. From the same reason, an infinite set under the discrete topology cannot be k-compact for $k \geqslant \aleph_0$.

[1] The terminology here used is due to G. Nöbeling. The first definition of compactness was given by M. Fréchet. The term "compact" is often used as a synonym for our "$\aleph_0$-compact," while "bicompact" and "quasi-compact" are used for our "completely compact." N. Bourbaki reserves the name "compact" exclusively for completely compact T_2-spaces.

11.III The order topology on a totally ordered set $(R, \leqslant)$ is completely compact if and only if R is conditionally complete and possesses an initial and terminal element: For under these assumptions every ultrafilter must be bounded and, by 6.1, also convergent. Conversely, if the initial or terminal element is lacking then the filter bases $\{[\leftarrow, x] : x \in R\}$ or $\{[x, \rightarrow] : x \in R\}$, respectively, generate a filter having no adherent point. If we want the order topology to be completely compact we must guarantee the existence of initial and terminal element and the underlying set must—by 6.1 again—be conditionally complete. Accordingly, every closed interval $[a, b]$ of the real line is completely compact.

11.3 (Bolzano-Weierstrass) *In an $\aleph_0$-compact space every infinite subset possesses at least one point of accumulation. Conversely, a T_1-space, wherein every infinite subset possesses at least one point of accumulation, must be $\aleph_0$-compact.*

Proof: Let M be an infinite subset of the $\aleph_0$-compact space (R, τ). Then there exists an infinite sequence $(a_n)_{n \in \mathbf{N}}$ composed of distinct points of M. Let $\mathfrak{a}$ be the Fréchet filter of this sequence. Then $\text{Rk}(\mathfrak{a}) = \aleph_0$ and hence $\mathfrak{a}$ possesses an adherent point p and thus $\tau p \wedge \mathfrak{a} \neq \mathfrak{o}$. Since the sequence $(a_n)_{n \in \mathbf{N}}$ is composed of distinct points, we have $\mathfrak{a} \leqslant \overline{M \cap \complement p}$ and consequently

$$\mathfrak{o} \neq \tau p \wedge \mathfrak{a} \leqslant \tau p \wedge \overline{M \cap \complement p}.$$

Hence p is a point of accumulation of M.

Conversely, let every infinite subset of the T_1-space (R, τ) possess a point of accumulation. In addition, let $\mathfrak{a}$ be a filter different from $\mathfrak{o}$ and such that $\text{Rk}(\mathfrak{a}) \leqslant \aleph_0$. If $\text{Rk}(\mathfrak{a}) < \aleph_0$ then $\mathfrak{a}$ must be a principal filter and must therefore possess an adherent point. We can therefore assume $\text{Rk}(\mathfrak{a}) = \aleph_0$. If $\{A_n : n \in \mathbf{N}\}$ denotes a basis of $\mathfrak{a}$, then there exists a sequence $(P_n)_{n \in \mathbf{N}}$ of distinct points such that $p_n \in A_1 \cap \ldots \cap A_n$. By hypothesis the set $\{p_n : n \in \mathbf{N}\}$ possesses a point of accumulation q. If $\tau q \wedge \mathfrak{a} = \mathfrak{o}$, then we would have $\tau q \wedge \overline{A_n} = \mathfrak{o}$ for a certain index n. This would mean that q is already a point of accumulation of the finite set $\{p_1, \ldots, p_{n-1}\}$, but this is impossible in a T_1-space. Hence $\tau q \wedge \mathfrak{a} \neq \mathfrak{o}$, and q is an adherent point of $\mathfrak{a}$. ∎

Axiom $(\mathbf{T}_1)$ is essential as a hypothesis in the latter part of the above theorem (cf. 11A).

11.4 *If τ_1 is a k-compact (completely compact) topology and $\tau_1 \leqslant \tau_2$ then τ_2 is likewise k-compact (completely compact).*

Proof: From $\tau_1 \leqslant \tau_2$ and $\mathfrak{a} \wedge \tau_1 p \neq \mathfrak{o}$ it follows that $\mathfrak{a} \wedge \tau_2 p \neq \mathfrak{o}$. If therefore p is adherent to $\mathfrak{a}$ under the τ_1-topology then it certainly is under the τ_2-topology. ∎

11.5 *Every completely compact T_2-topology is finer than any other completely compact topology but coarser than any other T_2-topology.*

Proof: Let τ be a completely compact T_2-topology of R, and suppose τ' is a completely compact topology finer than τ, i.e., $\tau' < \tau$. Then there exists a $p \in R$ such that $\tau' p < \tau p$ and hence there exists a filter $\mathfrak{v}$ such that $\mathfrak{v} \leqslant \tau p$ and $\mathfrak{v} \wedge \tau' p = \mathfrak{o}$. Since τ' is completely compact, there is a $q \in R$ such that $\mathfrak{v} \leqslant \tau' q$. Hence $\mathfrak{v} \leqslant \tau' q \wedge \tau p \leqslant \tau q \wedge \tau p$. By $(\mathbf{T}_2)$, this implies $p = q$ which, however, contradicts $\mathfrak{o} = \mathfrak{v} \wedge \tau' p = \mathfrak{v} \wedge \tau' q$. Hence τ is the most fine of completely compact topologies. Now let τ'' be a T_2-topology such that $\tau \leqslant \tau''$. By 11.4, τ'' is completely compact and therefore the finest completely compact topology. Hence $\tau = \tau''$ and τ is the coarsest T_2-topology. ∎

Definition 11c: *A subset M of (R, τ) is said to be completely compact (k-compact) whenever the subspace (M, τ_M) is completely compact (k-compact). M is said to be* completely compact (k-compact in R *whenever every filter $\mathfrak{a}$ such that $\mathfrak{o} < \mathfrak{a} \leqslant \overset{\sqcap}{M}$ (and* $\mathrm{Rk}(\mathfrak{a}) \leqslant k$) *possesses an adherent point in R.*

Some authors use the terminology *relative compact* rather than our "completely compact in R." M is manifestly completely compact in R if and only if every ultrafilter $\mathfrak{v} \leqslant \overset{\sqcap}{M}$ is convergent in R. Every subset of a set completely compact (k-compact) in R is likewise completely compact (k-compact) in R.

11.6 *Every closed subset of a completely compact (k-compact) space is itself completely compact (k-compact).*

Proof: Let M be a closed subset of the completely compact (k-compact) space (R, τ). Every filter $\mathfrak{a}$ such that $\mathfrak{o} < \mathfrak{a} \leqslant \overset{\sqcap}{M}$ (and $\mathrm{Rk}(\mathfrak{a}) \leqslant k$) possesses an adherent point p in R. Consequently $\tau p \wedge \overset{\sqcap}{M} \geqslant \tau p \wedge \mathfrak{a} \neq \mathfrak{o}$, and therefore $p \in \bar{M}$. But M is closed and hence $p \in M$.[2] ∎

As for the converse, a completely compact subset need not be closed (cf. 11D). It will, however, be closed if we assume the Hausdorff separation axiom.

11.7 *Every completely compact subset of a T_2-space is closed.*

Proof: Let M be a completely compact subset of the T_2-space (R, τ). From $p \in \bar{M}$ we deduce $\tau p \wedge \overset{\sqcap}{M} > \mathfrak{o}$, and then there exists an ultrafilter $\mathfrak{v}$ such that $\mathfrak{v} \leqslant \tau p \wedge \overset{\sqcap}{M}$. Since M is completely compact, by 11.2 $\mathfrak{v}$ also converges to a point $q \in M$. Since axiom $(\mathbf{T}_2)$ holds, $\mathfrak{v}$ converges to only one point. Thus $p = q$ and $p \in M$. ∎

[2] To be more precise in our proof we must start from a filter $\mathfrak{a}'$ of M. This fixes a unique filter $\mathfrak{a}$ of R such that $\mathrm{Tr}_M(\mathfrak{a}) = \mathfrak{a}'$. The above proof then tells us that

$$\mathfrak{a}' \leqslant \mathrm{Tr}_M(\tau p) = \tau_M p$$

(cf. p. 66.) These remarks apply, *mutatis mutandis*, to all further analogous situations.

11.8 *The union of a finite number of completely compact (k-compact) sets is itself completely compact (k-compact).*

In a T_2-space the intersection of any number of completely compact sets is again completely compact.

Proof: Let the sets $M_1, \dots, M_n$ be completely compact (k-compact). From $\mathfrak{o} < \mathfrak{a} \leqslant \ulcorner M_1 \cup \dots \cup M_n \urcorner$ we deduce

$$\mathfrak{o} < \mathfrak{a} \wedge \ulcorner M_1 \cup \dots \cup M_n \urcorner = (\mathfrak{a} \wedge \ulcorner M_1 \urcorner) \vee \dots \vee (\mathfrak{a} \wedge \ulcorner M_n \urcorner).$$

Hence there exists an index ν_0 such that $\mathfrak{b} = \mathfrak{a} \wedge \ulcorner M_{\nu_0} \urcorner \neq \mathfrak{o}$. Moreover since $\mathrm{Rk}(\mathfrak{a}) \leqslant k$ we surely have $\mathrm{Rk}(\mathfrak{b}) \leqslant k$. Hence $\mathfrak{b}$ and also $\mathfrak{a}$ each possess an adherent point in M_{ν_0} and therefore all the more so in $M_1 \cup \dots \cup M_n$.

Let $\{M_\iota : \iota \in I\}$ be a system of completely compact sets with intersection $D = \bigcap M_\iota$. If $\mathfrak{v}$ is an ultrafilter such that $\mathfrak{v} \leqslant \ulcorner D \urcorner$, then for every $\iota \in I$, $\mathfrak{v}$ converges to a point $p_\iota \in M_\iota$. But by ($\mathbf{T}_2$) the limit of $\mathfrak{v}$ must be unique. Hence all the p_ι are the same and the limit of $\mathfrak{v}$ lies in D. ∎

We might naturally suspect that the closure of a set completely compact in R is itself completely compact, but this is not so, even when R is a T_2-space (cf. 11E). However, as soon as R is a T_3-space, our suspicion is correct. We then have:

11.9 *Let (R, τ) be a T_3-space and M completely compact in R. Then $\bar{M}$ is completely compact.*

Proof: Let $\mathfrak{v}$ be an ultrafilter such that $\mathfrak{v} \leqslant \ulcorner \bar{M} \urcorner$. For every open set A such that $\mathfrak{v} \leqslant \ulcorner A \urcorner$ we have $A \cap M = \emptyset$. But the system of all such opens sets A is a filter basis, generating a filter $\mathfrak{a}$. We must have $\mathfrak{v} \leqslant \mathfrak{a}$ and $\mathfrak{b} = \mathfrak{a} \wedge \ulcorner M \urcorner \neq \mathfrak{o}$. Hence $\mathfrak{b}$ possesses an adherent point p. Since $\mathfrak{b} \leqslant \ulcorner M \urcorner$, $p \in \bar{M}$. Suppose $\tau p \wedge \mathfrak{v} = \mathfrak{o}$. By ($\mathbf{T}_3$) there would then exist a $U \in \tau p$ such that $\mathfrak{v} \leqslant \ulcorner \complement \bar{U} \urcorner$. We should then have $\complement \bar{U} \in \mathfrak{a}$ and $\tau p \wedge \mathfrak{b} \leqslant \tau p \wedge \mathfrak{a} \leqslant \tau p \wedge \ulcorner \complement \bar{U} \urcorner = \mathfrak{o}$, but this would contradict the fact that p is an adherent point to $\mathfrak{b}$. Hence $\tau p \wedge \mathfrak{v} \neq \mathfrak{o}$ and therefore $\mathfrak{v} \leqslant \tau p$. ∎

The definition we gave of compactness involved a convergence condition which guarantees a sufficient number of convergent filters, or limits. We can arrive at an equivalent definition by starting from coverings (10e) and then imposing a certain finiteness condition. This procedure leads to a group of results, known as the *Heine-Borel-Lebesque covering theorems.*

11.10 *A subset M of the space (R, τ) is completely compact (k-compact) in R if and only if every open covering $\mathfrak{H}$ of R (such that $|\mathfrak{H}| \leqslant k$) contains a finite subcovering of M.*

Proof: Let M be completely compact and H be an open covering of R, and thereby of M also. Let us assume that $\mathfrak{v}$ is an ultrafilter where $\mathfrak{v} \leqslant \ulcorner M \urcorner$. Then, by 11.2, $\mathfrak{v}$ must converge to a point $p \in R$. With p we can associate an $H \in \mathfrak{H}$

such that $p \in H$. Since H is open, we have $\mathfrak{v} \leqslant \tau p \leqslant \overset{\sqcap}{H}$. By 3.13, $\mathfrak{H}$ must therefore contain a finite subcover of M. If M is only assumed to be k-compact with $|\mathfrak{H}| \leqslant k$, then every filter $\mathfrak{a}$ such that $\mathfrak{o} < \mathfrak{a} \leqslant \overset{\sqcap}{M}$ possesses an adherent point p in R. Hence there again exists an $H \in \mathfrak{H}$ such that $p \in H$; consequently, $\mathfrak{a} \wedge \overset{\sqcap}{H} \geqslant \mathfrak{a} \wedge \tau p \neq \mathfrak{o}$ and once more, by 3.13, our assertion is proved.

Conversely, let $\mathfrak{a}$ be a filter such that $\mathfrak{o} < \mathfrak{a} \leqslant \overset{\sqcap}{M}$ (and such that $\mathrm{Rk}(\mathfrak{a}) \leqslant k$), but possessing no adherent point in R. In addition, let $\mathfrak{A}$ be a basis of $\mathfrak{a}$ (such that $|\mathfrak{A}| \leqslant k$). Then we have $\bigcap \{\bar{A} : A \in \mathfrak{A}\} = \emptyset$, and $\mathfrak{H} = \{\complement \bar{A} : A \in \mathfrak{A}\}$ is an open covering of R (such that $|\mathfrak{H}| \leqslant k$). Consequently $\mathfrak{a} \wedge \overset{\sqcap}{H} = \mathfrak{o}$ for all $H \in \mathfrak{H}$, and using 3.13 again, H cannot contain any finite subcover of M. ∎

In the special case where $M = R$, we have:

11.11 *A space (R, τ) is completely compact (k-compact) if and only if every open covering $\mathfrak{H}$ of R (with $|\mathfrak{H}| \leqslant k$) contains a finite subcovering of R.*

In the case of complete compactness we can sharpen the statements of the last two theorems by considering only those coverings whose members belong to a fixed subbasis.

11.12 (Alexander) *Let $\mathfrak{C}$ be a subbasis of (R, τ). If every covering of R whose members belong to $\mathfrak{C}$ contains a finite cover of M, then M is completely compact in R.*

Proof: Let $\mathfrak{v}$ be an ultrafilter such that $\mathfrak{v} \leqslant \overset{\sqcap}{M}$, and which we shall assume not convergent. For each $p \in R$, the sets of form $C_1 \cap \ldots \cap C_n$ where $C_\nu \in \mathfrak{C}$ and $p \in C_\nu$ for $\nu = 1, \ldots, n$, constitute a basis of τp. If $\mathfrak{v} \wedge \overset{\sqcap}{C} \neq \mathfrak{o}$, then $\mathfrak{v} \leqslant \overset{\sqcap}{C}$ for all $C \in \mathfrak{C}$ where $p \in C$. This implies $\mathfrak{v} \leqslant \tau p$, contradicting the assumption that $\mathfrak{v}$ is not convergent. Hence, given any $p \in R$ there exists a $C_p \in \mathfrak{C}$ such that $p \in C_p$ and $\mathfrak{v} \wedge \overset{\sqcap}{C_p} = \mathfrak{o}$. The subcover $\mathfrak{H}$ of $\mathfrak{C}$, where $\mathfrak{H} = \{C_p : p \in R\}$ is a covering of R which by 3.13 can not itself contain a finite subcover of M. ∎

For applications, complete compactness is the most important type of compactness. The usefulness of the notion of k-compactness stems from the fact that when certain additional conditions are satisfied, k-compactness guarantees complete compactness.

11.13 *Suppose that for a space (R, τ) we have either $|R| \leqslant k$ or* Bd $(R, \tau) \leqslant k$. *Then every set which is k-compact in R is completely compact in R. In particular, if the whole space is k-compact then it must be completely compact.*

Proof: Let $\mathfrak{H}$ be an open covering of R. For each $p \in R$ let H_p be a member of $\mathfrak{H}$ such that $p \in H_p$. Then $\mathfrak{H}' = H_p : p \in R\}$ is likewise an open covering of R and $\mathfrak{H}'$ is contained in $\mathfrak{H}$. If $|R| \leqslant k_1$, then $|\mathfrak{H}'| \leqslant k$. If Bd $(R, \tau) \leqslant k$, then by 10.9 we know there exists a covering $\mathfrak{H}'$ of R such that $\mathfrak{H}' \subseteq \mathfrak{H}$ and $|\mathfrak{H}'| \leqslant k$. If the set M is k-compact in R, then by 11.10 there exists a finite cover of R contained in $\mathfrak{H}'$ and therefore also in $\mathfrak{H}$. Hence, every open covering of R contains a finite covering of M and M is completely compact in R by 11.10. ∎

As a corollary to the above theorem, we have:

11.14 *Every $\aleph_0$-compact space possessing a countable basis is completely compact.*
We also immediately have (cf. p. 74):

11.15 *Every $\aleph_0$-compact Lindelof space is completely compact.*
We even have a similar result for pseudometric spaces.

11.16 *Every $\aleph_0$-compact pseudometric space (R, δ) is completely compact.*

Proof: For every $n \in \mathbf{N}$ let $\mathfrak{M}_n$ designate the family of all subsets M of R such that $\delta(p, q) \geqslant 1/n$ for all $p, q \in M$. Considering $\mathfrak{M}_n$ as ordered by the usual set-theoretic inclusion, let $\mathfrak{K}$ be a chain in $\mathfrak{M}_n$ and let $K = \cup \{M : M \in \mathfrak{K}\}$. If $p, q \in K$, then $p, q \in M$ for a certain $M \in \mathfrak{K}$ and hence $\delta(p, q) \geqslant 1/n$. This tells us that $K \in \mathfrak{M}_n$ and thus, applying Zorn's lemma, $\mathfrak{M}_n$ must contain a maximal element M_n. If M_n were infinite, it would have to possess an accumulation point because of 11.3 and the fact that R is $\aleph_0$-compact. But this contradicts $\delta(p, q) \geqslant 1/n$ for all $p, q \in M_n$, and hence the sets M_n are all finite. Thus $M = \cup\{M_n : n \in \mathbf{N}\}$ is at most countably infinite. Since M_n is maximal, for each $p \in R$ and for each $n \in \mathbf{N}$, we can find a $q \in M_n$ such that $\delta(p, q) < 1/n$. Hence M is dense. R is therefore separable and possesses, by 10.11, a countable basis. By 11.14 R must be completely compact. ∎

The condition of complete compactness allows us to link together more closely the different separation axioms.

11.17 *Every completely compact T_3-space is a T_4-space. Every completely compact T_2-space is normal.*

Proof: Let R be a completely compact T_3-space and A, B two disjoint closed subsets of R. Given any $p \in A$, we can, by $(\mathbf{T_3})$ find open sets U_p and V_p such that $p \in U_p$, $B \subseteq V_p$, and $U_p \cap V_p = \varnothing$ (8.5). The system $\{U_p : p \in A\}$ is an open covering of A. But since, by 11.6, A must be completely compact, we know that a finite number of the U_p already cover A (11.11). Let this finite subcover be composed of $U_{p_1}, \ldots, U_{p_n}$. The open sets

$$U = U_{p_1} \cup \ldots \cup U_{p_s} \qquad \text{and} \qquad V = V_{p_1} \cap \ldots \cap V_{p_s}$$

therefore satisfy the conditions $A \subseteq U$, $B \subseteq V$, and $U \cap V = \varnothing$, and hence R is a T_4-space.

Let us now suppose that R is a completely compact T_2-space. Let M be a closed subset of R and p be a point not in M. Given any point $q \in M$ we can find open neighborhoods $U_q \in \tau p$ and $V_q \in \tau p$ such that $U_q \cap V_q = \varnothing$. The system $\{V_q : q \in M\}$ is an open covering of M, and for the same reasons as above, M can already be covered by a finite number of these V_q—say by $V_{q_1}, \ldots, V_{q_n}$. Then for the open sets

$$U = U_{q_1} \cap \ldots \cap U_{q_n} \qquad \text{and} \qquad V = V_{q_1} \cup \ldots \cup V_{q_n}$$

we again have $p \in U$, $M \subseteq V$, and $U \cap V = \emptyset$. Then R is certainly a T_3-space and, by what we have just proved above, also a T_4-space. Since we have supposed that R is also a T_2-space, it must therefore be normal. ∎

The notion of compactness can also be characterized by means of set-theoretic intersections.

11.18 *A set M is completely compact (k-compact) in R if and only if, for every closed filter $\mathfrak{a}$ (such that* $\mathrm{Rk}(\mathfrak{a}) \leqslant k$*), the condition $\mathfrak{a} \wedge \ulcorner M \urcorner \neq \mathfrak{o}$ implies*

$$\bigcap \{A : A \in \mathfrak{a}\} \neq \emptyset.$$

Proof: Let M be completely compact (k-compact) in R. In addition, suppose $\mathfrak{a} = \bar{\mathfrak{a}}$ (and $\mathrm{Rk}(\mathfrak{a}) \leqslant k$). From $\mathfrak{b} = \mathfrak{a} \wedge \ulcorner M \urcorner \neq \mathfrak{o}$ it follows (for $\mathrm{Rk}(\mathfrak{b}) \leqslant \mathrm{Rk}(\mathfrak{a})$) that $\mathfrak{b}$ possesses an adherent point p. By (11.1) we have $\ulcorner p \urcorner \leqslant \bar{\mathfrak{b}} \leqslant \bar{\mathfrak{a}} = \mathfrak{a}$, hence $p \in \bigcap \{A : A \in \mathfrak{a}\}$. To prove the converse, suppose M satisfies the above intersection property. Then $\mathfrak{o} < \mathfrak{a} \leqslant \ulcorner M \urcorner$ implies $\bar{\mathfrak{a}} \wedge \ulcorner M \urcorner \neq \mathfrak{o}$ (and $\mathrm{Rk}(\mathfrak{a}) \leqslant k$ implies $\mathrm{Rk}(\bar{\mathfrak{a}}) \leqslant k$), hence $\bigcap \{A : A \in \bar{\mathfrak{a}}\} \neq \emptyset$. Thus there exists a $p \in R$ such that $\ulcorner p \urcorner \leqslant \bar{\mathfrak{a}}$; i.e., p is an adherent point to $\mathfrak{a}$. ∎

If in the above theorem we let $M = R$, we obtain the corollary:

11.19 *A space R is completely compact (k-compact) if and only if, for every closed filter $\mathfrak{a}$ (with* $\mathrm{Rk}(\mathfrak{a}) \leqslant k$*) the condition $\mathfrak{a} \neq \mathfrak{o}$ implies $\bigcap \{A : A \in \mathfrak{a}\} \neq \emptyset$.*

11.20 (G. Cantor's Intersection Theorem) *A space R is $\aleph_0$-compact if and only if every sequence $(A_n)_{n \in \mathbf{N}}$ of nonempty closed sets where $A_{n+1} \subseteq A_n$ possesses a nonempty intersection $D = \bigcap A_n$.*

Proof: Let R be $\aleph_0$-compact. The sequence of sets $(A_n)_{n \in \mathbf{N}}$ is the basis of a closed filter $\mathfrak{a}$. Since $A_n \neq \emptyset$ for all n we have $\mathfrak{a} \neq \mathfrak{o}$, and by 11.19 we have $D \neq \emptyset$. Conversely, suppose R is not $\aleph_0$-compact. By 11.19 again, there exists a closed filter $\mathfrak{b}$ such that $\mathrm{Rk}(\mathfrak{b}) \leqslant \aleph_0$, $\bigcap \{B \in \mathfrak{b}\} = \emptyset$ and $\mathfrak{b} \neq \mathfrak{o}$. If the sets $\{B_n : n \in \mathbf{N}\}$ form a basis composed of closed sets of $\mathfrak{b}$, then so do the sets $A_n = B_1 \cap \ldots \cap B_n$. Since $\mathfrak{b} \neq \mathfrak{o}$, the sets A_n are nonempty. Also $A_{n+1} \subset A_n$. However, $\bigcap A_n = \bigcap B_n = \emptyset$. ∎

Notes and Exercises

11A Let $\mathbf{R}^+$ be the set of all positive real numbers. To each $x \in \mathbf{R}^+$ let us associate the filter basis $\mathfrak{U}_x = \{]y, \rightarrow] : 0 < y < x\}$. These filter bases then satisfy the Haussdorf neighborhood axioms. Hence $\tau x = \ulcorner \mathfrak{U}_x \urcorner$ defines a topology τ on $\mathbf{R}^+$ (τ is not a T_1-topology.) The Fréchet filter associated with the sequence $(V_n)_{n \in \mathbf{N}}$ does not possess any adherent point in $(\mathbf{R}^+, \tau)$ and hence this space is not $\aleph_0$-compact. Nevertheless, every infinite subset and even every two-point subset of $\mathbf{R}^+$ possess a point of accumulation: for $p < q$ implies $\ulcorner q \urcorner \leqslant \tau p$; i.e.,

p is a point of accumulation of the set $\{p, q\}$. Hence we cannot drop axiom $(\mathbf{T}_1)$ from the second part of 11.3.

11B *Exercise:* On an uncountable set the topology $\tau^{\aleph_0}$ is completely compact, while the topology $\tau^{\aleph_1}$ is never $\aleph_0$-compact (cf. 4.III).

11C Let $(R, \leqslant)$ be a well-ordered set having no end points, under the order topology τ. If M is a subset of R, let $S(M)$ denote the set of all $x \in R$, for which we can find an $m \in M$ where $x \leqslant m$. Lastly, let k be the smallest cardinal number such that there exists a subset M of R with $|M| = k$ and $S(M) = R$.

Exercise: The filter generated by the filter basis $\{[x, \rightarrow] : x \in R\}$ possesses no adherent points in (R, τ). The space (R, τ) is therefore not k-compact. Nevertheless, (R, τ) is a k'-compact space for every cardinal number k' where $k' < k$ [34, p. 95].

11D Let τ denote the standard topology on the real line $\mathbf{R}$. For each integer $n \in \mathbf{N}$ let

$$U'_n(p) = \{q : |p - q| < 1/n \quad \text{or} \quad q > n\}.$$

Then $\mathfrak{U}'_p = \{U'_n(p) : n \in \mathbf{N}\}$ is a filter basis, and $\tau' p = \overbracket{\mathfrak{U}'_p}$ defines a new topology τ' on $\mathbf{R}$, for which we evidently have $\tau < \tau'$. Moreover, τ' is a T_1-topology, but not a T_2-topology. The set $M = \{p : p \geqslant 0\}$ is completely compact under the τ'-topology: For let $\mathfrak{v}$ be an ultrafilter such that $\mathfrak{v} \leqslant \overbracket{M}$. If $\mathfrak{v}$ contains an interval $[a, b]$, then $\mathfrak{v}$ is already τ-convergent, hence certainly τ'-convergent, to a point in M. If $\mathfrak{v}$ does not contain such an interval, $\mathfrak{v}$ is unbounded and τ'-convergent to every point of $\mathbf{R}$. The set M is nevertheless not τ'-closed; we have $h_{\tau'}(M) = \mathbf{R}$. Hence the hypothesis of $(\mathbf{T}_2)$ is essential to the statement of 11.7.

11E In the space $\mathbf{R}^2$ let τ be the topology defined by the Euclidean norm (cf. 6C). Let $K = \{x : \|x\| < 1\}$ (open unit disk), $S = \{x : \|x\| = 1\}$ (circumference of unit circle), and let p be a fixed point on S. The conditions

$$\tau' q = \begin{cases} \tau q & \text{for} \quad q \neq p \\ \overbracket{p} \vee (\tau p \wedge \overbracket{K}) & \phantom{\text{for}} \quad q = p \end{cases}$$

define a topology τ' on $\mathbf{R}^2$. With respect to this topology, K is completely compact in $\mathbf{R}^2$. The closure of K under this topology is the set $K \cup S$. This closure is not completely compact: for let $\mathfrak{a}$ be a filter convergent to p under the norm topology and such that $\mathfrak{a} \leqslant \overbracket{S \cap \complement p}$. Then $\mathfrak{a}$ admits of no adherent point under τ'.

12. Locally Compact Spaces

Complete compactness guarantees the convergence of all ultrafilters and this very fact is often quite instrumental in carrying out the proof of many theorems. But several spaces important for their applications are not even $\aleph_0$-compact—for instance, the real line is not $\aleph_0$-compact because the set of integers possesses no accumulation point. Nevertheless, in this case, and in many others, we still can put compactness to good use. Compactness does not hold globally, in the whole space, but it does hold locally, i.e., in a sufficiently small neighborhood of every point. The real line is locally compact because all closed intervals $[a, b]$ are completely compact.

Definition 12a: *A space R and its topology are both said to be* locally completely compact *if every point of R possesses a completely compact neighborhood. A subset M of a topological space is said to be locally completely compact whenever the subspace associated with M is locally completely compact. We shall often use the abbreviation "l.c.c." for "locally completely compact."*

Every completely compact space is manifestly l.c.c., for the whole space itself can serve as a neighborhood of each of its points. The concept of k-compactness can likewise be localized, but since locally k-compact spaces are not of major importance we shall not dwell upon them here (cf. 12A). In the literature of a few decades ago the space (R, τ) was often called locally completely compact whenever each point $p \in R$ possessed an open neighborhood $U \in \tau p$ whose closure $\bar{U}$ was completely compact. This definition is not equivalent to the one we have used (cf. 12B). Only when we assume that the separation axiom $(\mathbf{T}_2)$ or $(\mathbf{T}_3)$ holds, do the two definitions become equivalent.[3]

12.1 *A T_2- or T_3-space (R, τ) is l.c.c. if and only if every $p \in R$ possesses a (open) neighborhood $U \in \tau p$ whose closure $\bar{U}$ is completely compact.*

Proof: If U is a neighborhood of p, so is $\bar{U}$. The conditions of the hypothesis imply immediately that R is l.c.c., no extra axioms of separation being needed. To prove the converse, let (R, τ) be assumed l.c.c. and let U be a completely compact neighborhood belonging to τp. If τ is a T_2-topology, then by 11.7 U must be closed. The interior $V = \mathring{U}$ is then an open neighborhood belonging to τp, whose closure $\bar{V}$ is contained in U and is also completely compact by 11.6. If, on the other hand, τ is a T_3-topology, then there exists an open neighborhood $V \in \tau p$ such that $\bar{V} \subseteq U$ and for the very same reasons as above, $\bar{V}$ is completely compact. ∎

[3] For N. Bourbaki a space is locally completely compact whenever it is what we would call locally completely compact and T_2.

12.2 *A T_2- or T_3-space is l.c.c. if and only if it admits of a covering by open sets whose closures are completely compact.*

Proof: If the underlying space is l.c.c., by 12.1 each point p possesses an open neighborhood U_p whose closure is completely compact. Then $\{U_p : p \in R\}$ is a covering of the desired type. Conversely, for a covering of this type, each member-set is a neighborhood of all its points and has a completely compact closure. Using 12.1 again, this shows that the underlying space if l.c.c. ∎

12.3 *If the l.c.c. sets $M_1, ..., M_n$ are closed, then their union $M = M_1 \cup ... \cup M_n$ is also l.c.c.*

In a T_2-space the intersection of a finite number of l.c.c. sets is itself l.c.c.

Proof: If $p \in M$ and we renumber the sets M_ν appropriately, we must have $p \in M_1 \cap ... \cap M_r$ and $p \notin M_{r+1} \cup ... \cup M_n$ for a suitable r where $1 \leqslant r \leqslant n$. Then, for $\rho = 1, ..., r$ we can find a $U_p \in \tau p$ such that $V_\rho = M_\rho \cap U_\rho$ is completely compact. Let $V = V_1 \cup ... \cup V_r$ and

$$U = U_1 \cap ... \cap U_r \cap \complement(M_{r+1} \cup ... \cup M_n).$$

Because the sets $M_{r+1}, ..., M_r$ are closed, U must be a neighborhood belonging to τp. Hence

$$U \cap M = (U \cap M_1) \cup ... \cup (U \cap M_n) \subseteq (U_1 \cap M_1) \cup ... \cup (U_r \cap M_r) = V \subseteq \dot{M}.$$

Therefore V is a neighborhood belonging to $\tau_M p$ and, by 11.8, V is completely compact.

The second contention follows immediately from 11.8. ∎

12.4 *In an l.c.c. space R suppose $A \cap M$ is closed for every completely compact subset M of R. Then A itself must be closed.*

Proof: Suppose $p \in \bar{A}$. If U is a completely compact neighborhood belonging to τp, then p is an adherent point of $A \cap U$. But by hypothesis $A \cap U$ is closed and therefore $p \in A \cap U \subseteq A$. ∎

12.5 *Every closed subset of an l.c.c. space is itself l.c.c. In an l.c.c. T_2-space every open subset is l.c.c.*

Proof: In the l.c.c. space (R, τ) let A be a closed set. If U is a completely compact neighborhood of the point $p \in A$, then $V = U \cap A$ is closed in U and is a neighborhood belonging to $\tau_A p$. By 11.4, V is also completely compact. Now let τ_2 be a T_2-topology and B an open set. With $p \in B$ let us associate the completely compact neighborhood U belonging to τp. The subspace (U, τ_U) is, by 11.17, always regular. Hence there exists a $V \in \tau_U p$ such that

$$W = h_{\tau_U}(V) \subseteq B \cap U.$$

W is a closed subset of (U, τ_U) and hence, by 11.6, W is completely compact. Since $U \in \tau p$ we also have $W \in \tau_B p$. ■

12.6 *Let (R, τ) be a l.c.c. T_2-space. A subset M of R is l.c.c. if and only if there exists a closed set A and an open set B such that $M = A \cap B$.*

Proof: By 12.5, A and B are both l.c.c. spaces. Then, by 12.3, their intersection $M = A \cap B$ is also l.c.c. Conversely, let M be l.c.c. If we let $A = \bar{M}$ and $B = M \cup \complement\bar{M}$, then A is closed, and we have

$$A \cap B = \bar{M} \cap (M \cup \complement\bar{M}) = (\bar{M} \cap M) \cup (\bar{M} \cap \complement\bar{M}) = M.$$

Hence it only remains to show that B is open. From $p \in \complement\bar{M}$ it follows that $\tau p \leqslant \ulcorner\complement\bar{M}\urcorner \leqslant \ulcorner B\urcorner$. Therefore, $p \in M$ and there must exist a neighborhood $U \in \tau p$ such that $V = U \cap M$ is completely compact. By 11.7, V is also closed. If $q \in \mathring{U}$ and $q \notin M$ then $\complement V \in \tau q$ and $\tau q \wedge \ulcorner M\urcorner \leqslant \ulcorner\complement V\urcorner \wedge \ulcorner U\urcorner \wedge \ulcorner M\urcorner = \mathfrak{o}$, and hence, finally, $q \notin \bar{M}$. Thus $\mathring{U}$ is a neighborhood of p and contained in $B = M \cup \complement\bar{M}$. ■

Every completely compact space is also l.c.c. If we remove one point from a completely compact T_2-space R, then what remains, $C_R\{p\}$, is by 12.5 an l.c.c. space. Later we shall establish the converse: If we adjoin one point (at infinity) to any l.c.c. T_2-space we obtain an extension (R, τ) which can be viewed in a natural way as a completely compact T_2-space (cf. p. 168). If we have $\mathrm{Rk}(\tau, p) \leqslant \aleph_0$, i.e., if there exists a countable basis $\{U_n : n \in \mathbf{N}\}$ of τp where all the U_n are open, then the sets $\complement U_n$ are completely compact and constitute a cover of the space remaining after we remove p. Hence this residue space is a union of a countable number of completely compact subsets. Such considerations lead us to formulate the following concept.

Definition 12b: *A space R is said to be* countable at infinity *if R can be expressed as a union of a countable number of completely compact subsets.*

12.7 *Every l.c.c. T_2-space (R, τ) which possesses a countable basis is* countable at infinity.

Proof: Let $\mathfrak{B}$ be a countable basis of (R, τ), and let $\mathfrak{B}'$ denote the system of all sets $B \in \mathfrak{B}$ for which $\bar{B}$ is completely compact. Given any $p \in R$ we can find a completely compact neighborhood $U \in \tau p$. By 11.7, U is even closed. Since $\mathfrak{B}$ is a basis, there exists a $B \in \mathfrak{B}$ such that $p \in B$ and $B \subseteq U$, hence such that $\bar{B} \subseteq U$. By 11.6, $\bar{B}$ is completely compact. Consequently $R = \bigcup \{\bar{B} : B \in \mathfrak{B}'\}$ and since $|\mathfrak{B}'| \leqslant |\mathfrak{B}| \leqslant \aleph_0$, R must be countable at infinity. ■

Let us also note that local complete compactness can serve as a link between the separation axioms $(\mathbf{T}_2)$ and $(\mathbf{T}_3)$.

12.8 *Every l.c.c. T_2-space is regular.*

Proof: Let U be a neighborhood belonging to τp. Then $\tau p \leqslant \ulcorner\mathring{U}\urcorner$ and, by

12.5, $\mathring{U}$ is also l.c.c. Hence there exists a completely compact neighborhood $V \in \tau p$ such that $V \subseteq U$. By 11.7, V is closed. Hence ($\mathbf{T}_3$) is satisfied and our space is regular. ∎

Notes and Exercises

12A A space is said to be locally k-compact if each of its points possesses a k-compact neighborhood.

Exercise

1. Let (R, τ) be locally k-compact. If $|R| \leqslant k$ or Bd $(R, \tau) \leqslant k$ then (R, τ) must be l.c.c.

2. The union of a finite number of locally k-compact closed sets is likewise locally k-compact.

3. Every closed subset of a locally k-compact space is itself locally k-compact [34, p. 106].

12B *Exercise*

1. The space defined in 11D is l.c.c.

2. But this space is not $\aleph_0$-compact (the negative integers possess no point of accumulation).

3. The closure of any of its nonempty open sets is the whole space. Hence in this space no point can possess a neighborhood whose closure is completely compact.

12C *Exercise:* In a T_2-space every dense and l.c.c. subset is open [24, p. 163].

12D Let $\mathfrak{H}$ denote the collection of all open coverings of the space (R, τ). For each covering $\mathfrak{h} \in \mathfrak{H}$ let $\mathfrak{c}_{\mathfrak{h}}$ denote the associated complementary-filter (cf. p. 28).

Exercise

1. If (R, τ) is l.c.c., then the filter $\bigvee \{\mathfrak{c}_{\mathfrak{h}} : \mathfrak{h} \in \mathfrak{H}\}$ possesses no adherent point.

2. If (R, τ) is a T_3-space and $\bigvee \{\mathfrak{c}_{\mathfrak{h}} : \mathfrak{h} \in \mathfrak{H}\}$ possesses no adherent point, then (R, τ) must be l.c.c.

13. Paracompact Spaces

It is possible to generalize complete compactness in yet another way different from localization. This other generalization stems from the characterization of complete compactness by means of coverings, and achieves its objective through finiteness conditions imposed upon these coverings. We shall first need some preparatory definitions.

Definition 13a: *Let $\mathfrak{S}$ be a system of subsets of the space (R, τ). Then $\mathfrak{S}$ is said to be locally finite (discrete) if, given any $p \in R$ we can find a neighborhood $U \in \tau p$ such that U possesses a nonempty intersection with only a finite number (at most one) of the members of $\mathfrak{S}$.*

Evidently any finite system of sets and any discrete system are both locally finite.

13.1 *In the space (R, τ) let $\mathfrak{S}$ be a locally finite (discrete) system of sets. Then $\overline{\mathfrak{S}} = \{\bar{s} : s \in S\}$ is also locally finite (discrete). For every subsystem $\mathfrak{S}'$ of $\mathfrak{S}$ we have*

$$\overline{\bigcup_{S \in \mathfrak{S}'} S} = \bigcup_{S \in \mathfrak{S}'} \bar{S}.$$

Proof: Given any $p \in R$ we can find an open neighborhood $U \in \tau p$ such that U has a nonempty intersection with at most a finite number (with at most one) of the $S \in \mathfrak{S}$. Since U is open, $U \cap S = \emptyset$ implies $U \cap \bar{S} = \emptyset$. Hence U possesses the above property with respect to $\overline{\mathfrak{S}}$ also. Let $S_1, \ldots, S_n$ represent those sets belonging to $\mathfrak{S}'$ which possess a nonempty intersection with U. If

$$p \in \overline{\bigcup \{S : S \in \mathfrak{S}'\}}$$

it follows that

$$p \in \overline{S_1 \cup \ldots \cup S_n} = \bar{S}_1 \cup \ldots \cup \bar{S}_n \subseteq \bigcup \{\bar{S} : S \in \mathfrak{S}'\}.$$

Since $\bigcup \bar{S} \subseteq \overline{\bigcup S}$, the desired equality must hold. ∎

Definition 13b: *Let $\mathfrak{S}$ and $\mathfrak{T}$ be two systems of subsets of the fundamental set R. We say that $\mathfrak{T}$ is a* refinement *of $\mathfrak{S}$ and write $\mathfrak{T} \ll \mathfrak{S}$ if, given any $T \in \mathfrak{T}$ we can always find an $S \in \mathfrak{S}$ such that $T \subseteq S$.*

This relation between systems of sets is manifestly reflexive and transitive. We also see immediately that $\mathfrak{T} \subseteq \mathfrak{S}$ implies $\mathfrak{T} \ll \mathfrak{S}$. We are now in a position to give the promised generalization of complete compactness.

Definition 13c: *A space (R, τ) is said to be* paracompact *if, given any open covering $\mathfrak{H}$ of R we can find an open and locally finite covering $\mathfrak{K}$ of R such that $\mathfrak{K} \ll \mathfrak{H}$.*[4]

13.2 *Every completely compact space is paracompact.*

[4] Most authors demand that paracompact spaces be T_2- or T_3-spaces, in addition to all the stipulations listed in our definition. We shall, however, never assume any such separation conditions.

Proof: Given any open covering $\mathfrak{H}$ of a completely compact space, by 11.11, we can find a finite cover $\mathfrak{K}$ such that $\mathfrak{K} \subseteq \mathfrak{H}$. Then $\mathfrak{K}$ is open and hence $\mathfrak{K} \ll \mathfrak{H}$. Since $\mathfrak{K}$ is finite it must be locally finite. ∎

13.3 *Every paracompact T_3-space is a T_4-space. Every paracompact T_2-space is normal.*

Proof: Let (R, τ) be a paracompact T_3-space (T_2-space) and let A be a closed subset of R. If B is another closed set such that $A \cap B = \varnothing$ (and if p is a point in $\complement A$), then for each $q \in A$ we can find an open neighborhood $U_q \in \tau p$ such that $\bar{U}_q \cap B = \varnothing (p \notin \bar{U}_q)$. The system $\mathfrak{H} = \{U_q : q \in A\} \cup \{\complement A\}$ is then an open covering of R, for which there exists an open and locally finite covering $\mathfrak{K}$ of R such that $\mathfrak{K} \ll \mathfrak{H}$. Let $\mathfrak{K}' = \{K : K \in \mathfrak{K}, K \cap A \neq \varnothing\}$. Then, by 13.1, the set $M = \cup \{\bar{K} : K \in \mathfrak{K}'\}$ is closed. Hence $U = \cup \{K : K \in \mathfrak{K}'\}$ and $V = \complement M$ are open sets such that $A \subseteq U$ and $U \cap V = \varnothing$. Since $\mathfrak{K} \ll \mathfrak{H}$ it follows that $K \in \mathfrak{K}'$, whence $K \cap A \neq \varnothing$, which in turn implies that there exists a $q \in A$ such that $K \subseteq U_q$. We then have $B \subseteq \complement \bar{K}$ $(p \in \complement \bar{K})$ for all $K \in \mathfrak{K}'$ and hence $B \subseteq V$ $(p \in V)$. We thus see that condition $(\mathbf{T}_2)$ implies $(\mathbf{T}_3)$, while condition $(\mathbf{T}_3)$ implies $(\mathbf{T}_4)$. ∎

Let $\mathfrak{M}$ be a system of subsets of the underlying set R and let A be an arbitrary subset of R. We define the *star of A* over M by

$$\mathrm{St}_{\mathfrak{M}}(A) = \cup \{M : M \in \mathfrak{M}, M \cap A \neq \varnothing\}.$$

Rather than write $\mathrm{St}_{\mathfrak{M}}(\{p\})$ we simply write $\mathrm{St}_{\mathfrak{M}}(p)$. If $A \subseteq B$, then $\mathrm{St}_{\mathfrak{M}}(A) \subseteq \mathrm{St}_{\mathfrak{M}}(B)$, and if $\mathfrak{M} \ll \mathfrak{N}$, then $\mathrm{St}_{\mathfrak{M}}(A) \subseteq \mathrm{St}_{\mathfrak{N}}(A)$. Let us further define

$$\mathrm{St}(\mathfrak{M}) = \{\mathrm{St}_{\mathfrak{M}}(p) : p \in R\}$$

and

$$\mathrm{St}^*(\mathfrak{M}) = \{\mathrm{St}_{\mathfrak{M}}(M) : M \in \mathfrak{M}\}$$

Suppose $\mathfrak{H}$ is an open covering of the space R. We say that $\mathfrak{K}$ is a *star refinement* of $\mathfrak{H}$ whenever $\mathfrak{K}$ is an open covering of R such that $\mathrm{St}(\mathfrak{K}) \ll \mathfrak{H}$. If we have $\mathrm{St}^*(\mathfrak{K}) \ll \mathfrak{H}$, we say that $\mathfrak{K}$ is a *strong star refinement* of $\mathfrak{H}$. Every strong star refinement is a star refinement of $\mathfrak{H}$: Given any $p \in R$ there exists a $K \in \mathfrak{K}$ such that $p \in K$, and consequently $\mathrm{St}_{\mathfrak{K}}(p) \subseteq \mathrm{St}_{\mathfrak{K}}(K)$, which implies

$$\mathrm{St}(\mathfrak{K}) \ll \mathrm{St}^*(\mathfrak{K}) \ll \mathfrak{H}.$$

Definition 13d: *A space (R, τ) is said to be* completely normal *if every open covering of R admits of a star refinement.*

13.4 *The space R is completely normal if and only if every open covering of R possesses a strong star refinement.*

Proof: Let R be completely normal. Given an open covering $\mathfrak{H}$ of R we can find a star refinement $\mathfrak{K}$ of $\mathfrak{H}$ and likewise a star refinement $\mathfrak{L}$ of $\mathfrak{K}$. Let L be a fixed member set of $\mathfrak{L}$. Since $\mathrm{St}(\mathfrak{L}) \ll \mathfrak{K}$, given a $p \in L$ we can find a $K_p \in \mathfrak{K}$ such that $\mathrm{St}_{\mathfrak{L}}(p) \subseteq K_p$, and consesquently $L \subseteq \mathrm{St}_{\mathfrak{L}}(p) \subseteq K_p$. Let q be a fixed point in L. Then for all $p \in L$ we have $q \in K_p$ and thus

$$\mathrm{St}_{\mathfrak{L}}(L) = \cup \{\mathrm{St}_{\mathfrak{L}}(p) : p \in L\} \subseteq \cup \{K_p : p \in L\} \subseteq \mathrm{St}_{\mathfrak{K}}(q).$$

Hence, $\mathrm{St}^*(\mathfrak{L}) \ll \mathrm{St}(\mathfrak{K}) \ll \mathfrak{H}$. The converse is trivial, for a strong star refinement is always a star refinement. ∎

13.5 *Let $\mathfrak{H}$ be an open covering of the arbitrary space R and let K be a closed and locally finite covering of R such that $\mathfrak{K} \ll \mathfrak{H}$. Then $\mathfrak{H}$ surely possesses a star refinement.*

Proof: Since $\mathfrak{K} \ll \mathfrak{H}$, given any $K \in \mathfrak{K}$ we can find an $H_K \in \mathfrak{H}$ such that $K \subseteq H_K$. For each $p \in R$, let

$$L_p = \bigcap_{\substack{K \in \mathfrak{K} \\ p \in K}} H_K \cap \mathbf{C} \bigcup_{\substack{K \in \mathfrak{K} \\ p \notin K}} K.$$

Since $\mathfrak{K}$ is locally finite there can only be a finite number of $K \in \mathfrak{K}$ such that $p \in K$. This, together with 13.1, implies that the L_p are open. Since in addition $p \in L_p$, $\mathfrak{L} = \{L_p : p \in R\}$ must be an open covering of R. Let q be a fixed point and let us suppose that, for the point p, we have $q \in L_p$. If $q \in K$, then $p \in K$ and hence $L_p \subseteq H_K$, and finally $\mathrm{St}_{\mathfrak{L}}(q) \subseteq H_K$. Therefore $\mathrm{St}(\mathfrak{L}) \ll \mathfrak{H}$. ∎

13.6 *A T_3-space (R, τ) is completely normal if, given any open covering $\mathfrak{H}$, we can find a locally finite covering $\mathfrak{K}$ (not necessarily open or closed) such that $\mathfrak{K} \ll \mathfrak{H}$.*

Proof: Let $\mathfrak{G}$ be an open covering of R. Given any $p \in R$ we can find a $G_p \in \mathfrak{G}$ such that $p \in G_p$, and, in addition, an open neighborhood $H_p \in \tau p$ such that $\bar{H}_p \subseteq G_p$. The system $\mathfrak{H} = \{H_p : p \in R\}$ is an open covering of R, for which we can find, by hypothesis, a locally finite covering $\mathfrak{K}$ such that $\mathfrak{K} \ll \mathfrak{H}$. From $K \in \mathfrak{K}$ we deduce $K \subseteq H_p$, and thus $\bar{K} \subseteq \bar{H}_p \subseteq G_p$ for a $p \in R$. Hence $\bar{\mathfrak{K}} \ll \mathfrak{G}$. By 13.1, $\bar{\mathfrak{K}}$ must be a locally finite and closed covering, and therefore, by 13.5, $\mathfrak{G}$ admits of a star refinement. ∎

The above theorem immediately implies:

13.7 *Every paracompact T_3-space is completely normal.*

In what follows we shall give further characterizations of paracompact T_3-spaces and, at the same time, establish results useful in attacking the converse to 13.7.

13.8 *In the completely normal space (R, τ) let $\mathfrak{S}$ denote a system of open sets and $\mathfrak{T}$ a locally finite (discrete) system of sets such that $\mathfrak{T} \ll \mathfrak{S}$. Then there exists a locally finite (discrete) system $\mathfrak{B}$ of open sets such that $\mathfrak{T} \ll \mathfrak{B} \ll \mathfrak{S}$.*

Proof: Given any point $p \in R$ there exists an open neighborhood $U_p \in \tau p$ such that U_p possesses a nonempty intersection with only a finite number (with only one) of the sets of $\mathfrak{T}$. By 13.4 the open covering $\mathfrak{U} = \{U_p : p \in R\}$ possesses a strong star refinement $\mathfrak{H}$. The system $\mathfrak{K} = \{\mathrm{St}_{\mathfrak{H}}(T) : T \in \mathfrak{T}\}$ is then composed of open sets, and we have $\mathfrak{T} \leqslant \mathfrak{K}$. Given any $q \in R$ we can find an $H_q \in \mathfrak{H}$ such that $q \in H_q$. From $H_q \cap \mathrm{St}_{\mathfrak{H}}(T) \neq \emptyset$ it follows that $\mathrm{St}_{\mathfrak{H}}(H_q) \cap T \neq \emptyset$. Since $\mathrm{St}^*(\mathfrak{H}) \leqslant \mathfrak{U}$, there exists a $p \in R$ such that $\mathrm{St}_{\mathfrak{H}}(H_q) \subseteq U_p$, and then we surely have $U_p \cap T \neq \emptyset$. Since this holds for only a finite number (at most one) of the $T \in \mathfrak{T}$, the neighborhood $H_q \in \tau q$ also possesses a nonempty intersection with only a finite number (at most one) of the sets in $\mathfrak{K}$, i.e., $\mathfrak{K}$ is likewise locally finite (discrete). Since $\mathfrak{T} \leqslant \mathfrak{S}$, given any $T \in \mathfrak{T}$ can find an $S_T \in \mathfrak{S}$ such that $T \subseteq S_T$. Then the system $\mathfrak{B} = \{\mathrm{St}_{\mathfrak{H}}(T) \cap S_T : T \in \mathfrak{T}\}$ verifies $\mathfrak{T} \leqslant \mathfrak{B}$, $\mathfrak{B} \leqslant \mathfrak{S}$, and $\mathfrak{B} \leqslant \mathfrak{K}$. Moreover, all the members of $\mathfrak{B}$ are open sets. From the very definition of $\mathfrak{B}$, and since $\mathfrak{K}$ is locally finite (discrete), the same must hold for $\mathfrak{B}$. ∎

Definition 13e: *Let k be a transfinite cardinal number. A system of sets $\mathfrak{S}$ is said to be k-*locally finite (*k*-discrete) *if* $\mathfrak{S} = \bigcup \{\mathfrak{S}_\iota : \iota \in I\}$ *where the systems* $\mathfrak{S}_\iota$ *are locally finite (discrete) and* $|I| \leqslant k$.

"$\aleph_0$-locally finite" and "$\aleph_0$-discrete" spaces are often called, respectively, "σ-locally finite" and "σ-discrete."

A system is thus $\aleph_0$-locally finite ($\aleph_0$-discrete) if and only if it is the union of a countable number of locally finite (discrete) systems. Since every system containing one member is discrete, we see that every countable system is $\aleph_0$-discrete.

13.9 *In a completely normal space (R, τ), given any open covering $\mathfrak{H}$ of R we can find an $\aleph_0$-discrete open covering $\mathfrak{B}$ such that $\mathfrak{B} \leqslant \mathfrak{H}$.*

Proof: Let $\mathfrak{H}_0$ be a star refinement of $\mathfrak{H}$ and let $\mathfrak{H}_n$ be a strong star refinement of $\mathfrak{H}_{n-1}$ for each $n \in \mathbf{N}$ (13.4). Suppose $H_\nu \in \mathfrak{H}_\nu$ for $\nu = 1, \ldots, n$ and

$$H_1 \cap H_2 \neq \emptyset, \ldots, H_{n-1} \cap H_n \neq \emptyset.$$

Since $\mathfrak{H}_n \leqslant \mathfrak{H}_{n-1}$, we first deduce that $H_n \cup H_{n-1} \subseteq \mathrm{St}_{\mathfrak{H}_{n-1}}(H_{n-1})$ and since $\mathrm{St}^*(\mathfrak{H}_{n-1}) \leqslant \mathfrak{H}_{n-2}$, we finally deduce that $H_n \cup H_{n-1} \subseteq H'_{n-2}$ for a certain $H'_{n-2} \in \mathfrak{H}_{n-2}$. Hence $H'_{n-2} \cap H_{n-2} \neq \emptyset$, and thus, proceeding similarly, we get $H_n \cup H_{n-1} \cup H_{n-2} \subseteq H'_{n-2} \cup H_{n-2} \subseteq \mathrm{St}_{\mathfrak{H}_{n-2}}(H_{n-2})$. Proceeding in this way we ultimately have $H_n \cup \ldots \cup H_1 \subseteq \mathrm{St}_{\mathfrak{H}_1}(H_1) \subseteq H_0$ for a certain $H_0 \in \mathfrak{H}_0$. If we therefore put

$$\mathfrak{K}_1 = \mathfrak{H}_1 \qquad \text{and} \qquad \mathfrak{K}_{n+1} = \{\mathrm{St}_{\mathfrak{H}_{n+1}}(K) : K \in \mathfrak{K}_n\},$$

then we have $\mathfrak{K}_n \leqslant \mathfrak{H}_0$ for all $n \in \mathbf{N}$.

Now let $\leqslant$ be a fixed well-ordering of R (well ordering theorem!). For each $n \in \mathbf{N}$ and $p \in R$ suppose

$$L_n(p) = \mathrm{St}_{\mathfrak{R}_n}(p) \cap \complement \cap \{\mathrm{St}_{\mathfrak{R}_{n+1}}(q) : q < p\}$$

$$\mathfrak{L}_n = \{L_n(p) : p \in R\} \qquad \text{and} \qquad \mathfrak{L} = \cup \{\mathfrak{L}_n : n \in \mathbf{N}\}.$$

Since $L_n(p) \subseteq \mathrm{St}_{\mathfrak{R}_n}(p)$ and $\mathfrak{R}_n \ll \mathfrak{H}_0$, we have $\mathfrak{L}_n \ll \mathrm{St}\,(\mathfrak{R}_n) \ll \mathrm{St}\,(\mathfrak{H}_0) \ll \mathfrak{H}$ for all $n \in \mathbf{N}$.

Let x be a fixed point in R and let p be the first point (with respect to the well-ordering of R) such that $x \in \cup\,(\mathrm{St}_{\mathfrak{R}_n}(p) : n \in \mathbf{N}\}$. We are assured of the existence of such a p because $x \in \mathrm{St}_{\mathfrak{H}_1}(x) = \mathrm{St}_{\mathfrak{R}_1}(x)$. We then have $x \in \mathrm{St}_{\mathfrak{R}_n}(p)$ for a certain n and $x \notin \mathrm{St}_{\mathfrak{R}_{n+1}}(q)$ for all $q < p$. Consequently $x \in L_n(p)$ and $\mathfrak{L}$ is a covering of R.

Each of the systems $\mathfrak{L}_n$ is discrete: Given any $x \in R$ we can find an $H \in \mathfrak{H}_{n+1}$ with $H \in \tau x$. Suppose $H \cap \mathfrak{L}_n(p) \neq \emptyset$, $H \cap \mathfrak{L}_n(q) \neq \emptyset$ and, say, $q \leqslant p$. From $H \cap L_n(q) \neq \emptyset$ we deduce that $H \cap K \neq \emptyset$ for a certain $K \in \mathfrak{R}_n$ where $q \in K$. Hence $q \in H \cup K \subseteq K'$ for a $K' \in \mathfrak{R}_{n+1}$, and therefore $H \subseteq \mathrm{St}_{\mathfrak{R}_{n+1}}(q)$. If we had $q < p$, then we would have as a consequence $H \cap L_n(p) = \emptyset$, which is contrary to hypothesis. Hence $q = p$, and the neighborhood H of x possesses a nonempty intersection with at most one of the members of $\mathfrak{L}_n$.

Since $\mathfrak{L}_n \ll \mathfrak{H}$, there exist, by 13.8, a discrete system $\mathfrak{B}_n$ of open sets such that $\mathfrak{L}_n \ll \mathfrak{B}_n \ll \mathfrak{H}$. Then $\mathfrak{B} = \cup \{\mathfrak{B}_n : n \in \mathbf{N}\}$ is an $\aleph_0$-discrete system such that $\mathfrak{B} \ll \mathfrak{H}$, and since $\mathfrak{L} \ll \mathfrak{B}$, $\mathfrak{B}$ is also a covering of R. ∎

13.10 *Given any $\aleph_0$-locally-finite open covering $\mathfrak{H}$ of any space (R, τ) we can find a locally finite covering $\mathfrak{R}$ such that $\mathfrak{R} \ll \mathfrak{H}$.*

Proof: Suppose $\mathfrak{H} = \cup\, \mathfrak{H}_n$ where the $\mathfrak{H}_n$ are locally finite open systems of sets. For each $H \in \mathfrak{H}$ we have $H \in \mathfrak{H}_n$ for some first index n. Let then

$$K_H = H \cap \complement \cup \{H' : H' \in \mathfrak{H}_m\,,\, m < n\}.$$

Since $K_H \subseteq H$, $\mathfrak{R} = \{K_H : H \in \mathfrak{H}\}$ is a refinement of $\mathfrak{H}$. For arbitrary $p \in R$ let n be the smallest index for which there exists an $H_p \in \mathfrak{H}_n$ such that $p \in H_p$. Consequently $p \in K_{H_p}$, and $\mathfrak{R}$ is a covering of R. The neighborhood H_p of p is disjoint from all sets K_H where $H \in \mathfrak{H}_m$ and $m > n$. If then U_ν, for $\nu = 1, \ldots, n$ denotes a neighborhood of p which has a nonempty intersection with at most a finite number of members of $\mathfrak{H}_\nu$, then $V = H_p \cap U_1 \cap \ldots \cap U_n$ must be a neighborhood belonging to τp which has a nonempty intersection with only a finite number of members of $\mathfrak{R}$. Hence $\mathfrak{R}$ is locally finite covering of R and $\mathfrak{R} \ll \mathfrak{H}$. ∎

To summarize, we have the following characterizations:

13.11 *In a T_3-space R, the following conditions are pairwise equivalent:*

(1) *R is paracompact.*

(2) *R is completely normal.*

(3) *Given any open covering $\mathfrak{H}$ of R we can find an $\aleph_0$-discrete open covering $\mathfrak{K}$ such that $\mathfrak{K} \ll \mathfrak{H}$.*

(4) *Given any open covering $\mathfrak{H}$ of R we can find an $\aleph_0$-locally-finite open covering $\mathfrak{K}$ such that $\mathfrak{K} \ll \mathfrak{H}$.*

(5) *Given any open covering $\mathfrak{H}$ of R we can find a (not necessarily open) locally finite covering $\mathfrak{K}$ such that $\mathfrak{K} \ll \mathfrak{H}$.*

Proof: (1) → (2)—was proved in 13.7, while (2) → (3) was proved in 13.9. Since every $\aleph_0$-discrete system is, all the more so, $\aleph_0$-locally finite, (3) → (4). By 13.10, (4) → (5). By 13.6, (5) implies initially that R is completely normal. Suppose, in addition, that $\mathfrak{H}$ is an open covering and $\mathfrak{K}$ a locally finite covering of R with $\mathfrak{K} \ll \mathfrak{H}$. Then, by 13.8, there must exist a locally finite system $\mathfrak{B}$ of open sets such that $\mathfrak{K} \ll \mathfrak{B} \ll \mathfrak{H}$. Since $\mathfrak{K}$ is a covering and $\mathfrak{K} \ll \mathfrak{B}$, $\mathfrak{B}$ must certainly also be a covering. Hence (5) implies (1). ∎

13.12 *Every T_3-Lindelof space is paracompact. Every T_3-space which is countable at infinity is paracompact* (12b).

Proof: In a Lindelöf space every open covering contains a countable covering. And every countable covering is $\aleph_0$-discrete. Hence our first assertion follows from 13.11. Furthermore, let R be the union of countably many completely compact subspaces R_n. If then $\mathfrak{H}$ is an open covering of R, then every subspace R_n is already covered by a finite number of sets belonging to $\mathfrak{H}$ and hence all of R is covered by countably many such sets. Hence R is a Lindelöf space and the second assertion is a consequence of the first assertion. ∎

13.13 *Every pseudometric space is paracompact.*

Proof: Let $\mathfrak{H}$ be an open covering of the pseudometric space (R, δ). To each $p \in R$ let us associate an $H_p \in \mathfrak{H}$ such that $p \in H_p$. Finally, let $n(p)$ be a natural number chosen so that $(n(p))^{-1} \leqslant \delta(p, \complement H_p)$, $\epsilon(p) = (3n(p))^{-1}$ and $U_p = U^{\delta}_{\epsilon(p)}(p)$. Then $\mathfrak{K} = \{U_p : p \in R\}$ is an open covering of R. Let now p be a fixed point and $n_0 = \min\{n(q) : p \in U_q\}$. Let this minimum be attained at the point q_0. From $p \in U_q$ and $r \in U_q$ we deduce

$$\delta(r, q_0) \leqslant \delta(r, q) + \delta(q, p) + \delta(p, q_0) < 3\epsilon(q_0) < \delta(q_0, \complement H_{q_0}).$$

Hence $\mathrm{St}_{\mathfrak{K}}(p) = \cup\{U_q : p \in U_q\} \subseteq H_{q_0}$, and thus $\mathfrak{K}$ is even a star refinement of $\mathfrak{H}$. Hence the space (R, δ) is completely normal. Since, by 8.12, (R, δ) is also a T_3-space, we see, by 13.11, that it is paracompact. ∎

The following concepts are akin to those of locally finite covering and paracompact space.

Definition 13f: *A covering* $\mathfrak{H}$ *of the space* R *is said to be* point finite *if every point of* R *is contained in at most a finite number of sets belonging to* $\mathfrak{H}$. *A topological space is called* metacompact *whenever, given any open covering* $\mathfrak{H}$ *of* R, *we can find a point finite open covering* $\mathfrak{K}$ *such that* $\mathfrak{K} \ll \mathfrak{H}$.

Every finite and every locally finite covering is evidently also point-finite. Every paracompact space is metacompact.

Definition 13g: *Let* $\mathfrak{H}$ *be an open covering of the space* R. *If to each* $H \in \mathfrak{H}$ *we can associate an open set* K_H *such that* $\bar{K}_H \subseteq H$ *and such that* $\mathfrak{K} = \{K_H : H \in \mathfrak{H}\}$ *is a covering of* R, *then* $\mathfrak{K}$ *is said to be a* contraction *of* $\mathfrak{H}$.

13.14 *In a* T_4*-space* R *every open and point-finite covering* $\mathfrak{H}$ *of* R *admits of a contraction.*

Proof: Let φ be a mapping from a subsystem $\mathfrak{H}_\varphi$ of $\mathfrak{H}$ into the system of all open sets of R possessing the following properties:

(a) $H \in \mathfrak{H}_\varphi$ implies $\overline{\varphi(H)} \subseteq H$.

(b) We have $R = \cup \{\varphi(H) : H \in \mathfrak{H}_\varphi\} \cup U \{H \in \mathfrak{H}, H \notin \mathfrak{H}_\varphi\}$.

One such mapping always exists: Let H_0 be a fixed set belonging to $\mathfrak{H}$. Then $A = \complement \cup \{H : H \in \mathfrak{H}, H \neq H_0\}$ is a closed set such that $A \subseteq H_0$ and hence $A \cap \complement H_0 = \emptyset$. By ($\mathbf{T}_4$) there exist open sets U and V such that $A \subseteq U$, $\complement H_0 \subseteq V$ and $U \cap V = \emptyset$. Hence $\bar{U} \subseteq H_0$. If we put $H_\varphi = \{H_0\}$ and $\varphi(H_0) = U$, then conditions (a) and (b) are satisfied. Let us denote by Σ the set of all such mappings. For $\varphi_1, \varphi_2 \in \Sigma$ let us define $\varphi_1 \leqslant \varphi_2$ by

$$\mathfrak{H}_{\varphi_1} \subseteq \mathfrak{H}_{\varphi_2} \quad \text{and} \quad \varphi_1(H) = \varphi_2(H) \quad \text{for all} \quad H \in \mathfrak{H}_{\varphi_1}.$$

This relation is an order on Σ. Now let $\mathbf{K}$ be a chain in the ordered set Σ. By the prescription $\mathfrak{H}_\psi = \cup\{\mathfrak{H}_\varphi : \varphi \in \mathbf{K}\}$ and $\psi(H) = \varphi(H)$ for $H \in \mathfrak{H}_\varphi$, $\varphi \in \mathbf{K}$ we define a mapping ψ belonging to Σ. Property (a) is a direct consequence of the definition of ψ. To prove (b), let p be a fixed point in R. Then p is contained in only a finite number of sets of $\mathfrak{H}$, say $H_1, \ldots, H_n$. Suppose $H_\nu \in \mathfrak{H}_\psi$ and $p \notin \psi(H_\nu)$ for $\nu = 1, \ldots, n_n$ then we could find a $\varphi \in \mathbf{K}$ such that $H_\nu \in \mathfrak{H}_\varphi$ and $p \notin \psi(H_\nu)$ for $\nu = 1, \ldots, n$. But this would be contradictory to the fact that φ possesses property (b). By Zorn's lemma, Σ contains a maximal element φ_0. For this element, we have $\mathfrak{H}_{\varphi_0} = \mathfrak{H}$; for if there did exist an $H_0 \in \mathfrak{H}$ such that $H_0 \notin \mathfrak{H}_{\varphi_0}$, then, with the aid of the previously constructed set U, we could define a mapping $\psi \in \Sigma$ by $\mathfrak{H}_\psi = \mathfrak{H}_{\varphi_0} \cup \{H_0\}$, $\psi(H) = \varphi_0(H)$ for $H \in \mathfrak{H}_{\varphi_0}$ and $\psi(H_0) = U$. We would then have $\varphi_0 < \psi$, contradicting the maximality of φ_0. Hence the sets $K_H = \varphi_0(H)$ constitute a contraction of $\mathfrak{H}$. ∎

For paracompact T_3-spaces we can prove directly and simply that arbitrary open coverings admit of a contraction.

13.15 *In a paracompact T_3-space every open covering admits of a contraction.*

Proof: Let $\mathfrak{H}$ be an open covering of the paracompact T_3-space (R, δ). Given any $p \in R$ we can find an open neighborhood $U_p \in \tau p$ such that $\bar{U}_p \subseteq H$ for a certain $H \in \mathfrak{H}$. Then $\mathfrak{U} = \{U_p : p \in R\}$ is an open covering of R, and for the system $\bar{\mathfrak{U}} = \{\bar{U}_p : p \in R\}$ we have $\bar{\mathfrak{U}} \ll \mathfrak{H}$. Suppose also that $\mathfrak{K}$ is an open and locally finite covering of R such that $\mathfrak{K} \ll \mathfrak{U}$, and for each $H \in \mathfrak{H}$ let $\mathfrak{K}_H$ be the system of all $K \in \mathfrak{K}$ for which we can find a $U \in \mathfrak{U}$ such that $K \subseteq U$ and $\bar{U} \subseteq H$. Let us now define

$$L_H = \cup \{K : K \in \mathfrak{K}_H\}.$$

Then L_H is open, and by 13.1 we have:

$$\overline{L_H} = \cup \{\bar{K} : K \in \mathfrak{K}_H\} \subseteq H.$$

For each $p \in R$ there exists a $K \in \mathfrak{K}$ such that $p \in K$. Since $\mathfrak{K} \ll \mathfrak{U}$ there also exists a $U \in \mathfrak{U}$ such that $K \subseteq U$, and since $\bar{U} = \mathfrak{H}$ we can also find an $H \in \mathfrak{H}$ with $\bar{U} \subseteq H$. Hence $p \in L_H$ and $\mathfrak{L} = \{L_H : H \in \mathfrak{H}\}$ is a contraction of $\mathfrak{H}$. ∎

The notion of contraction can also be used to characterize T_4-spaces.

13.16 *A space R is a T_4-space if and only if every finite open cover of R admits of a* contraction.

Proof: Let A and B be two disjoint closed sets. Then $\{\complement A, \complement B\}$ is a finite open covering of R. If this covering admits of the contraction $\{H, K\}$, i.e., if $\bar{H} \subseteq \complement A$, $\bar{K} \subseteq \complement B$, and $H \cup K = R$, then $U = \complement \bar{H}$ and $V = \complement \bar{K}$ are open sets with $A \subseteq U, B \subseteq V$ and $U \cap V = \emptyset$. Since every finite covering is also point finite, the converse assertion follows from 13.14. ∎

13.17 *Every point-finite covering $\mathfrak{H}$ of R contains a cover $\mathfrak{M}$ of R, which is minimal in the following sense: No proper subsystem of $\mathfrak{M}$ is a covering of R.*

Proof: Let Σ denote the system of all coverings of R contained in $\mathfrak{H}$. Let $\mathbf{K}$ be a chain in Σ, ordered with respect to set-theoretic inclusion. Then $\mathfrak{D} = \cap \{\mathfrak{K} : \mathfrak{K} \in \mathbf{K}\}$ is also a covering of Σ: We certainly have $\mathfrak{D} \subseteq \mathfrak{H}$. Given any $p \in R$, there are only a finite number of sets belonging to $\mathfrak{H}$ and containing p. At least one of these sets must belong to each $\mathfrak{K} \in \mathbf{K}$, and hence to $\mathfrak{D}$. Thus $\mathfrak{D}$ is also a covering of R, and our contention follows from Zorn's lemma. ∎

13.18 *If a metacompact T_1-space (R, τ) is, in addition, $\aleph_0$-compact, then it is completely compact.*

Proof: Let $\mathfrak{H}$ be an open covering of (R, τ). Then there exists a point finite open covering $\mathfrak{K}$ of (R, τ) where $\mathfrak{K} \ll \mathfrak{H}$. By 13.17, $\mathfrak{K}$ contains a minimal covering $\mathfrak{L}$ of R. Then $\mathfrak{L} \ll \mathfrak{K} \ll \mathfrak{H}$ and, by 11.11, all that remains to be proved is that $\mathfrak{L}$ is finite. Since $\mathfrak{L}$ is minimal, in each $L \in \mathfrak{L}$ we can find a point p_L contained in no other member set of $\mathfrak{L}$. If $\mathfrak{L}$ were infinite, then $A = \{p_L : L \in \mathfrak{L}\}$ would be an infinite subset of R and hence, by 11.3, A would possess a point of accumulation q. We would have $q \in L'$ for a certain $L' \in \mathfrak{L}$. By $(\mathbf{T}_1)$ and the fact that L' is open, the set $A \cap L'$ would contain two distinct points, contradicting the definition of the point p_L. ∎

Totally ordered sets are not necessarily paracompact under their order topology. Paracompact spaces can also possess subspaces which need not be paracompact (cf. 13A). It is only in the case of special subspaces that paracompactness is preserved, as the following theorem shows.

13.19 *In a paracompact space (R, τ), let M be a set expressible as the union of a countable number of closed sets.*[5] *If the subspace (M, τ_M) is a T_3-space*[6] *then it is paracompact. If M itself is closed, then without assuming that (M, τ_M) is a T_3-space, it still follows that it is paracompact.*

Proof: Suppose $M = \cup \{M_n : n \in \mathbf{N}\}$ where the M_n are closed. Suppose also that $\mathfrak{H}$ is an open covering of the subspace (M, τ_M). Given any $H \in \mathfrak{H}$ we can find a τ-open subset U_H of R where $H = U_H \cap M$. For every n the systems of sets $\mathfrak{U}_n = \{\complement M_n\} \cup \{U_H : H \in \mathfrak{H}\}$ form an open covering of R. Hence there exist locally finite open coverings $\mathfrak{B}_n$ of R such that $\mathfrak{B}_n \ll \mathfrak{U}_n$. The subsystem $\mathfrak{B}_n' = \{V : V \in \mathfrak{B}_n, V \cap M_n \neq \emptyset\}$ of $\mathfrak{B}_n$ is likewise locally finite and a covering of M_n. Hence we finally see that

$$\mathfrak{W} = \cup \mathfrak{W}_n \qquad \text{where} \qquad \mathfrak{W}_n = \{V' \cap M : V' \in \mathfrak{B}_n'\}$$

is an $\aleph_0$-locally finite open covering of (M, τ_M). From $V \in \mathfrak{B}_n'$ we see that $V' \cap M_n = \emptyset$ and since $\mathfrak{B}_n' \ll \mathfrak{B}_n \ll \mathfrak{U}_n$, we must have $V' \subseteq U_H$ and $V' \cap M \subseteq U_H \cap M = H$ for a certain $H \in \mathfrak{H}$. Hence $\mathfrak{W} \ll \mathfrak{H}$. If now we assume (M, τ_M) to be a T_3-space, the desired contention follows, using 13.11. If M is already closed, put $M = M_1$. The above proof immediately furnishes us with a locally finite open covering $\mathfrak{W}_1$ of (M, τ_M) where $\mathfrak{W}_1 \ll \mathfrak{H}$. ∎

For further properties and characterizations of paracompact spaces, cf. 18.13, 22.8, and 31.7.

Notes and Exercises

13A In a nondenumerable well-ordered set let p be the first point such that the interval $[\leftarrow, p]$ has the power $\aleph_1$. Let $R = [\leftarrow, p]$, $R' = [\leftarrow, p[$, and let τ denote the standard topology of R, τ' that of R'. Then (R', τ') is a subspace of (R, τ). The well-ordered set R is conditionally complete. By 11.III (R, τ)

[5] Such a set is known as an F_σ-set.

[6] By 9.4 this condition is surely fulfilled if (R, τ) is itself a T_3-space.

is therefore completely compact, and *a fortiori* paracompact. But the subspace (R', τ') is nevertheless not paracompact: $\mathfrak{H} = \{[\leftarrow, x[: x \in R'\}$ is an open covering of R. For the open covering $\mathfrak{K}$ of R', suppose $\mathfrak{K} \ll \mathfrak{H}$. For each $K \in \mathfrak{K}$ we have $\sup\{y : y \in K\} = s(K) < x$ for an $x \in R'$. Hence there exist in $\mathfrak{K}$ a countable infinity of sets $K_1, K_2, \ldots$ such that $s(K_1) < s(K_2) < \ldots$. Since $s(K_n) < p$ the sets $[\leftarrow, s(K_n)]$ and $V = U\{[\leftarrow, s(K_n)] : n \in \mathbf{N}\}$ are countable, and $s = \sup\{x : x \in V\}$ is a point in R'. But every τ'-neighborhood of s is disjoint from an infinity of the K_n [34, p. 172].

13B *Exercise:* Let us assume that the totally ordered set $(R, \leqslant)$ is conditionally convergent under its order topology τ. The space (R, τ) is then paracompact if and only if it is representable as a union of countably many bounded closed intervals (or equivalently, if and only if (R, τ) is countable at infinity).

13C *Exercise:* Let M be a subset of a paracompact space and let M satisfy the following condition: Given any open set U with $M \subseteq U$ we can find an F_σ-set A (cf. footnote 5, p. 100) such that $M \subseteq A \subseteq U$. Then the subspace (M, τ_M) is also paracompact.

13D *Exercise:* If in a paracompact space every open subspace is paracompact, then so is every subspace. (Let $\mathfrak{H}$ be an open covering of the arbitrary subspace M. For each $H \in \mathfrak{H}$ there exists a set U_H, open in the whole space, such that $H = U_H \cap M$. Consider the open subspace generated by $U = \cup\{U_H : H \in \mathfrak{H}\}$.

14. Connectedness

A system of sets $\mathfrak{Z}$ is said to be a *decomposition* of the set R if $R = \cup\{Z : Z \in \mathfrak{Z}\}$ and if the members Z of $\mathfrak{Z}$ are all nonempty and pairwise disjoint sets. Whenever $\mathfrak{Z}$ is a decomposition of R we also say that R has been decomposed into the sets of $\mathfrak{Z}$. Suppose the space (R, τ) has been decomposed into two open sets A and B. Then the neighborhoods filter τp of R is completely determined by its traces on A and B. This means that no relation or connection exists between the behavior of the underlying topology on A and its behavior on B —we say in this case that the space is *disconnected.*

Definition 14a: *A space (R, τ) and its topology are both said to be* connected *if R cannot be decomposed into two open sets. A subset M of R is said to be connected whenever the subspace (M, τ_M) is connected.*

14.1 *The following assertions are pairwise equivalent:*

(1) *R is connected.*

(2) *R cannot be decomposed into two closed sets.*

(3) *R cannot be decomposed into two separated sets.*[7]

(4) *The only sets in R that are simultaneously open and closed are $\emptyset$ and R itself.*

Proof: Let $\mathfrak{Z} = \{A, B\}$ be a decomposition of R. Then $A = \complement B$ and $B = \complement A$. Hence A and B are both open if and only if they are both closed—or equivalently, if and only if they are each simultaneously open and closed. If A and B are closed then they are separated. If, conversely, they are separated, then $\bar{A} \cap \complement A = \bar{A} \cap B = \varphi$ and $\bar{B} \cap \complement B = \bar{B} \cap A = \emptyset$. This implies $\bar{A} = A$ and $\bar{B} = B$. ∎

14.2 *A subset M of R is connected if and only if M can be decomposed (in R) into two separated subsets.*

Proof: For any subsets A and B of M, we have, by 9.2,

$$h_{\tau_M}(A) \cap B = h_\tau(A) \cap M \cap B = h_\tau(A) \cap B.$$

Hence A and B are separated in (M, τ_M) if and only if they are separated in R. The theorem then follows from 14.1. ∎

Examples

14.I The trivial topology on any set is connected.

14.II The discrete topology on any set R is disconnected whenever R contains at least two points.

14.III The order topology on a totally ordered set is connected if and only if the order is gap-free and conditionally complete. For by 6.2 the whole space and the empty set are the only sets simultaneously open and closed and conversely. In particular, the ordinary topology on the real line is connected.

14.3 *If $\tau_1 \leqslant \tau_2$ then every τ_1-connected set is also τ_2-connected.*

Proof: Any τ_2-separated set is also τ_1-separated. Hence if M can be decomposed into two τ_2-separated sets it can also be decomposed into two τ_1-separated sets. ∎

14.4 *Let M be a connected subset of R and suppose $M \subseteq N \subseteq \bar{M}$. Then N is also connected.*

Proof: Suppose N is disconnected, and let (A, B) be a decomposition of N into separated sets. Then $A \cap M = \emptyset$, for otherwise we would have $M \subseteq B$ and consequently

$$A = A \cap N \subseteq A \cap \bar{M} \subseteq A \cap \bar{B} = \emptyset,$$

which would be a contradiction. Likewise $B \cap M \neq \emptyset$. Hence $\{A \cap M, B \cap M\}$

[7] Cf. p. 61.

is a decomposition of M into nontrivial separated sets and M is therefore disconnected. ∎

14.5 *Let $\mathfrak{S}$ be a system of connected subsets of the space R possessing the following property: Whenever $\{\mathfrak{S}_1, \mathfrak{S}_2\}$ is a decomposition of $\mathfrak{S}$, then*

$$V_1 = \cup \{S : S \in \mathfrak{S}_1\} \qquad \text{and} \qquad V_2 = \cup (S : S \in \mathfrak{S}_2\}$$

are not separated. Whenever this condition is satisfied, $V = \cup\{S : S \in \mathfrak{S}\}$ is also connected.

Note: The above condition is always satisfied whenever the intersection of all the member sets of $\mathfrak{S}$ is not empty or whenever there is a member S of $\mathfrak{S}$ such that all the other member sets of $\mathfrak{S}$ have a nonempty intersection with S.

Proof: Suppose V is disconnected and (A, B) a decomposition of V into separated sets. From $S \in \mathfrak{S}$ we deduce $S \subseteq A$ or $S \subseteq B$: For if not, then $\{A \cap S, B \cap S\}$ would be a decomposition of S into separated sets, contradicting the fact that S is connected. Let $\mathfrak{S}_1 = \{S : S \subseteq A\}$ and $\mathfrak{S}_2 = \{S : S \subseteq B\}$. Then $\{\mathfrak{S}_1, \mathfrak{S}_2\}$ is a decomposition of $\mathfrak{S}$ such that $V_1 \subseteq A$ and $V_2 \subseteq B$. This means that V_1 and V_2 are separated, contradicting the hypothesis. ∎

14.6 *Let the T_4-space (R, τ) be completely compact (k-compact). If the filter $\mathfrak{a}$ of R (where $\text{Rk}(\mathfrak{a}) \leqslant k$) possesses a basis whose members are all connected, then $D = \cap \{\bar{A} : A \in \mathfrak{a}\}$ is also a connected set.*

Proof: Let (M, N) be a decomposition of D, wherein M and N are closed in D. Since D is itself closed, M and N must also be closed in R (9.3). By $(\mathbf{T}_4)$, there exist disjoint open sets U, V such that $M \subseteq U$ and $N \subseteq V$. Let $\mathfrak{b} = \mathfrak{a} \wedge \overline{\mathbf{C}(U \cup V)}$. (We then have $\text{Rk}(\mathfrak{b}) \leqslant \text{Rk}(\mathfrak{a}) \leqslant k$.) Let $F = \cap \{\bar{B} : B \in \mathfrak{b}\}$. Since $\mathfrak{b} \leqslant \mathfrak{a}$ we have, on the one hand, $F \subseteq D = M \cup N \subseteq U \cup V$, and since $\mathfrak{b} \leqslant \overline{\mathbf{C}(U \cup V)}$ we have, on the other hand, $F \subseteq \mathbf{C}(U \cup V)$. Hence $F = \varnothing$. By 11.19 we have $\mathfrak{b} = \mathfrak{o}$ and consequently there exists a connected set $A \in \mathfrak{a}$ such that $A \subseteq U \cup V$. Since U and V are separated, we must have $A \subseteq U$ or $A \subseteq V$, hence either $D \subseteq \bar{A} \subseteq \mathbf{C}V$ or $D \subseteq \bar{A} \subseteq \mathbf{C}V$. Since $M \subseteq D \cap U$ and $N \subseteq D \cap V$ we finally have $M = \varnothing$ or $N = \varnothing$, but this contradicts the assumption that $\{M, N\}$ is a decomposition. ∎

14.7 *Every nonempty connected subset M of the space R is contained in exactly one maximal connected subset of R.*

Proof: By 14.5 (cf. Note) the union of all connected sets S such that $M \subseteq S$ is also connected. ∎

The empty set is always connected, as are all one-point sets. Hence in any (nonempty) space we are sure that there exists at least one (nonempty) connected subset.

Definition 14b: *The maximal connected subsets of a space are called its* (*connected*) components.

14.8 (1) *Every component is a closed set.*

(2) *The collection of all the components of a nonempty space forms a decomposition of the space.*

(3) *A space is connected if and only if it has only one component.*

(4) *A space is connected if and only if, given any two distinct points, there exists a connected subset M containing both of them.*

Proof: (1) If K is a component, then by 14.4 $\bar{K}$ is also connected. Since K is maximal, $K = \bar{K}$.

(2) Since every one point set is connected in any nonempty space, we see that the components are nonempty and each point is contained in some component. If K_1 and K_2 are distinct components, we have $K_1 \cap K_2 = \emptyset$: For, otherwise, using 14.5 (cf. Note) we should have $K_1 \cup K_2$ connected, contradicting the maximality of K_1 and K_2.

(3) If R is connected, it is the unique component. Conversely, if $R \neq \emptyset$ and possesses only one component, then, by (2) above, this component equals R and hence R is connected by definition.

(4) If R is connected, we can take $M = R$. Conversely, let R be disconnected. Then, by (3) above, there exist at least two components K_1 and K_2 and hence two points, $p \in K_1$ and $q \in K_2$ cannot then lie in the same connected component. ∎

The components of a subspace generally are not the intersections of the components of the whole space and the subspace. The components of a closed subspace of R are themselves closed in R. However, the components of a space are not generally open. Nevertheless, if a space possesses only a finite number of components, say $K_1, \ldots, K_n$ then they must be open because $K_{\nu_0} = \complement \cup \{K_\nu : \nu = 1, \ldots, n; \nu \neq \nu_0\}$.

14.9 *In the space (R, τ) let K be the component containing the point p and let $\mathfrak{S}$ be the system of the sets S containing p which are simultaneously open and closed. Then $K \subseteq \cap \{S : S \in \mathfrak{S}\}$. This inequality becomes an equality when (R, τ) is a completely compact T_4-space.*

Proof: Since K is connected, we either have $K \cap S = \emptyset$ or $K \cap S = K$ for all $S \in \mathfrak{S}$. From $p \in K \cap S$ we see that $K \cap S = K$ and hence $K \subseteq S$. Let us now suppose that τ is a completely compact T_4-topology. In order to prove the second contention above, we need only prove that $D = \cap \{S : S \in \mathfrak{S}\}$ is connected. Suppose there exists a decomposition $D = \{A, B\}$ of D into two sets closed in D. Without loss of generality we may suppose that $p \in A$. Since D itself is closed, the sets A and B are closed in R and, by ($\mathbf{T}_4$), they possess open neighborhoods U and V such that $U \cap V = \emptyset$. The system $\mathfrak{S}$ is a filter basis, and $\{\complement S : S \in \mathfrak{S}\}$ is an open covering of the closed (and therefore completely compact) set $\complement U \cap \complement V$. Hence $\complement U \cap \complement V \subseteq \complement S$ for a certain $S \in \mathfrak{S}$, and consequently $S \cap U = S \cap \complement V$. This implies that the set $S' = S \cap U$ is both open and closed. Since $p \in A \subseteq D \cap U \subseteq S \cap U = S'$, we see that $S' \in \mathfrak{S}$, but this leads to the contradiction $B \subseteq D \subseteq S' \subseteq U$. ∎

By 11.17 and the above theorem we see that in a completely compact T_2-space the intersection of all the neighborhoods of p which are simultaneously open and closed is the component containing p.

Definition 14c: *A space and its topology are said to be* totally disconnected *if all its components are one-point sets.*

14.10 *A totally disconnected pseudometric space must be a metric space. Every countable metric space is totally disconnected. The order topology on any countable totally ordered set is totally disconnected.*

Proof: Let the pseudometric space (R, δ) be totally disconnected. For each $p \in R$ the set $\{p\}$ is connected, and, by 14.4, so is the set $\{\bar{p}\}$. Since $\{p\}$ is also a component, we must have $\{p\} = \{\bar{p}\}$. By 8.1, (R, δ) is a T_1-space and by 8.12, δ is a metric.

Every one-point set is manifestly both connected and totally disconnected —hence let us assume $2 \leqslant |R| \leqslant \aleph_0$. Let δ be a metric on R and M a subset of R containing at least two points. Let p be a fixed point of M and let $T = \{\delta(p, q) : q \in M\}$. Then T is a subset of $\mathbf{R}$ with $0 \in T$ and $2 \leqslant |T| \leqslant \aleph_0$. If a is a positive number belonging to T, because of the nondenumerability of the set of reals between 0 and a, there must exist a real number b such that $0 < b < a$ and $b \notin T$. The sets

$$A = \{q : q \in M, \delta(p, q) < b\} \quad \text{and} \quad B = \{q : q \in M, \delta(p, q) > b\}$$

are separated and together form a decomposition of M. Hence every subset containing more than one point is disconnected and the space (R, δ) is totally disconnected.

Finally, let $(R, \leqslant)$ be a totally ordered countable set with order topology τ and let M be a connected subset. By 14.III, M is gap-free and conditionally complete. If M were infinite, then M would have to be order-isomorphic to an interval of rational numbers, by 2.2. But since such an interval can never be conditionally complete, M must be finite. By 8.11, τ is a T_1-topology and the one-point subsets of R are therefore closed. Hence if M has but one point it must be connected. ∎

The set of rational numbers with the usual topology furnishes us with an example of a totally disconnected space. The following example is even more important for its applications.

Consider the real line with the usual metric δ. To every interval $J = [a, b]$ let us associate the set

$$J' = \left[a, a + \frac{b-a}{3}\right] \cup \left[a + 2\frac{b-a}{3}, b\right].$$

Let M designate the union of a finite number of such intervals $J_1, ..., J_k$ and let $M' = J_1' \cup ... \cup J_k'$. Then M is also the union of a finite number of closed intervals, and we have $M' \subseteq M$. We can inductively define a monotonically decreasing sequence of sets $(T_n)_{n \in \mathbf{N}}$ by

$$T_0 = [0, 1] \quad \text{and} \quad T_{n+1} = T_n'.$$

Let now $D = \cap\{T_n : n \in \mathbf{N}\}$. This subspace (D, δ) of $\mathbf{R}$ is known as the *Cantor Discontinuism.*[8] From the very definition of the Cantor Discontinuum: The set T_n is the union of 2^n closed and pairwise disjoint intervals $J_1^n, ..., J_{2^n}^n$. Hence the sets T_n are themselves closed. Each of the intervals J_ν^n has a diameter $\delta(J_\nu^n) = 3^{-n}$, and any two such intervals are at a distance $\geqslant 3^{-n}$.

14.11 *The Cantor Discontinuum D is a completely compact, completely normal, totally disconnected space. It has a countable basis and its cardinality is uncountable.*

Proof: Since it is a metric space, D is completely normal. Since the sets T_n are closed, D is a closed subset of the completely compact interval $T_0 = [0, 1]$, and hence completely compact. If p and q are two distinct points in the subset M of D, then $\delta(p, q) > 3^{-n}$ for an integer n large enough. Hence p and q cannot belong to the same interval J_ν^n. Without any loss of generality, we may suppose $p < q$. Hence there exists a real number x such that $p < x < q$ and $x \notin D$. The sets

$$A = M \cap [\leftarrow, x[\quad \text{and} \quad B = M \cap]x, \rightarrow]$$

are then both open in M and $\{A, B\}$ is a decomposition of M. Hence every subset of D containing at least two points is disconnected and D itself is totally disconnected. Since D is a subspace of the real line, by 10.4 it must possess a countable basis. With a view towards future applications, we have given an explicit description of a countable basis of D: the intervals J_ν^n are open in T_n. Hence $\mathfrak{B} = \{J_\nu^n \cap D : n \in \mathbf{N}, \nu = 1, ..., 2^n\}$ is a countable system of sets open in D. For a set M open in D and the point $p \in M$ we have $\delta(p, \complement M) = \epsilon > 0$. Suppose $3^{-n} < \epsilon$. If $p \in J_\nu^n$, then $J_\nu^n \subseteq U_\epsilon(p)$, hence $J_\nu^n \cap D \subseteq U_\epsilon(p) \cap D \subseteq M$. Hence $\mathfrak{B}$ is even a basis of the Cantor Discontinuum. Every real number x of the form

$$x = \sum_{\nu=1}^{\infty} \frac{x_\nu}{3^\nu} \quad \text{where} \quad x_\nu = 0 \quad \text{or} \quad x_\nu = 2$$

[8] By a *continuum* we mean any closed connected set. A *connected open set* is called a *domain.*

lies in D: Every partial sum of this series is the end-point of an interval J_ν^n, hence a point in D. Since D is completely compact, the limit x of this series must also lie in D. But the set of all such numbers x has the cardinality $2^{\aleph_0} = c > \aleph_0$. ∎

Notes and Exercises

14A *Exercise:* The topologies τ^k (cf. 4.III) are connected.

14B In the plane $\mathbf{R}^2$ let M_n be the set of all points (x, y) where $x = 1/n$ and $0 \leqslant y \leqslant 1$. Let M denote the union of the points (0, 0) and (0, 1) and of all the sets $M_n (n \in \mathbf{N})$.

Exercise: Describe all the components of the subspace M. Every subset, both open and closed in M, contains either both (0, 0) and (0, 1) or else neither of them [24, p. 60].

14C In a pseudometric space (R, δ) let p be a fixed point, and let M be the set of all points q verifying the following condition: Given any $\epsilon > 0$, there exists a finite number of points $a_0, \ldots, a_n$ with $a_0 = p$, $a_n = q$, and $\delta(a_\nu, a_{\nu+1}) < \epsilon$. The points of M are said to be ***linked to p in chain.***

Exercise

1. M is both an open and closed set.

2. If (R, δ) is connected, then any two of its points are linked to each other in chain.

3. The converse of the above results is false—cf. the set theoretic union of the graph of a hyperbola and its asymptotes.

4. But if (R, δ) is completely compact the converse of (2) above is valid [34, p. 115].

14D *Exercise:* If a T_2-space is such that the closure of any open set is also open, then the space must be totally disconnected [34, p. 115].

A completely compact T_2-space where the closure of every open set is itself open, is called a *Stone Space.*

15. Local Connectedness

As was the case for compactness, the localization of the notion of connectedness is of great importance.

Definition 15a: *A space R and its topology τ are said to be* locally connected *if, for every $p \in R$, the neighborhoods filter τp possesses a basis all of whose member sets are connected.*

The localization of connectedness differs appreciably from that of compactness: In the case of compactness we only demanded that every point p possess a completely compact, or k-compact, neighborhood. In the case of connectedness we demand that every point p possess a neighborhoods basis, each of whose members has to be a connected set. This difference means that local connectedness is not a true generalization of connectedness, as the following example makes clear.

15.I On the real line let J_n denote the open interval $]n, n + \frac{1}{2}[$ and let $A_n = \cup \{J_\nu : \nu \geqslant n\}$. Then $\mathfrak{A} = A_n : n \in \mathbf{N}\}$ is a filter basis. In addition, let τ denote the standard topology on R. By the rule

$$\tau'p = \tau p \quad \text{for} \quad p \neq 0 \qquad \text{and} \qquad \tau'0 = \tau 0 \vee \overline{\mathfrak{A}}$$

we can define a new topology τ' on R such that $\tau' > \tau$. By 14.3, since τ is connected, so is τ'. Nevertheless τ' is not locally connected, for every sufficiently small neighborhood $U' \in \tau'0$ contains an interval which is both open and closed in U'.

On the other hand, a locally connected space need not be connected: The subspace (M, τ_M) of $\mathbf{R}$ where $M = J_1 \cup J_2$ is locally connected but not connected.

15.II Consider a totally order set $(R, \leqslant)$ under the order topology. For such a space, connectedness implies local connectedness. For if R is gap-free and conditionally complete, so is every interval of R (cf. 14.III).

As we remarked in the preceding section, the components of a space need not be open. But in locally connected spaces we do have the following result.

15.1 *A space is locally connected if and only if the components of all open subsets are themselves open (by the components of a subset we mean, of course, the components of the subspace associated with this subset).*

Proof: Let (R, τ) be locally connected, let M be an open subset of R, and K a component of M. Every point $p \in K$ then possesses a connected neighborhood U where $U \subseteq M$. Since K is the largest connected subset of M containing p, we see that $U \subseteq K$ and K is thereby open.

Conversely, let the components of open subsets of R be assumed also open. If then U is an open neighborhood belonging to τp, then that component of U which contains p is likewise an open neighborhood of p. Hence τp possesses a basis made up of connected sets. ∎

The above proof even tells us that in a locally connected space every neighborhoods filter possesses a basis made up of connected *open* sets. The possible types of bases of a space can be used to characterize local connectedness.

15.2 *A space is locally connected if it possesses a basis made up of connected sets.*

Proof: Since in a locally connected space every neighborhoods filter possesses a basis composed of open, connected sets, the system of all connected, open sets constitutes a basis of the whole space. Conversely, suppose the whole space possesses a basis all of whose sets are connected. This basis certainly contains a basis for the neighborhoods filter of any point of the space. ∎

15.3 *Let Σ be a system of locally connected topologies on R. Then the sup topology $\tau = \bigvee \{\tau' : \tau' \in \Sigma\}$ is locally connected.*

Proof: By 15.1, our task is to show that, given a τ-open subset M of R, then every one of its components K is itself τ-open. For each $\tau' \in \Sigma$, M is a τ'-open set (7.1). Hence $p \in K$ implies the existence of a τ'-connected neighborhood $U' \in \tau' p$ such that $U' \subseteq M$. By 14.3, U' is also τ-connected and thereby a subset of K. This means that K is a τ'-open set for all $\tau' \in \Sigma$, and hence, by 7.4, also τ-open. ∎

The discrete topology is always locally connected. Hence the above theorem implies that, given any topology τ, there exists a coarsest locally connected topology τ' such that $\tau' \leqslant \tau$.

15.4 *Let (R, τ) be a T_3-space. If, given any finite open covering $\mathfrak{H}$ of R we can find a finite closed covering $\mathfrak{K}$ where $\mathfrak{K} \lll \mathfrak{H}$ and all the member sets of $\mathfrak{K}$ are connected, then (R, τ) must be locally connected.*

Conversely, in any completely compact, locally connected T_3-space, given an open covering $\mathfrak{H}$ we can always find a finite closed covering $\mathfrak{K}$ such that $\mathfrak{K} \lll \mathfrak{H}$ and all the members of $\mathfrak{K}$ are connected sets.

Proof: Let U be a neighborhood belonging to τp. By $(\mathbf{T_3})$ there exists a $V \in \tau p$ such that $\bar{V} \subseteq U$. Then $\mathfrak{H} = \{U, \complement\bar{V}\}$ is an open covering of R. By hypothesis there exists a closed covering $\mathfrak{K} = \{K_1, \ldots, K_n\}$ where $\mathfrak{K} \lll \mathfrak{H}$ and all the sets belonging to $\mathfrak{K}$ are connected. Suppose $p \in K_1 \cap \ldots \cap K_s$ but $p \notin K_{s+1} \cup \ldots \cup K_n$. Let $W = K_1 \cup \ldots \cup K_s$. Then by 14.5 (cf. Note), W is connected. Since $\mathfrak{K} \lll \mathfrak{H}$, $p \in K_\sigma$ impl $K_\sigma \subseteq U$. Hence $W \subseteq U$. Finally we see that $\complement(K_{s+1} \cup \ldots \cup K_n)$ is an open set contained in W and containing the point p, whence $W \in \tau p$.

Let us now suppose that (R, τ) is a completely compact, locally connected T_3-space and that $\mathfrak{H}$ is an open covering of R. For each $p \in R$ we can find a set $H_p \in \mathfrak{H}$ such that $p \in H_p$. By $(\mathbf{T_3})$ and by the local connectedness of R, there exists a connected open neighborhood $V_p \in \tau p$ where $\bar{V}_p \subseteq H_p$. Since the whole space is completely compact the covering $\mathfrak{K}' = \{V_p : p \in R\}$ contains a finite covering $\mathfrak{K}'' = \{V_p : p \in R\}$. Then $\mathfrak{K} = \{\bar{K} : K \in \mathfrak{K}''\}$ is a finite closed covering and $\mathfrak{K} \lll \mathfrak{H}$. Since the sets belonging to $\mathfrak{K}''$ are connected by construction, so are the sets belonging to $\mathfrak{K}$. ∎

The following theorem gives a characterization of connected order topologies.

15.5 *The topology of a connected space (R, τ) is in reality nothing other than the order topology on a totally ordered R, if and only if it is a locally connected T_1-topology such that, given any three distinct, nonempty, connected proper subsets of R, then there are always two of them which do not form a covering of R.*

Proof: Let us first suppose that $(R, \leqslant)$ is a totally ordered set with connected order topology τ. Then τ is certainly a T_1-topology (8.11). By 15.II, τ is also locally connected. Let us suppose that M is a connected subset of R, not containing the point p. Then we must have $]p, \rightarrow] \cap M = \emptyset$ or $[\leftarrow, p[\cap M = \emptyset$, for otherwise these separated sets would form a decomposition of M. Hence a proper, connected subset of R can not contain both an interval of the form $[\leftarrow, x]$ and an interval of form $[y, \rightarrow]$. Hence, given any three such sets, there are always two of them which do not form a covering of R.

We shall use the following lemmas to prove the converse.

Lemma 1: *For every point $p \in R$ and every component K of $\complement p$ we have $\bar{K} - K \cup \{p\}$.*

By $(\mathbf{T}_1)$, $\complement p$ is an open subset of R, and so is K, by 15.1. K is closed in $\complement p$, by 14.8. Hence $\bar{K} \subseteq K \cup \{p\}$. If $p \notin \bar{K}$, then it would follow that K is closed in R, which contradicts the fact that R is connected.

Lemma 2: *For each $p \in R$, $\complement p$ possesses at most two components.*

By 14.4, 14.5, and Lemma 1, the union of the closures of any number of components of $\complement p$ is itself connected. If $\complement p$ possesses more than two components, let us denote any two of them by K_1 and K_2 and let V denote the union of the closures of the remaining components. Consider

$$M_1 = \bar{K}_1 \cup V, \qquad M_2 = \bar{K}_2 \cup V, \qquad M_3 = \bar{K}_1 \cup \bar{K}_2.$$

These sets are nonempty, connected, proper subsets of R. In contradiction to the hypothesis, any two of them form a covering of R.

Lemma 3: *There exist at most two points in R having connected complements.*

For if the complements of three distinct points p were connected, then they would give an example of three sets so constituted as to contradict the hypothesis.

If p is a point whose complement is decomposable into two components, let us denote these components by K_p and K'_p. But if $\complement p$ is connected, then K_p and K'_p must denote the sets $\complement_p$ and $\emptyset$. (Just which of K_p and K'_p shall denote the empty set will be decided later.)

Lemma 4: *If $p \neq q$ it follows that: Exactly one of the sets K_p, K'_p is contained in K_q or K'_q.*

If $\complement p$ is connected, then, since $q \in \complement p$, $\complement p$ is contained in neither K_q nor K'_q, while $\emptyset$ is, of course, contained in both K_q and K'_q. If $\complement p$ is disconnected we may assume, without loss of generality, that $p \in K_q$ and $q \in K_p$. Since $q \in K_p$, we see that K_p is contained in neither K_q nor K'_q. Since $q \notin K'_p$,

$$K'_p = (K'_p \cap K_q) \cup (K'_p \cap K'_q).$$

The intersections on the right side of this equation are disjoint open sets, and since K'_p is connected, one of these intersections must be empty, which immediately establishes our contention.

A finite T_1-space is connected if and only if it contains one point. We, therefore, assume $|R| \geqslant \aleph_0$. Hence, by Lemma 3, we can choose a fixed point $x \in R$ such that $\complement x$ is disconnected. We now normalize our notations for the sets K_p, K'_p as follows: By lemma 4, one of the sets K_p, K'_p at most is contained in K_x. If one of these sets is effectively contained in K_x we shall denote this set by K_p. Otherwise, K_p or K'_p is contained in K'_x. But then K'_x is contained in neither K_p nor K'_p (for K'_x is nonempty). Again, by Lemma 4, we must therefore have $K_x \subseteq K_p$ or $K_x \subseteq K'_p$. In this eventuality we choose our notation so that $K_x \subseteq K'_p$. With these notational conventions, exactly one of the cases, $K_p \subseteq K_x$ or $K_x \subseteq K_p$, respectively, can occur.

Lemma 5: *For points $p, q \in R$ let us define the relation $p \leqslant q$ if $K_p \subseteq K_q$. Then $\leqslant$ is an order relation of R.*

Reflexivity and transitivity follow immediately from the definition. If $p \leqslant q$ and $q \leqslant p$, then $K_p = K_q$, and hence $\overline{K_p} = \overline{K_q}$. By Lemma 1 we must have $K_p \cup \{p\} = K_p \cup \{q\}$, and since $p \notin K_p$, it follows that $p = q$. Lastly, let us suppose that neither $p \leqslant q$ nor $q \leqslant p$. Then the sets

$$M_1 = K_p \cup \overline{K'_q} \qquad \text{and} \qquad M_2 = \overline{K'_p} \cup K_q$$

are, in any case, nonempty. Since $K_p \subseteq K_q$, we have $K_p \cap \overline{K'_q} \neq \emptyset$. Hence M_1 is also connected. Similarly, M_2 is connected. Finally M_1 and M_2 are both proper subsets of R—for if $M_1 = R$ then $K_q = \complement\overline{K'_q} \subseteq K_p$, contradicting the hypothesis. There are now two possibilities: (a) $K_p \subseteq K_q$. In this case we then also have $K_q \subseteq K_x$, for otherwise $K_x \subseteq K_q$ and this would imply $K_p \subseteq K_q$, contrary to our assumption. Hence $\overline{K'_x} \subseteq \overline{K'_p} \cap \overline{K'_q}$ and the sets M_1, M_2, and K_x serve as a contradiction to the hypothesis of our theorem: Whenever case (a) fails to hold, case (b) must hold, i.e., $K_x \subseteq K_p$. Just as above, one deduces that $K_x \subseteq K_q$ and now the sets M_1, M_2, and $\overline{K'_x}$ furnish a contradiction to the hypothesis of our theorem. Consequently we always have $p \leqslant q$ or $q \leqslant p$.

If $\complement p$ is connected, then p is manifestly the initial or terminal point of the ordered set R.

Lemma 6: *The given topology τ on R coincides with the order topology on the totally ordered set R.*

Let us first suppose that p is neither the initial nor terminal point of R; i.e., let us suppose $\complement p$ is connected. If a and b are two points in R with $a < p < b$, then the interval $]a, b[= K'_a \cap K_b$ is an open neighborhood of p under the order topology. Since R is locally connected, the sets K'_a and K_b must be open, for they are components of open sets. This implies that $]a, b[$ is an open neighborhood of p in the topology τ. Conversely, let U be an open neighborhood belonging to τp. Without loss of generality, we may assume U to be connected. By Lemma 1, $p \in \overline{K_p}$ and $p \in \overline{K'_p}$. Hence there exist points $a, b \in U$ with $a < p < b$. If $]a, b[$ were not contained in U, then there would exist a $q \in]a, b[$ such that $q \notin U$. Then we would have $U = (U \cap K_q) \cup (U \cap K'_q)$. But this would contradict the connectedness of U, for since $a \in U \cap K_q$ and $b \in U \cap K'_q$ these two sets cannot be empty. Hence U is also a neighborhood of p under the order topology. If p is the initial or end point of R, reasoning analogous to that above leads us to the same conclusion. ∎

Definition 15b: *A finite system of sets* $\mathfrak{B} = \{V_1, ..., V_n\}$ *is said to be a* connection *of the points* p *and* q, *if the following conditions are verified:*

(a) *The sets belonging to* $\mathfrak{B}$ *are open and connected.*

(b) *We have* $p \in V_1$ *and* $q \in V_n$.

(c) *For* $\nu = 1, ..., n$ *we have* $V_\nu \cap V_{\nu+1} \neq \emptyset$.

A connection $\mathfrak{B}$ *is called a* minimal connection *of* p *and* q *whenever no proper subsystem of* $\mathfrak{B}$ *is a connection of* p *and* q.

Evidently every connection of p and q contains at least one minimal connection of p and q. By 14.5, the union of all the member sets of a connection is itself open and connected.

15.6 *A space* R *is connected if and only if, given any two of its points* p *and* q, *we can find a connection of* p *and* q.

Proof: If R is connected, then the system $\{R\}$ is itself a connection of any two points of R. Conversely, if we can associate a connection $\mathfrak{B}$ with any two points p, q of R, then $M = \cup \{V : V \in \mathfrak{B}\}$ is a connected set containing p and q. Hence, by 14.8, R must be connected. ∎

15.7 *Let* M *be a connected subset of the locally connected space* (R, τ) *and let* $\mathfrak{H}$ *be an open covering of* M. *Then, given any two points* $p, q \in M$ *we can find a connection* $\mathfrak{B}$ *of these points such that* $\mathfrak{B} \ll \mathfrak{H}$.

Proof: Let p be a fixed point in M, and A the set of all points $a \in M$, for which we can find a connection $\mathfrak{B}_a$ of p and a such that $\mathfrak{B}_a \ll \mathfrak{H}$. If H is a member set of the family $\mathfrak{H}$ such that $p \in H$, then, by the local connectedness of R, there exists a connected open neighborhood $U \in \tau p$ such that $U \subseteq H$.

Then the system $\{U\}$ is a connection of p to p, for which $\{U\} \ll \mathfrak{H}$. Hence $p \in A$ without any doubt whatsoever. Let us further suppose that for $a \in A$ the connection $\mathfrak{B}_a = \{V_1, ..., V_n\}$. Then $V_n \cap M$ is a neighborhood in M of a, and $\mathfrak{B}_a$ is also a connection of p and a' for all $a' \in V_n \cap M$. Hence the set A is open in M. Finally, let q be a point in $\bar{A} \cap M$. Then we can again find a set $H \in \mathfrak{H}$ such that $q \in H$ and again find an open connected neighborhood $U \in \tau q$ with $U \subseteq H$. Since $q \in \bar{A}$ there exists an $a \in A$ such that $a \in U$. Then $\mathfrak{B} = \mathfrak{B}_a \cup \{U\}$ is a connection of p and q such that $\mathfrak{B} = \mathfrak{H}$. Consequently $q \in A$ and therefore A is closed in M. Since A is nonempty and both open and closed in M, while M itself is connected, we must have $A = M$. ∎

Notes and Exercises

15A *Exercise:* The topologies τ^k (cf. 4.III) are locally connected.

15B *Exercise:* The spaces $\mathbf{R}^n$ and $\mathbf{C}^n$ and also the Hilbert Space (cf. 6D) are each both connected and locally connected spaces.

15C In the plane $\mathbf{R}^2$ let the set M be composed of all points on the curve $y = \sin(1/x)$ $(x \neq 0)$ and all points $(0, y)$ where $-1 \leqslant y \leqslant +1$.

Exercise: The above subspace M of $\mathbf{R}^2$ is connected, but not locally connected [34, p. 119].

IV

Mappings

We can discover and establish the relationships between two topological spaces (R, τ) and (R', τ') by examining maps from R into R'. Since (R, τ) is not to be thought of as a set but rather as a set carrying a certain topology, it behooves us to only consider maps from R into R', which are, in a certain sense, compatible with the topologies τ and τ'. The most important class of such maps is the class of continuous maps. In the second half of this chapter and with the aid of the theory of continuous maps, we shall establish general principles for creating new topological spaces out of old ones.

A special subclass of continuous maps will allow us to define an appropriate notion of "isomorphism" between topological spaces. In general, an isomorphism between two sets is a 1-1 mapping between these sets "preserving" whatever structure we are interested in studying. In algebra, this preservation of structure is achieved by demanding that the map in question commute with the underlying algebraic structures. Likewise, an order isomorphism commutes with the order relations on the underlying sets. Topological isomorphism can also be characterized by appropriate commutativity conditions.

16. Topological Maps, Homeomorphisms

Let (R, τ) and (R', τ') be two topological spaces and let φ be a 1-1 mapping of R onto R'. As we have seen in Section 3, φ also induces a 1-1 mapping of the lattice of filters $F(R)$ onto the lattice of filters $F(R')$. Since τ maps R into $F(R)$ while τ' maps R' into $F(R')$, we see that the composite maps $\varphi\tau$ and $\tau'\varphi$ are both mappings of R into $F(R')$. If these two maps are equal, then $\varphi\tau p = \tau'\varphi p$ for all $p \in R$. This commutativity property means that φ not only maps points of R onto points of R', but also maps the neighborhoods filter of a point p of R into the neighborhoods filter of the point φp of R'.

Definition 16a: *A mapping* $\varphi : R \to R'$ *is said to be a* topological mapping *or a* homeomorphism *of the space* (R, τ) *onto the space* (R', τ') *whenever* φ *is a 1-1 mapping of* R *onto* R' *obeying the commutativity condition,* $\varphi\tau = \tau'\varphi$.

The spaces (R, τ) *and* (R', τ') *are said to be homeomorphic or* topologically equivalent *if there exists a topological map or homeomorphism of* (R, τ) *onto* (R', τ'). *We symbolize such a homeomorphism by* $(R, \tau) \cong (R', \tau')$.

The identity map of any space onto itself is a homeomorphism. If φ is a homeomorphism, so is φ^{-1}. The multiplication, or composition, of homeomorphisms yields again a homeomorphism. The relation of homeomorphism between spaces is therefore an equivalence relation in the class of all topological spaces—equivalent spaces are said to belong to the same homeomorphism class.

Examples

16.I The real line is homeomorphic to the interval $]-1, +1[$: The contraction map

$$\varphi x = \frac{x}{1 + |x|}$$

will carry out this homeomorphism.

16.II Whenever the underlying set R is uncountable, the topological spaces $(R, \tau^{\aleph_1})$ and $(R, \tau^{\aleph_0})$ cannot be homeomorphic (cf. 4.III): A set U where $p \in U$ and $|\complement U| = \aleph_0$ is a typical neighborhood belonging to $\tau^{\aleph_1}p$. Since 1-1 mappings cannot alter cardinalities, such a map cannot send U onto a neighborhood belonging to $\tau^{\aleph_0}p$.

By topological statements, concepts, properties, etc. we mean all such which remain invariant under homeomorphism. Concepts like "cardinality," "filter," "ultrafilter," "sup," and "inf" of sets or filters are invariant under 1-1 mappings. Statements containing such concepts and otherwise containing only purely logical components, or only mention of the topology τ at hand, must therefore be topological statements—for homeomorphisms send neighborhoods filters into neighborhoods filters. All of the notions we have previously examined (save for those of Section 7), such as open set, closed set, closure, interior, basis, separation, compactness, connectedness are topological concepts. With respect to their topological properties we can say that homeomorphic spaces are really identical and we can think of them as being realizations of one and the same abstract structure.

16.1 *Let* φ *be a 1-1 mapping of* R *onto* R'. *Then the following statements are pairwise equivalent:*

(1) φ *is a homeomorphism of* (R, τ) *onto* (R', τ').

(2) M *is* τ*-open* (τ*-closed*) *if and only if* φM *is* τ'*-open* (τ'*-closed*).

(3) *We have* $\varphi(\mathring{M}) = \widehat{\varphi M}^{\circ}$ ($\varphi\bar{M} = \overline{\varphi M}$).

(4) *φ maps every basis (subbasis) of (R, τ) onto a basis (subbasis of (R', τ')).*

(5) *There exists at least one basis (subbasis) of (R, τ) mapped by φ onto a basis (subbasis of (R', τ')).*

Proof: From $\varphi\tau = \tau'\varphi$ we deduce the equivalence of the conditions $\tau p \leqslant \overline{M}$ and $\tau'\varphi p \leqslant \overline{\varphi M}$. Hence M is τ-open if and only if φM is τ'-open. This tells us immediately $\varphi \mathring{M} = \widehat{\varphi M}^{\circ}$, whence

$$\varphi\tau p = \varphi\{M : p \in \mathring{M}\} = \{\varphi M : \varphi p \in \widehat{\varphi M}^{\circ}\} = \tau'\varphi p.$$

Since the parenthetical contentions in (2) and (3) above follow from the non-parenthetical contentions by taking complements, we see that (1), (2), and (3) are pairwise equivalent. If $\mathfrak{B}$ is a basis of (R, τ) then, by (2), $\varphi\mathfrak{B}$ must contain only τ'-open sets. If M' is a τ'-open set, then $M = \varphi^{-1}M'$ is a τ-open set, and we have $M = \cup \{B : B \in \mathfrak{B}, B \subseteq M\}$, and therefore

$$M' = \varphi M = \cup \{\varphi B : \varphi B \in \varphi\mathfrak{B}, \varphi B \subseteq M'\}.$$

Hence $\varphi\mathfrak{B}$ is a basis of (R', τ'). If $\mathfrak{B}$ is only a subbasis, we can still make conclusions analogous to those above, because the intersection of sets belonging to $\mathfrak{B}$ is mapped onto the intersection of the images of the sets in question. Hence (2) implies (4), while (5) follows immediately from (4). Finally, let $\mathfrak{B}$ be a basis of (R, τ) and $\varphi\mathfrak{B}$ a basis of (R', τ'). Then $\mathfrak{U}_p = \{B : B \in \mathfrak{B}, p \in B\}$ is a basis of τp and $\varphi\mathfrak{U}_p = \{\varphi B : B \in \mathfrak{B}, \varphi p \in \varphi B\}$ is a basis of $\tau'\varphi p$; i.e., $\varphi\tau p = \tau'\varphi p$. In the case of subbases one comes to a similar conclusion. Hence (1) follows from (5). ∎

Let R and R' be two totally ordered sets. A 1-1 mapping of R onto R' is said to be an antiorder isomorphism if $a < b$ is equivalent to $\varphi a > \varphi b$.

16.2 *Let τ and τ' be the order topologies on two totally ordered sets R and R'. Then every order isomorphism, as well as every antiorder isomorphism of R onto R' is a homeomorphism of (R, τ) onto (R', τ').*

Proof: The subbasis $\{[\leftarrow, x[: x \in R\} \cup \{]x, \rightarrow] : x \in R\}$ of (R, τ) gets mapped by φ onto the corresponding subbasis of (R', τ'). Hence our contention follows from 16.1. ∎

Note that a homeomorphism between totally ordered sets need be neither an order nor an antiorder isomorphism. For instance, consider the natural numbers $\mathbf{N}$, ordered as usual. Then the order topology is identical with the discrete topology on $\mathbf{N}$, yet every permutation of $\mathbf{N}$ is a homeomorphism. If, however, we stipulate that the order topology must be connected, then the converse to 16.2 will hold.

16.3 *Let τ and τ' be the order topologies on the totally ordered sets R, respectively R', and let φ be a homeomorphism of (R, τ) onto (R', τ'). If τ is connected then φ must be either an order isomorphism or an order anti-isomorphism.*

Proof: Let us suppose $x, y \in R$ and $x < y$. Since $[\leftarrow, x]$, $[x, y]$ and $[y, \rightarrow]$ are connected sets, so must their images under φ be connected, i.e., intervals. Hence $\varphi x < \varphi y$ implies

$$\varphi[\leftarrow, x] = [\leftarrow, \varphi x], \qquad \varphi[x, y] = [\varphi x, \varphi y], \qquad \varphi[y, \rightarrow] = [\varphi y, \rightarrow]$$

and likewise, $\varphi y < \varphi x$ implies

$$\varphi[\leftarrow, x] = [\varphi x, \rightarrow], \qquad \varphi[x, y] = [\varphi y, \varphi x], \qquad \varphi[y, \rightarrow] = [\leftarrow, \varphi y].$$

For any four points $x, y, u, v \in R$ where $u < x < v < y$ we must therefore have $\varphi u < \varphi x < \varphi v < \varphi y$ or else $\varphi u > \varphi x > \varphi v > \varphi y$. We can tell just which of these two possibilities actually occurs by comparing the order of any two of the images of our four points. For instance, if $a < b$ and $\varphi a < \varphi b$, then φ must be order-preserving and hence an order isomorphism. If however $a < b$ and $\varphi a > \varphi b$, then φ is order-reversing and hence an antiorder isomorphism. ∎

16.4 *Let R be a totally ordered infinite set with order topology τ. The space (R, τ) is homeomorphic to one of the intervals* $[0, 1]$, $]0, 1]$, or $]0, 1[$ *on the real line if and only if τ is both connected and separable.*

Proof: If τ is connected, then the order of R must be gap-free and conditionally complete (14.III). By 10.12 we see that (R, τ) must be order-isomorphic to one of the intervals $[0, 1]$, $]0, 1]$, $[0, 1[$, or $]0, 1[$, and by 16.2, even homeomorphic. Note that the intervals $]0, 1]$ and $[0, 1[$ are homeomorphic to each other: the map $\varphi x = 1 - x$ carries out this homeomorphism. Conversely, any interval on the real line possesses the properties listed above. ∎

The set of all homeomorphisms of the space (R, τ) onto itself forms a group, called the *homeomorphisms group* of this space. Different topologies τ and τ' on the same underlying set can lead to the same homeomorphisms group. For example, consider the space (R, τ^k) discussed in 4.III. The homeomorphisms group of this space is identical with the totality of all 1-1 mappings of R onto itself, no matter what the value of k may be. A space is said to be *homogeneous* if its homeomorphisms group is transitive, i.e., if, given any two points p, q, there always exists a homeomorphism φ such that $\varphi p = q$. For instance, n-dimensional space R^n is homogeneous, for its homeomorphisms group certainly contains all translations.

Notes and Exercises

16A The set $S^n = \{x : x \in \mathbf{R}^{n+1}, \| x \| = 1\}$ is called the n-dimensional *sphere.* If we remove one single point from S^n, the remaining, so-called punctured sphere, is homeomorphic to the n-dimensional affine space $\mathbf{R}^n$. A mapping establishing this homeomorphism is the familiar *stereographic projection.*

16B The n-dimensional (real) projective space P^n can be thought of as a topological space as follows. Remove an arbitrary but fixed hyperplane from P^n. As we know, the remaining space can be thought of as an n-dimensional affine space $\mathbf{R}^n$. Let the system $\mathfrak{B}$ be made up of all subsets B of P^n for which $B \cap \complement H$, for all hyperplanes H, is an open subset of $\mathbf{R}^n$. Then $\mathfrak{B}$ is a basis for a topology on P^n (usual topology on P^n).

Exercise: If H denotes a hyperplane, then the subspace $\complement H$ of P^n is homeomorphic to R^n. Every projective mapping (collineation) of P^n onto itself is also a homeomorphism. The space P^n is homogeneous, connected, and completely compact.

16C Let τ be the standard topology on the real line. We can create a second topology τ' on $\mathbf{R}$ by proceeding as follows: Given any point $p \in \mathbf{R}$ we shall take as neighborhoods of p all sets of the form $U \cup \complement A$, where the following conditions determine U and A: U must be a member of τp and A must be such that $| A | \leqslant \aleph_0$, $A \subseteq [\leftarrow, p]$, and in $(\mathbf{R}, \tau)$ A must possess at most the point p as accumulation point.

Exercise

1. We have $\tau \wedge \tau^{\aleph_1} < \tau' < \tau$ (cf. 4.III).

2. The spaces $(\mathbf{R}, \tau)$ and $(\mathbf{R}, \tau \wedge \tau^{\aleph_1})$ possess the same homeomorphisms group and this group contains as a proper subgroup the homeomorphisms group of $(\mathbf{R}, \tau')$.

16D The formation of sups and infs in the lattice of topologies on a fundamental set R is not a topological operation. From the homeomorphisms $(R, \tau_1) \simeq (R', \tau_1')$ and $(R, \tau_2) \simeq (R', \tau_2')$ we cannot deduce the homeomorphism $(R, \tau_1 \wedge \tau_2) \simeq (R', \tau_1' \wedge \tau_2')$. Suppose, for example, that p and q are two points of R. Let the topologies τ_1 and τ_2 of R be defined by

$$\tau_1 x = \begin{cases} \overline{x} & \text{for} \quad x = p \\ \overline{R} & \phantom{\text{for}} \quad x \neq p \end{cases} \qquad \text{and} \qquad \tau_2 x = \begin{cases} \overline{x} & \text{for} \quad x = q \\ \overline{R} & \phantom{\text{for}} \quad x \neq q. \end{cases}$$

Then we have the homeomorphisms $(R, \tau_1) \simeq (R, \tau_1)$ and $(R, \tau_1) \simeq (R, \tau_2)$. But the spaces (R, τ_1) and $(R, \tau_1 \wedge \tau_2)$ are not homeomorphic.

17. Continuous Mappings

In real analysis a function φ is said to be continuous at the point p if, whenever a sequence $(a_n)_{n\in\mathbf{N}}$ converges to p it follows that the sequence of values $(\varphi a_n)_{n\in\mathbf{N}}$ must converge to the value φp. If we transcribe this situation to the general case of mappings between topological spaces, we arrive at the following generalization: Let (R, τ) and (R', τ') be two topological spaces and φ be a mapping from R into R'. The mapping φ will be said to be continuous if, whenever a filter $\mathfrak{a}$ is τ-convergent to p it follows that the filter $\varphi\mathfrak{a}$ is τ'-convergent to φp; i.e., $\mathfrak{a} \leqslant \tau p$ will always imply $\varphi\mathfrak{a} \leqslant \tau'\varphi p$. This is manifestly equivalent to the condition $\varphi\tau p \leqslant \tau'\varphi p$, in which form we can now see the generalized continuity condition as a weakening of the commutativity condition of real analysis.

Definition 17a: *A mapping* $\varphi : R \to R'$ *is called a* continuous *mapping of* (R, τ) *into* (R', τ') *whenever we have* $\varphi\tau p \leqslant \tau'\varphi p$ *for all* $p \in R$; *it is said to be an* open mapping *of* (R, τ) *into* (R', τ') *whenever the reverse inequality,* $\varphi\tau p \geqslant \tau'\varphi p$, *holds for all* $p \in R$.

Hence a 1-1 mapping φ of R onto R' is continuous (open) if and only if its inverse map φ^{-1} is open (continuous). Such a mapping φ is therefore a homeomorphism if and only if it is both continuous and open, or, equivalently, if and only if φ and φ^{-1} are both continuous (or both open). The composition of two continuous (open) maps gives as product once again a continuous (open) map. Since continuous maps occur so frequently and are so important we shall use the abbreviation "$\varphi : (R, \tau) \to (R', \tau')$" for "$\varphi$ is a continuous mapping of (R, τ) into (R', τ')." We shall also say that φ is *continuous* (*open*) at the fixed point $p \in R$ whenever $\varphi\tau p \leqslant \tau'\varphi p$ ($\varphi\tau p \geqslant \tau'\varphi p$).

We can reword the continuity condition $\varphi\tau p \leqslant \tau'\varphi p$ as follows: Given any neighborhood $U' \in \tau'\varphi p$, i.e., neighborhood of the image φp, then we can find a neighborhood $U \in \tau p$ such that $\varphi U \subseteq U'$. Likewise, the condition of openness, $\varphi\tau p \geqslant \tau'\varphi p$, of the mapping φ is equivalent to the following statement: Given any $U \in \tau p$ there exists a $U' \in \tau'\varphi p$ such that $\varphi U \supseteq U'$.

Examples

17.I Whenever τ is the discrete, or τ' the trivial topology, then every mapping from (R, τ) into (R', τ') is continuous. Conversely, if τ is now the trivial, or τ' the discrete topology, then every mapping from (R, τ) into (R', τ') is open.

17.II Let (R, δ) and (R', δ') be two pseudometric spaces. The continuity of $\varphi : R \to R'$ can in this case be reformulated as follows: Given any $p \in R$ and any $\epsilon > 0$, we can find an $\eta > 0$ such that for all $q \in R$, the condition $\delta(p, q) < \eta$ always implies $\delta'(\varphi p, \varphi q) < \epsilon$. In the particular case where $\varphi : \mathbf{R} \to \mathbf{R}$ we have the familiar definition of continuity from real analysis: Given any $p \in \mathbf{R}$ and any $\epsilon > 0$, we can find an $\eta > 0$ such that for all $q \in \mathbf{R}$, the condition $|p - q| < \eta$ always implies $|\varphi p - \varphi q| < \epsilon$.

17.III The mapping φ defined by

$$\varphi x = \begin{cases} x & \text{for } x \geqq 0 \\ |x| + 1 & \phantom{\text{for }} x < 0 \end{cases}$$

is open but not continuous, from $\mathbf{R}$ onto the subspace $[0, \rightarrow]$.

17.IV The mapping $\varphi : \mathbf{R} \rightarrow [0, \rightarrow]$ defined by $\varphi x = \max\{0, x - 1\}$ is continuous, but not open.

The theorems that follow contain further characterizations of continuous and open mappings.

17.1 *The following statements are pairwise equivalent.*

(1) *φ is a continuous mapping from (R, τ) into (R', τ').*

(2) *The inverse images of open (closed) sets are open (closed): If M' is an open (closed) subset of (R', τ'), then $\varphi^{-1}M'$ is also an open (closed) subset of (R, τ).*

(3) *For subsets M' of R' we have $\varphi^{-1}\overset{\circ}{\widehat{M'}} \subseteq \overset{\circ}{\widehat{\varphi^{-1}M'}}$.*

(4) *For subsets M of R we have $\varphi\bar{M} \subseteq \overline{\varphi M}$.*

(5) *There exists at least one subbasis of (R', τ') such that, under the mapping φ, all the members of this subbasis have inverse images which are open.*

Proof

(1) $\rightarrow$ (2): Let M' be τ'-open. If $p \in \varphi^{-1}M'$, then $\varphi p \in M'$, hence

$$\varphi\tau p \leqslant \tau'\varphi p \leqslant \overbracket{M'}.$$

Then we have $\tau p \leqslant \varphi^{-1}\varphi\tau p \leqslant \varphi^{-1}\overbracket{M'}$. Hence $\varphi^{-1}M'$ is a τ-open set. Since $\varphi^{-1}\mathbf{C}M' = \mathbf{C}\varphi^{-1}M'$, the statements in parentheses in (2) must be true by duality.

(2) $\rightarrow$ (5): The member sets of any subbasis of (R', τ') are τ'-open, and hence by (2) they possess open inverse images.

(5) $\rightarrow$ (1): Let $\mathfrak{B}'$ be a subbasis of (R', τ') whose member sets possess open inverse images. The sets of the form $U' = B_1' \cap \ldots \cap B_n'$, where $\varphi p \in B'$ and $B_\nu' \in \mathfrak{B}'$ for $\nu = 1, \ldots, n$, constitute a basis of $\tau'\varphi p$. Since

$$\varphi^{-1}U = \varphi^{-1}B_1' \cap \ldots \cap \varphi^{-1}B_n',$$

we see that $\varphi^{-1}U'$ is a τ-open set such that $p \in \varphi^{-1}U'$. This implies $\tau p \leqslant \overbracket{\varphi^{-1}U'}$, hence $\varphi\tau p \leqslant \overbracket{U'}$ for all the sets U', and consequently $\varphi\tau p \leqslant \tau'\varphi p$.

(2) → (3): Evidently $\varphi^{-1}(\overset{\circ}{\widehat{M'}}) \subseteq \varphi^{-1}M'$. Like $\overset{\circ}{\widehat{M'}}$, $\varphi^{-1}(\overset{\circ}{\widehat{M'}})$ is also open and hence contained in the interior of $\varphi^{-1}M'$, i.e., in $\overset{\circ}{\widehat{\varphi^{-1}(M')}}$. Conversely, let M' be open. With the aid of (3) above we see $\overset{\circ}{\widehat{\varphi^{-1}M'}} \supseteq \varphi^{-1}(\overset{\circ}{\widehat{M'}}) = \varphi^{-1}M'$, and hence $\varphi^{-1}M'$ is open.

(2) → (4): Since $\overline{\varphi M}$ is closed, $\varphi^{-1}\overline{\varphi M}$ must also be closed. Hence $\bar{M} \subseteq \varphi^{-1}\overline{\varphi M}$ and $\varphi\bar{M} \subseteq \overline{\varphi M}$. Conversely, let M' be a τ'-closed set. By (4) above we see that $\varphi\overline{\varphi^{-1}M'} \subseteq \overline{\varphi\varphi^{-1}M'} \subseteq \overline{M'} = M'$, hence $\overline{\varphi^{-1}M'} \subseteq \varphi^{-1}M'$. Consequently $\varphi^{-1}M'$ is closed. ∎

17.2 *The following statements are pairwise equivalent:*

(1) *φ is an open mapping of (R, τ) into (R', τ').*

(2) *The images of open sets are open: If M is open in R, then φM is open in R'.*

(3) *For subsets M of R we have $\varphi\mathring{M} \subseteq \overset{\circ}{\widehat{\varphi M}}$.*

(4) *There exists at least one basis of (R, τ) such that, under the mapping φ, all the members of this basis have open images.*

Proof

(1) → (2): Let M be open. Given any $p' \in \varphi M$ there exists a $p \in M$ such that $\varphi p = p'$. We see that $\tau'p' = \tau'\varphi p \leqslant \varphi\tau p \leqslant \ulcorner\varphi M\urcorner$. Hence φM is open.

(2) → (4): Since the member sets of every basis are open, so must their images be open.

(4) → (1): Let $\mathfrak{B}$ be a basis of (R, τ) and let us suppose that the images of all member sets of $\mathfrak{B}$ are open. If $B \in \mathfrak{B}$ and $p \in B$ we then have $\tau'\varphi p \leqslant \varphi\tau p \leqslant \ulcorner\varphi B\urcorner$. Since $\{B : B \in \mathfrak{B}, p \in B\}$ forms a basis of τp, we see that $\tau'\varphi p \leqslant \varphi\tau p$.

(2) → (3): $\varphi(\mathring{M})$, like $\mathring{M}$, is an open set and $\varphi\mathring{M}$ is therefore contained in both φM and $\overset{\circ}{\widehat{\varphi M}}$. If, conversely, M is open, then (3) implies $\overset{\circ}{\widehat{\varphi M}} \subseteq \varphi\mathring{M} = \varphi M$, which tells us that φM is open. ∎

It is, in general, not possible to characterize open mappings by means of closed sets, closures, or subbases. Only by adding further hypotheses can we hope to achieve any such characterizations.

17.3 (1) *If φ is an open mapping from (R, τ) into (R', τ'), then we have, for every subset M' of R',*

$$(*) \qquad \varphi^{-1}\overline{M'} \subseteq \overline{\varphi^{-1}M'}.$$

(2) *Let φ be a mapping from R onto R'. Then* (*) *implies that φ must be an open mapping and that M' is closed whenever $\varphi^{-1}M'$ is closed.*

Proof

(1) From $p \in \varphi^{-1}\overline{M'}$ we deduce that $\tau'\varphi p \wedge \ulcorner M' \urcorner \neq \mathfrak{o}$. By 3.15 we have $\varphi(\tau p \wedge \ulcorner \varphi^{-1}M' \urcorner) = \varphi\tau p \wedge \ulcorner M' \urcorner \geqslant \tau'\varphi p \wedge \ulcorner M' \urcorner \neq \mathfrak{o}$; hence we also have

$$\tau p \wedge \ulcorner \varphi^{-1}M' \urcorner \neq \mathfrak{o}$$

and thus $p \in \overline{\varphi^{-1}M'}$.

(2) If φ is a mapping onto R', then $\varphi\varphi^{-1}M' = M'$ for all subsets M' of R'. By (*) above we then have

$$\varphi\overset{\circ}{M} \subseteq \varphi(\widehat{\varphi^{-1}\varphi M}) = \varphi\mathbf{C}\overline{\varphi^{-1}\mathbf{C}\varphi M} \subseteq \varphi\mathbf{C}\varphi^{-1}\overline{\mathbf{C}\varphi M} = \mathbf{C}\overline{\mathbf{C}\varphi M} = (\widehat{\varphi M})$$

and therefore, by 17.2, φ must be open. If $\varphi^{-1}M'$ is closed, we see, further, that $\varphi^{-1}\overline{M'} \subseteq \overline{\varphi^{-1}M'} = \varphi^{-1}M'$ and hence $\overline{M'} \subseteq M'$. ∎

17.4 *Suppose $\varphi : R \to R'$ and $M' = \varphi R$. Then $\varphi : (R, \tau) \to (R', \tau')$ is equivalent to $\varphi : (R, \tau) \to (M', \tau'_{M'})$.*

Proof: If $\varphi\tau p \leqslant \tau'_{M'}\varphi p$, then all the more so, $\varphi\tau p \leqslant \tau'\varphi p$. But since $\varphi\tau p \leqslant \ulcorner M' \urcorner$ we see that, conversely, $\varphi\tau p \leqslant \tau'\varphi p$ implies $\varphi\tau p \leqslant \tau'\varphi p \wedge \ulcorner M' \urcorner$, whence $\varphi\tau p \leqslant \tau'_{M'}\varphi p$. ∎

Since, in the above case, φ is a mapping of R onto M', when considering continuous mappings, we can always think of them as being "onto" mappings. A similar theorem will not hold for open maps. For example, $\varphi x = |x|$ defines an open map of $\mathbf{R}$ onto $[0, \to]$, but this mapping is not an open mapping from $\mathbf{R}$ into $\mathbf{R}$.

17.5 *Let $(R', \leqslant)$ be a totally ordered set with order topology τ' and let φ and ψ be two continuous mappings from (R, τ) into (R', τ'). Then the mappings σ_1 and σ_2 defined by*

$$\sigma_1 p = \min\{\varphi p, \psi p\} \qquad \textit{and} \qquad \sigma_2 p = \max\{\varphi p, \psi p\}$$

are continuous.

Proof: If $\varphi p = \psi p$, we have $\sigma_i\tau p \leqslant \varphi\tau p \vee \psi\tau p \leqslant \tau'\varphi p \vee \tau'\psi p = \tau'\sigma_i p$ for $i = 1, 2$. If $\varphi p \neq \psi p$, suppose $\varphi p < \psi p$. By 18.11, τ' is, in both these cases, a T_2-topology. Hence there exist neighborhoods $U' \in \tau'\varphi p$ and $V' \in \tau'\psi p$ where $u' < v'$ for all $u' \in U'$ and $v' \in V'$. Because of the continuity of φ and ψ there exists a neighborhood $W \in \tau p$ such that $\varphi W \subseteq U'$ and $\psi W \subseteq V'$. The mapping

σ_1 is identical to φ on W while σ_2 is identical to ψ on W, and both maps σ_1 and σ_2 are continuous at p since φ and ψ are continuous. ∎

Note that φ and ψ may be open mappings, yet σ_1 and σ_2 need not be open, as example 17.IV shows.

17.6 (1) *The conditions* $\tau_2 \leqslant \tau_1$, $\tau_1' \leqslant \tau_2'$ *and* $\varphi : (R, \tau_1) \to (R', \tau_1')$ *imply* $\varphi : (R, \tau_2) \to (R', \tau_2')$.

(2) *Let* $\tau = \bigvee \{\tau_\iota : \iota \in I\}$. *Then if* $\varphi : (R_1 \tau_\iota) \to (R', \tau')$ *for all* $\iota \in I$, *we also have* $\varphi : (R, \tau) \to (R', \tau')$.

(3) *Let* $\tau' = \bigwedge \{\tau_\iota' : \iota \in I\}$. *Then if* $\varphi : (R, \tau) \to (R', \tau_\iota')$ *for all* $\iota \in I$, *we also have* $\varphi : (R, \tau) \to (R', \tau')$.

Proof

(1) The hypotheses immediately imply $\varphi\tau_2 p \leqslant \varphi\tau_1 p \leqslant \tau_1'\varphi p \leqslant \tau_2'\varphi p$.

(2) Let M' be a τ'-open set. Then, by 17.1, $\varphi^{-1}M'$ is a τ_ι-open set for all $\iota \in I$. By 7.4, $\varphi^{-1}M'$ must also be τ-open, whence our contention follows, using 17.1 again. ∎

17.7 (1) *Let* φ *be an open mapping from* (R, τ_1) *into* (R', τ_1'). *If* $\tau_2 \geqslant \tau_1$ *and* $\tau_1' \geqslant \tau_2'$, *then* φ *is also an open mapping from* (R, τ_2) *into* (R', τ_2').

(2) *Let* $\tau' = \bigvee \{\tau_\iota' : \iota \in I\}$. *If, for all* $\iota \in I$, φ *is an open mapping of* (R, τ) *into* (R', τ_ι'), *then* φ *is also an open mapping of* (R, τ) *into* (R', τ').

Proof

(1) We have $\varphi\tau_2 p \geqslant \varphi\tau_1 p \geqslant \tau_1'\varphi p \geqslant \tau_2'\varphi p$.

(2) Let M be a τ-open set. By 17.2, φM is a τ_ι'-open set for all $\iota \in I$ and by 7.4, φM must therefore also be τ'-open. Our contention again follows immediately from 17.2. ∎

Similar results for the inf of topologies on R do not, in general, hold for open mappings.

Under an open mapping the image of a closed set need not be closed. It therefore is natural to introduce a third category of mappings.

Definition 17b: *A mapping* $\varphi : R \to R'$ *is said to be a* closed mapping *of* (R, τ) *into* (R', τ') *whenever the images under* φ *of closed subsets of* (R, τ) *are* τ'*-closed in* (R', τ').

17.8 *The following statements are pairwise equivalent:*

(1) φ *is a closed mapping from* (R, τ) *into* (R', τ').

(2) *For all* $p' \in R'$ *we have* $\varphi^{-1}\tau' p' \leqslant \bigvee \{\tau q : \varphi q = p'\}$.

(3) *For all subsets* M *of* R *we have* $\varphi\bar{M} \supseteq \overline{\varphi M}$.

Proof

(1) → (2): Suppose statement (2) does not hold. Then, given any $q \in R$ where $\varphi q = p'$, we can find an open neighborhood $U_q \in \tau q$ possessing the following property: For the set $M = \complement \cup \{U_q : \varphi q = p'\}$ we have $\varphi^{-1}\tau' p' \wedge \ulcorner M \urcorner \neq \mathfrak{o}$, hence $\tau' p' \wedge \ulcorner \varphi M \urcorner \neq \mathfrak{o}$. (3.15). Since M is closed, (1) above tells us that φM is also closed and therefore $p' \in \varphi M$. Thus there exists a $q \in M$ such that $\varphi q = p'$. But by the very definition of M, $q \in M$ implies $\varphi q \neq p'$.

(2) → (3): From $p' \in \overline{\varphi M}$ it follows that $\tau' p' \wedge \ulcorner \varphi M \urcorner \neq \mathfrak{o}$, whence

$$\mathfrak{o} \neq \ulcorner M \urcorner \wedge \varphi^{-1}\tau' p' \leqq \ulcorner M \urcorner \wedge \bigvee_{\varphi q = p'} \tau q = \bigvee_{\varphi q = p'} (\ulcorner M \urcorner \wedge \tau q).$$

Therefore $\tau q \wedge \ulcorner M \urcorner \neq \mathfrak{o}$ and $q \in \bar{M}$ for at least point q with $\varphi q = p'$, and hence $p' \in \varphi \bar{M}$.

(3) → (1): If M is closed, then, since $\overline{\varphi M} \subseteq \varphi \bar{M} = \varphi M$, we see that φM is also closed. ∎

17.9 *Let φ be a closed mapping from R onto R'. If then $\varphi^{-1}M'$ is open, so is M'.*

Proof: $\complement \varphi^{-1} M' = \varphi^{-1} \complement M'$ is closed. Consequently $\complement M' = \varphi\varphi^{-1} \complement M'$ is closed and M' itself is open. ∎

17.10 *Let φ be a* 1-1 *mapping from R onto R'. Then φ is an open mapping if and only if it is a closed mapping.*

Proof: φ is closed if and only closed sets of R have closed inverse images under the mapping φ^{-1} from R' onto R, where φ^{-1} is the inverse map to φ. But by 17.1 this can only happen if φ^{-1} is continuous, and hence φ is open. ∎

17.11 (1) *If $\tau_2 \geqslant \tau_1$ and $\tau_1' \geqslant \tau_2'$, then every closed mapping φ from (R, τ_1) into (R', τ_1') is also a closed mapping from (R, τ_2) into (R', τ_2').*

(2) *Suppose $\tau' = \bigvee \{\tau_\iota' : \iota \in I\}$. If φ is a closed mapping from (R, τ) into (R', τ_ι') for all $\iota \in I$, then φ is also a closed mapping of (R, τ) into (R', τ').*

Proof

(1) This contention follows from 17.8:

$$\varphi^{-1}\tau_2' p' \leqq \varphi^{-1}\tau_1' p' \leqq \bigvee_{q\, p = p'} \tau_1 p \leqq \bigvee_{q\, p = p'} \tau_2 p.$$

(2) If M is closed, then φM is a τ_ι'-closed set for all $\iota \in I$, and hence it is also τ'-closed. ∎

Notes and Exercises

17A *Exercise:* The mapping $\varphi :]0, \rightarrow] \rightarrow [-1, +1]$ defined by $\varphi x = \sin(1/x)$ is continuous, but neither open or closed.

17B Let the topology τ on R be connected and let $(R', \leqslant)$ be a totally ordered set with order topology τ', and let φ be a continuous mapping from (R, τ) into (R', τ').

Exercise: For the points $p, q \in R$ suppose $\varphi p < y < \varphi q$. Then there exists an $x \in R$ such that $\varphi x = y$ (*Intermediate-Value Property*).

17C Let $R = [-1, +1]$. The systems of sets

$$\mathfrak{B}_1 = \{[-1, +1], [-1, 0]\} \quad \text{and} \quad \mathfrak{B}_2 = \{[-1, +1], [0, +1]\}$$

are then bases for the topologies τ_1 and τ_2, respectively. In addition, let (R', τ') represent the interval $[0, +1]$ under the usual topology.

Exercise: The mapping $\varphi : R \rightarrow R'$ defined by $\varphi x = | x |$ is an open mapping from (R, τ_1) onto (R', τ') and, likewise, an open mapping from (R, τ_2) into (R', τ'). But φ is not an open mapping from $(R, \tau_1 \wedge \tau_2)$ onto (R', τ').

17D The existence of continuous 1-1 mappings from (R, τ) onto (R', τ') and back from (R', τ') onto (R, τ) does not imply that these two spaces are homeomorphic, as the follow counter-example shows.

Let $R_0, R_1, R_2, \ldots$ be pairwise disjoint sets such that $| R_n | = \aleph_n$, and suppose $R = \cup \{R_n : n = 0, 1, 2, \ldots\}$. Let us choose a point p_n from each of the R_n. Then we can define two topologies τ_1 and τ_2 on R, as follows:

$$\tau_1 x = \tau_2 x = \{M : x \in M, | \complement M | < \aleph_0\} \qquad \text{for } x \in R_0,$$

$$\tau_1 p_n = \{M : p_n \in M, | \complement M | < \aleph_{2n+1}\} \quad \text{and} \quad \tau_2 p_n = \{M : p_n \in M, | \complement M | < \aleph_{2n}\}$$

$$\text{for } n = 1, 2, \ldots;$$

$$\text{other wise, } \tau_1 x = \tau_2 x = \ulcorner x \urcorner.$$

Now the identity mapping is continuous and 1-1 from (R, τ_1) onto (R, τ_2). Suppose the mapping φ to be constructed as follows: Let φ map R_0 in a 1-1 way onto the complement of p_0 in R_0. For $n = 1, 2, \ldots$ let $\varphi p_n = p_{n-1}$, and for all the other points of R let $\varphi x = x$. Then φ is a continuous 1-1 mapping of (R, τ_2) onto (R, τ_1). But the spaces (R, τ_1) and (R, τ_2) are not homeomorphic [24, p. 102].

18. Completely Regular Spaces

Given two topological spaces (R, τ) and (R', τ'), it is always possible to find continuous mappings of the one into the other—for example, those mappings which, say, send all of R onto the single point $p' \in R'$, must always be continuous. Under certain conditions these are the only possible type of continuous map from (R, τ) into (R', τ'). Suppose, for instance, that τ is the trivial topology while τ' is a T_1-topology. Then, for every continuous φ: $(R, \tau) \to (R', \tau')$ we must have $\varphi \overline{R} = \varphi \tau p$ and hence $\varphi R = \{\varphi p\}$; i.e., all of R must be mapped onto the single point p. In many instances, however, it behooves us to know that there exist "sufficiently many continuous maps" of one space into another. The most varied of hypotheses can give us this guarantee. Consider the following: If φ is continuous, then the convergence of the filter $\mathfrak{a}$ to the point p implies the convergence of the image filter $\varphi\mathfrak{a}$ to the image point φp. If we take the converse situation, we can formulate the following additional condition to impose on continuous maps: Whenever $\mathfrak{a}$ does not conver to p, we can demand that there exist a continuous map φ, for which $\varphi\mathfrak{a}$ does not converge to φp—imposing this condition will allow us to avoid trivial, constant (continuous) maps.

Definition 18a: *A collection Φ of continuous mappings from the space (R, τ) into the space (R', τ') is said to be a* probing system *of (R, τ) by means of (R', τ') whenever, for any filter $\mathfrak{a}$ and any point p of R, the condition $\varphi\mathfrak{a} \leqslant \tau'\varphi p$ for all $\varphi \in \Phi$ implies $\mathfrak{a} \leqslant \tau p$. A space (R, τ) is said to be probed by (R', τ') if the set of all continuous mapping of (R, τ) into (R', τ') is a probing system [of (R, τ) by means of (R', τ')].*

If Φ is a probing system of (R, τ) by means of (R', τ'), then τ is the coarsest topology on R with respect to which all the mappings belonging to Φ are continuous: Suppose $\varphi : (R, \tau_0) \to (R', \tau')$ for all $\varphi \in \Phi$. Then for any $p \in R$ and any $\varphi \in \Phi$ we have, first, $\varphi\tau_0 p \leqslant \tau'\varphi p$ and then $\tau_0 p \leqslant \tau p$; i.e., $\tau_0 \leqslant \tau$.

18.1 *The following statements are pairwise equivalent:*

(1) *Φ is a probing system of (R, τ) by means of (R', τ').*

(2) $\tau p = \bigwedge \{\varphi^{-1}\tau'\varphi p : \varphi \in \Phi\}$.

(3) *If $\mathfrak{B}'$ is a subbasis of (R', τ') then $\mathfrak{B} = \{\varphi^{-1}B' : B' \in \mathfrak{B}', \varphi \in \Phi\}$ is a subbasis of (R, τ).*

(4) *Given any ultrafilter $\mathfrak{v}$ of R such that $\mathfrak{v} \wedge \tau p = \mathfrak{o}$, we can find a $\varphi \in \Phi$ such that $\varphi\mathfrak{v} \wedge \tau'\varphi p = \mathfrak{o}$.*

Proof

(1) → (2): For every $\varphi \in \Phi$ we have, by continuity, $\tau p \leqslant \varphi^{-1}\tau'\varphi p$, and hence $\tau p \leqslant \bigwedge \{\varphi^{-1}\tau'\varphi p : \varphi \in \Phi\}$. Conversely, $\varphi_0 \in \Phi$ implies

$$\varphi_0 \bigwedge \{\varphi^{-1}\tau'\varphi p : \varphi \in \Phi\} \leqslant \varphi_0\varphi_0^{-1}\tau'\varphi_0 p \leqslant \tau'\varphi_0 p.$$

But since Φ is a probing system, we finally have $\bigwedge \{\varphi^{-1}\tau\varphi p : \varphi \in \Phi\} \leqslant \tau p$.

(2) → (3): Since the maps belonging to Φ are continuous, the sets $\varphi^{-1}B'$ are open. By (2), every $U \in \tau p$ is of the form $U = \varphi_1^{-1}U_1' \cap \ldots \cap \varphi_n^{-1}U_n'$ where $\varphi_\nu \in \Phi$ and $U_\nu' \in \tau'\varphi_\nu p$. Now every U_ν' itself contains a neighborhood of $\varphi_\nu p$ and this neighborhood can be expressed as the intersection of a finite number of member sets of $\mathfrak{B}'$. Hence the sets containing p and expressible as finite intersections of members of $\mathfrak{B}$ must constitute a basis of τp.

(3) → (4): From $\mathfrak{v} \wedge \tau p = \mathfrak{o}$ we deduce, by (3), that $\mathfrak{v} \wedge \varphi_1^{-1}\widehat{B_1'} \wedge \ldots \wedge \varphi_n^{-1}\widehat{B_n'} = \mathfrak{o}$ for appropriate $\varphi_\nu \in \Phi$ and $B_\nu' \in \mathfrak{B}'$ where $\varphi_\nu p \in B_\nu'$. Hence $\mathfrak{v} \wedge \varphi_\nu^{-1}\widehat{B_\nu'} = \mathfrak{o}$, and therefore $\varphi_\nu\mathfrak{v} \wedge \tau'\varphi_\nu p \leqslant \varphi_\nu\mathfrak{v} \wedge \widehat{B_\nu'} = \mathfrak{o}$ for at least one index ν.

(4) → (1): Suppose $\mathfrak{a} \leqslant \tau p$ is false. Then there exists a $U \in \tau p$ and an ultrafilter $\mathfrak{v}$ such that $\mathfrak{v} \leqslant \mathfrak{a} \wedge \widehat{\complement U}$. Hence $\mathfrak{v} \leqslant \mathfrak{a}$ and $\mathfrak{v} \wedge \tau p = \mathfrak{o}$. This means that there must exist a $\varphi \in \Phi$ with $\varphi\mathfrak{v} \wedge \tau'\varphi p = \mathfrak{o}$, which implies that $\varphi\mathfrak{a} \leqslant \tau'\varphi p$ must also be false. ∎

18.2 *Let (R, τ) be probed by (R', τ').*

(1) *If τ is a T_1- and τ' a T_2-topology, then τ must also be a T_2-topology.*

(2) *If τ' is a T_3-topology, so is τ.*

Proof

(1) From $\tau p \wedge \tau q = \mathfrak{o}$ it follows that, for an arbitrary continuous mapping φ we have $\mathfrak{o} \neq \varphi(\tau p \wedge \tau q) \leqslant \varphi\tau p \wedge \varphi\tau q \leqslant \tau'\varphi p \wedge \tau'\varphi q$, hence $\varphi p = \varphi q$ and then surely $\widehat{\varphi p} \leqslant \tau'\varphi q$. Since (R, τ) is probed by (R', τ'), we must have $\widehat{p} \leqslant \tau q$ and hence, by $(\mathbf{T}_1)$, $p = q$.

(2) By 17.1 (4) we have, for all continuous mappings φ,

$$\varphi\overline{\tau p} \leqslant \overline{\varphi\tau p} \leqslant \overline{\tau'\varphi p} = \tau'\varphi p,$$

and since (R, τ) is probed by (R', τ'), $\overline{\tau p} \leqslant \tau p$. ∎

18.3 *Suppose (R, τ) can be probed by (R', τ'). Then there exists a probing system Φ such that*

$$|\,\Phi\,| \leqslant \operatorname{Bd}(R, \tau) \leqslant |\,\Phi\,| \cdot \operatorname{Bd}(R', \tau'),$$

provided only that $\operatorname{Bd}(R, \tau)$ *is not finite. The second of the above inequalities holds for every probing system.*

Proof: Let Φ be a probing system, and $\mathfrak{B}'$ be a minimal basis of (R', τ'). By 18.1, $\mathfrak{B} = \{\varphi^{-1}B' : B' \in \mathfrak{B}', \varphi \in \Phi\}$ is a subbasis of (R, τ). Since Bd (R, τ) is not finite, the basis generated by $\mathfrak{B}$ has the same cardinality as $\mathfrak{B}$, and consequently Bd $(R, \tau) \leqslant |\mathfrak{B}| \leqslant |\Phi| \cdot |\mathfrak{B}'| = |\Phi| \cdot$ Bd (R', τ').

Suppose $\{B_\iota : \iota \in I\}$ is a minimal basis of (R, τ) and let Π denote the set of all ordered couples of indices (ι_1, ι_2) to which we can associate a finite number of mappings $\varphi_{\iota_1, \iota_2, \rho} (\rho = 1, ..., r_{\iota_1, \iota_2})$ and open subsets $V'_{\iota_1, \iota_2, \rho}$ of (R', τ') such that

$$B_{\iota_1} \subseteq \cap \{\varphi^{-1}_{\iota_1, \iota_2, \rho} V'_{\iota_1, \iota_2, \rho} : \rho = 1, ..., r_{\iota_1, \iota_2}\} \subseteq B_{\iota_2}.$$

Let

$$\Phi = \{\varphi_{\iota_1, \iota_2, \rho} : (\iota_1, \iota_2) \in \Pi, \rho = 1, ..., r_{\iota_1, \iota_2}\}.$$

For each $\iota_2 \in I$ and each $p \in B_{\iota_2}$, there exist continuous mappings $\psi_1, ..., \psi_n$ and open neighborhoods $U'_\nu \in \tau'\varphi_\nu p$ such that $\cap \{\psi_\nu^{-1} U'_\nu : \nu = 1, ..., n\} \subseteq B_{\iota_2}$. The left-side of this inclusion is an open neighborhood belonging to τp, and hence there exists an $\iota_1 \in I$ such that $p \in B_{\iota_1} \subseteq \cap \psi_\nu^{-1} U'_\nu$. Consequently $(\iota_1, \iota_2) \in \Pi$ and $p \in \cap \varphi_{\iota_1, \iota_2, \rho} V'_{\iota_1, \iota_2, \rho} \subseteq B_{\iota_2}$. This means that Φ is a probing system. Since $|I| =$ Bd $(R, \tau) \geqslant \aleph_0$ we see, in conclusion, that

$$|\Phi| \leqslant \aleph_0 \cdot |\Pi| \leqslant \aleph_0 \cdot |I|^2 = |I| = \text{Bd}\,(R, \tau). \blacksquare$$

Of special importance is the case of probing by means of the real line. This leads to a separation axiom which stands between $(\mathbf{T}_3)$ and $(\mathbf{T}_4)$.

Definition 18b: *A space (R, τ) is said to be a T_{3a}-space and τ a T_{3a}-topology whenever (R, τ) can be probed by means of the real line* **R** *under the usual topology. A T_{3a}-space which is also a T_1-space is said to be a* completely regular *space.*

18.4 *Every T_{3a}-space is a T_3-space; every completely regular space is regular.*

Proof: Since **R** is a T_3-space, our contention follows from 18.2. ∎

18.5 (R, τ) *is a T_{3a}-space if and only if, given any closed subset A and any point $p \in \complement A$, we can find a continuous mapping $\varphi : (R, \tau) \to [0, 1]$ such that $\varphi p = 0$ and $\varphi A = \{1\}$.*

Proof: If (R, τ) is a T_{3a}-space, then, by 18.1 (2) we can find continuous mappings ψ_ν $(\nu = 1, ..., n)$ of (R, τ) into **R** and neighborhoods $U_{\epsilon_\nu}(\psi_\nu p)$ such that $\psi_1^{-1} U_{\epsilon_1}(\psi, p) \cap ... \cap \psi_n^{-1} U_{\epsilon_n}(\psi_n p) \subseteq \complement A$. By 17.5, the mappings

$$\varphi_\nu x = \min \{1, \epsilon_\nu^{-1} |\psi_\nu x - \psi_\nu p|\} \qquad \text{and} \qquad \varphi x = \max \{\varphi_1 x, ..., \varphi_n x\}$$

are continuous. We have $\varphi p = 0$ and $0 \leqslant \varphi x \leqslant 1$ for all $x \in R$. If $\varphi x < 1$, then $|\psi_\nu x - \psi_\nu p| < \epsilon$ for $\nu = 1, ..., n$ and hence $x \in \complement A$.

Conversely, let the condition of the theorem be satisfied. If then $\mathfrak{v}$ is an ultrafilter of R with $\mathfrak{v} \wedge \tau p = \mathfrak{v}$, then there exists an open neighborhood $U \in \tau p$ such that $\mathfrak{v} \leqslant \overline{\complement U}$. Since $\complement U$ is a closed set, by hypothesis there exists a continous mapping $\varphi : (R, \tau) \to [0, 1]$ such that $\varphi p = 0$ and $\varphi \complement U = \{1\}$. Consequently $\varphi \mathfrak{v} = \overline{1}$, and thus, by 18.1, (R, τ) is probed by [0, 1] and, all the more so, by **R**. ∎

18.6 *A space (R, τ) is a T_4-space if and only if, given any two closed subsets A and B, we can find a continuous mapping φ: $(R, \tau) \to [0, 1]$ such that $\varphi A = \{0\}$ and $\varphi B = \{1\}$.*

Proof: If the above condition holds, then, by 17.1, the sets $U = \varphi^{-1}[0, \frac{1}{2}[$ and $V = \varphi^{-1}]\frac{1}{2}, 1]$ are open subsets of (R, τ) with $A \subseteq U, B \subseteq V$ and $U \cap V = \emptyset$.

Conversely, let (R, τ) be a T_4-space. We shall first show that, given any rational number r of the form $r = k/2^n$ $(n = 0, 1, 2, ...; k = 0, ..., 2^n)$ we can find a subset M_r of R such that the following conditions hold:

(a) $M_0 = A$ and $M_1 = \complement B$.

(b) If $r < r'$, then $\bar{M}_r \subseteq \overset{\circ}{\widehat{M_{r'}}}$. The proof is by induction on n with (a) as the first step of the induction. Suppose $r = k/2^n$ with $k \not\equiv 0 \bmod 2$, $s = r - (\frac{1}{2})^n$, and $t = r + (\frac{1}{2})^n$. By the inductive hypothesis, we suppose that M_s and M_t have already been constructed so that $\bar{M}_s \subseteq \overset{\circ}{\widehat{M_t}}$. Then by $(\mathbf{T}_4)$ there exist open sets U and V with $\bar{M}_s \subseteq U$, $\complement \overset{\circ}{\widehat{M_t}} \subseteq V$, and $U \cap V = \emptyset$. If we put $M_r = U$, then condition (b) is satisfied. We define a mapping $\varphi : R \to [0, 1]$ by

$$\varphi x = \begin{cases} \inf \{r : x \in \overline{M_r}\} & \text{for} \quad x \notin B \\ 1 & \phantom{\text{for}} \quad x \in B. \end{cases}$$

Then $\varphi A = \{0\}$ and $\varphi B = \{1\}$. It remains to prove that φ is continuous at the point x. Let us first assume that $0 < \varphi x < 1$. Then, given any $\epsilon > 0$, we can find in $U_\epsilon(\varphi(x))$ two rational numbers r and s of the form $k/2^n$ such that $r < \varphi x < s$. Then $x \in \complement \bar{M}_r \cap \overset{\circ}{\widehat{M_s}} = V$ and $\varphi V \subseteq U_\epsilon(\varphi x)$. Moreover, V is open and a neighborhood belonging to τx. If $\varphi x = 0$ or $\varphi x = 1$, the above proof that φ is continuous will still be valid if we only omit all reference to r, or s, respectively. ∎

18.7 *A T_4-space is a T_{3a}-space if and only if it is a T_3-space. Every normal space is completely regular.*

Proof: Let (R, τ) be a T_4-space. By 18.4, every T_{3a}-topology is also a T_3-topology. Conversely, let τ be a T_3-topology. If then A is a closed set and p a point in $\complement A$, there exists a $U \in \tau p$ with $\bar{U} \cap A = \emptyset$. By 18.6 there exists a continuous mapping $\varphi : (R, \tau) \to [0, 1]$ such that $\varphi A = \{1\}$ and $\varphi \bar{U} = \{0\}$. Since $p \in U$ we must have $\varphi p = 0$, and by 18.5, τ is a T_{3a}-topology. The second contention above is a special case of the first, for every normal space is both a T_3-space and a T_1-space. ∎

18.8 *If (R, τ) can be probed by means of (R', τ'), then so can every subspace of (R, τ). Every subspace of a T_{3a}-space (a completely regular space) is likewise a T_{3a}-space (a completely regular space).*

Proof: Let p be a point in the subset M of R, and let $\mathfrak{v}$ be an ultrafilter with $\mathfrak{v} \leqslant \overset{\sqcap}{M}$ and $\mathfrak{v} \wedge \tau_M p = \mathfrak{o}$. Then $\mathfrak{v} \wedge \tau p = \mathfrak{o}$, and there exists a continuous mapping $\varphi : (R, \tau) \to (R', \tau')$ such that $\varphi \mathfrak{v} \wedge \tau' \varphi p = \mathfrak{o}$. The restriction φ_M of φ to M is a continuous mapping of (M, τ_M) into (R', τ') such that $\varphi_M \mathfrak{v} \wedge \tau' \varphi_M p = \mathfrak{o}$. By 18.1, the subspace (M, τ_M) can be probed by (R', τ'). ∎

The above two theorems, taken together, imply:

18.9 *Every subspace of a normal space is completely regular.*

Later (cf. Section 27) we shall prove the converse result—that every completely regular space is a subspace of a normal space. The next theorem points up the transitive nature of the probing property.

18.10 *Suppose (R, τ) can be probed by (R', τ') and (R', τ') can be probed by (R'', τ''). Then (R, τ) can also be probed by (R'', τ''). Every space which admits of a probing by a T_{3a}-space must itself be a T_{3a}-space.*

Proof: For the ultrafilter $\mathfrak{v}$ of R let us suppose $\mathfrak{v} \wedge \tau p = \mathfrak{o}$. Then there exists a continuous mapping $\varphi : (R, \tau) \to (R', \tau')$ such that $\varphi \mathfrak{v} \wedge \tau' \varphi p = \mathfrak{o}$. Hence there exists a continuous mapping $\psi : (R', \tau') \to (R'', \tau'')$ such that $\psi \varphi \mathfrak{v} \wedge \tau'' \psi \varphi p = \mathfrak{o}$. Since $\psi \varphi : (R, \tau) \to (R'', \tau'')$ our contention must therefore follow from 18.1. ∎

Let φ be a mapping from the topological space R into the real line and suppose $T(\varphi) = \{p : p \in R, \varphi p \neq 0\}$. Then the closure $\overline{(T\varphi)}$ is called the *carrier* or *support* of the mapping φ. If $\overline{T(\varphi)}$ is completely compact, we say that is a mapping with *compact carrier* or *compact support*.

18.11 *Let (R, τ) be an l.c.c. T_2- or T_3-space. Then, given any completely compact subset K of R and any neighborhood U of K* (cf. p. 44), *we can find a continuous mapping $\varphi : (R, \tau) \to [0, 1]$ such that $\varphi K = \{1\}$ and also such that the carrier of φ is a subset of U.*

Proof: If τ is a T_2-topology, then by 12.8, τ must also be a T_3-topology. Hence we need only consider the case where τ is a T_3-topology.

Given any point $p \in K$ we can find an open neighborhood $V_p \in \tau p$ whose closure $\bar{V}_p$ is completely compact and contained in U (12.1). The family $\{V_p : p \in k\}$ is an open covering of K containing a finite family $\{V_1, \ldots, V_n\}$

which still covers K. Then $V = V_1 \cup \ldots \cup V_n$ is an open set with $K \subseteq V$. The closed set $K' = \bar{V} = \bar{V}_1 \cup \ldots \cup \bar{V}_n$ is completely compact by 11.8, and $K' \subseteq V$.[1] For the very same reasons, there exists an open set W with $K' \subseteq W \subseteq \bar{W} \subseteq U$, such that its closure $L = \bar{W}$ is completely compact. By 11.7, the subspace (L, τ_L) is a T_4-space, and by 18.6 there must exist a continuous mapping $\psi : (L, \tau_L) \rightarrow [0, 1]$ with $\psi K' = \{1\}$ and $\psi(L \cap \complement W\} = \{0\}$.[2]
Let us define the mapping $\varphi : R \rightarrow [0, 1]$ by

$$\varphi x = \begin{cases} \psi x & \text{for} \quad x \in L \\ 0 & \quad x \notin L. \end{cases}$$

Then $\varphi K = \{1\}$ and $T(\varphi) \subseteq L \subseteq U$. Since $\overline{T(\varphi)} \subseteq \bar{L} = L$ and since L is completely compact, φ is a mapping with compact carrier. Lastly, φ is continuous: At the points of W, the continuity of φ follows from the facts that W is open and that ψ is continuous. At the points of $\complement L$, continuity of φ follows from the openness of $\complement L$ and from the fact that φ is constant on $\complement L$. Finally, continuity, at the points of $L \cap \complement W$ follows for both the above reasons. ∎

18.12 *Every l.c.c. T_3-space is a T_{3a}-space. Every l.c.c. T_2-space is completely regular.*

Proof: Let A be a closed set and p a point in $\complement A$. Since every one-point set is completely compact, we can let $K = \{p\}$ and $U = \complement A$ in the statement of 18.11. We are therefore assured of the existence of a continuous mapping $\psi : (R, \tau) \rightarrow [0, 1]$ such that $\psi p = 1$ and $\psi A = \{0\}$. The map $\varphi x = 1 - \psi x$ then satisfies the conditions of 18.5. ∎

Definition 18c: *A collection Φ of continuous mappings $\varphi : (R, \tau) \rightarrow [0, 1]$ is said to be a* partition of unity *if it possesses the following properties:*

(1) *Given any $p \in R$ we can find a $U \in \tau p$ such that $\varphi U \neq \{0\}$ for at most a finite number of $\varphi \in \Phi$.*

(2) *For each $p \in R$ we have $\Sigma\{\varphi p : \varphi \in \Phi\} = 1$.*[3]

If Φ is a partition of unity of (R, τ), then for every $p \in R$ there exists, by (2), at least one $\varphi \in \Phi$ such that $\varphi p \neq 0$ and hence, such that $p \in T(\varphi)$. The sets $T(\varphi)$ are open because $T(\varphi) = \varphi^{-1}]0, 1[$, and therefore the system

$$T(\Phi) = \{T(\varphi) : \varphi \in \Phi\}$$

is an open covering of (R, τ). If $\mathfrak{H}$ is an arbitrary open covering of (R, τ), whenever $T(\Phi) \ll \mathfrak{H}$ we say that the partition Φ is *subordinate* to the covering $\mathfrak{H}$.

[1] K itself need not be closed! But if τ is a T_2-topology, then K is closed by 11.7, and we may omit these few steps in the above proof.

[2] The set $L \cap \complement W$ can be empty.

[3] $\Sigma\{\varphi p : \varphi \in \Phi\}$ denotes the sum of the numbers φp where $\varphi \in \Phi$. This seemingly infinite sum is really a finite sum, for by (1) only a finite number of the φp are different from 0.

18.13 *A T_2- or T_3-space (R, τ) is paracompact if, and only if, given any open covering $\mathfrak{H}$ of R we can find a partition of unity subordinate to $\mathfrak{H}$.*

Proof: Let (R, τ) be paracompact. With $\mathfrak{H}$ we can then associate an open and locally finite covering $\mathfrak{K}$ of R such that $\mathfrak{K} \ll \mathfrak{H}$(13C). By 13.3, τ is a T_4-topology and by 13.5 $\mathfrak{K}$ admits of a contraction $\mathfrak{L}$. To each $K \in \mathfrak{K}$ there corresponds an $L_K \in \mathfrak{L}$ such that $\overline{L_K} \subseteq K$ and also a continuous mapping $\psi_K : (R, \tau) \to [0, 1]$ with $\psi_K \complement K = \{0\}$ and $\psi_K \overline{L_K} = \{1\}$—the existence of such a ψ_K being guaranted by 18.6. If U is a neighborhood belonging to τp which admits of a nonempty intersection with only a finite number of sets of $\mathfrak{K}$, then $\psi_K U \neq \{0\}$ for at least one $K \in \mathfrak{K}$ but for only a finite number of $K \in \mathfrak{K}$. Therefore the sum $\Sigma\{\psi_K p : K \in \mathfrak{K}\}$ is finite and positive and the mapping $\varphi_K : (R, \tau) \to [0, 1]$ defined by

$$\varphi_K p = \frac{\psi_K p}{\Sigma\{\psi_{K'} p : K' \in \mathfrak{K}\}}$$

is continuous for each $K \in \mathfrak{K}$. Then $\Phi = \{\varphi_K : K \in \mathfrak{K}\}$ is a partition of unity: Since $\varphi_K p = 0$ and $\psi_K p = 0$ are equivalent, condition (1) must hold; while (2) follows directly from the definition of φ_K. Moreover, $T(\varphi_K) = T(\psi_K) \subseteq K$, whence the partition Φ is subordinate to the covering $\mathfrak{K}$ and therefore all the more subordinate to $\mathfrak{H}$.

Conversely, let Φ be a partition of the unity of (R, τ). Then the system $T(\Phi)$ is an open covering of R, which by (1) must be locally finite. If (R, τ) fulfills the conditions of the theorem, then, given any open covering $\mathfrak{H}$ of R, we can find an open and locally finite covering which is a refinement of $\mathfrak{H}$. ∎

18.14 *Suppose $\tau = \bigwedge \tau_\iota$ and that for each $\iota \in I$, (R, τ_ι) can be probed by means of (R', τ'). Then (R, τ) can also be probed by means of (R', τ'). The inf of any family of T_{3a}-topologies is again a T_{3a}-topology.*

Proof: Let us suppose that the ultrafilter $\mathfrak{v}$ satisfies $\mathfrak{v} \wedge \tau p = \mathfrak{o}$. Then, for a certain $\iota \in I$, we must have $\mathfrak{v} \wedge \tau_\iota p = \mathfrak{o}$. Hence there exists a continuous mapping $\varphi : (R, \tau_\iota) \to (R', \tau')$ such that $\varphi\mathfrak{v} \wedge \tau'\varphi p = \mathfrak{o}$ (18.1). But since $\tau \leqslant \tau_\iota$, φ must be a continuous mapping from (R, τ) into (R', τ'). ∎

Notes and Exercises

18A *Exercise:* If there exists a continuous 1-1 mapping of (R, τ) into (R', τ'), and if τ' is a T_1- or T_2-topology, then so is τ.

18B Let $C(R, \tau)$ denote the set of all continuous mappings from the space (R, τ) into the real line. We can define addition and multiplication between the maps of $C(R, \tau)$ in the usual way by $(\varphi + \psi)\, x = \varphi x + \psi x$ and $(\varphi \cdot \psi) x = (\varphi x) \cdot (\psi x)$.

With respect to these algebraic operations $C(R, \tau)$ is a commutative ring with unit element. An ideal J of the ring $C(R, \tau)$ is said to be *fixed* or *localized* at the point $p \in R$ whenever $J = \{\varphi : \varphi \in C(R, \tau), \varphi p = 0\}$.

Exercise

1. Every localized ideal of $C(R, \tau)$ is a proper maximal ideal.

2. Let (R, τ) be a completely compact T_2-space. Then every (proper) maximal ideal of $C(R, \tau)$ is localized at exactly one point. There exists a 1-1 correspondence between points of R and maximal ideals of $C(R, \tau)$.

3. Two completely compact T_2-spaces (R, τ) and (R', τ') are homeomorphic if and only if the rings $C(R, \tau)$ and $C(R', \tau')$ are isomorphic.

4. The contentions 1-3 above remain valid if we replace $C(R, \tau)$ by the ring $B(R, \tau)$, defined to be the ring of all continuous bounded mappings φ from (R, τ) into the real line (i.e., φR is a bounded subset of $\mathbf{R}$) (cf. 27A).

19. Properties of Continuous Mappings

The early theorems of this section serve as a link between continuity of mappings and compactness or connectedness properties of spaces.

19.1 *Let the space (R, τ) be completely compact (k-compact). If there exists a continuous mapping φ of (R, τ) onto (R', τ'), then (R', τ') must also be completely compact (k-compact).*

Proof: Let $\mathfrak{a}'$ be a filter distinct from $\mathfrak{o}$ belonging to R' (with $\operatorname{Rk}(\mathfrak{a}') \leqslant k$). Then $\varphi^{-1}\mathfrak{a}' \neq \mathfrak{o}$ (and $\operatorname{Rk}(\varphi^{-1}\mathfrak{a}') \leqslant k$), whence the filter $\varphi^{-1}\mathfrak{a}'$ possesses an adherent point p. We have $\mathfrak{o} \neq \tau p \wedge \varphi^{-1}\mathfrak{a}'$ and hence

$$\mathfrak{o} \neq \varphi(\tau p \wedge \varphi^{-1}\mathfrak{a}') \leqslant \varphi\tau p \wedge \varphi\varphi^{-1}\mathfrak{a}' \leqslant \tau'\varphi p \wedge \mathfrak{a}'.$$

Thus φp is an adherent point of $\mathfrak{a}'$. ∎

19.2 *Every continuous mapping φ of a completely compact space (R, τ) into a T_2-space (R', τ') is a closed mapping. If φ is a 1-1 mapping of R onto R', then φ is a homeomorphism.*

Proof: Let M be a closed subset of (R, τ). Then M must be completely compact (11.6), and by 19.1 φM must also be completely compact. Since τ' is a T_2-topology, φM must also be closed (11.7). Hence φ is a closed mapping. If in addition φ is a 1-1 mapping of R onto R', then, by 17.10, φ is also open. Thus φ is a 1-1 mapping, both continuous and open, and therefore a homeomorphism. ∎

19.3 *Let φ be a continuous mapping from (R, τ) into (R', τ'). If M is a connected subset of R, then the image φM is also connected.*

Proof: Let $\{A', B'\}$ be a decomposition of φM into two separated sets. Then the sets $A = \varphi^{-1}A'$ and $B = \varphi^{-1}B'$ constitute a decomposition of M. We have

$$\varphi(\bar{A} \cap B) \subseteq \varphi\bar{A} \cap \varphi B \subseteq \overline{\varphi A} \cap \varphi B = \bar{A}' \cap B' = \emptyset,$$

hence $\bar{A} \cap B = \emptyset$. Likewise we have $A \cap \bar{B} = \emptyset$. But this contradicts the hypothesis that M is connected. ∎

Definition 19a: *A subset M of R is said to be* stable *under the mapping $\varphi : R \to R'$, if $M = \varphi^{-1}\varphi M$.*

Manifestly, M is stable under the mapping $\varphi : R \to R'$ if and only if there exists a subset M' of R' such that $M = \varphi^{-1}M'$.

19.4 *Let φ be a continuous mapping of (R, τ) onto (R', τ') which sends open stable sets onto open sets. Then if (R, τ) is locally connected, so is (R', τ').*

Proof: Let M' be an open subset of (R',τ') and K' a component of M'. Then $K = \varphi^{-1}K'$ is a stable set and K is contained in the open set $M = \varphi^{-1}M'$. For each $p \in M$ let K_p denote the component of M containing the point p. From $p \in K$ we see that $\varphi p \in K' \cap \varphi K_p$. Since φK_p is connected by 19.3, we must also have $K' \cup \varphi K_p$ connected, by the note to 14.5. Hence $\varphi K_p \subseteq K'$ and therefore $K_p \subseteq K$. But K_p is a component of the open set M and hence, by 15.1, K_p is itself open and thereby a neighborhood of p. Hence K is open and stable and $K' = \varphi K$ is, by hypothesis, open. The components of the open subsets of (R', τ') are themselves open, whence, by 15.1, (R', τ') is locally connected. ∎

A continuous mapping $\varphi : (R, \tau) \to (R', \tau')$ is said to be *quasi-compact* if it sends closed stable sets onto closed sets. For instance, every closed mapping is quasi-compact. If φ is a mapping onto R, then $\varphi\, \complement M = \complement\varphi M$ for every stable subset M of R, and the closed stable subsets are identical with the complements of open stable subsets. In the above theorem we can therefore replace the hypothesis on φ by the completely equivalent hypothesis "*let φ be a quasi-compact mapping of* (R, τ) *onto* (R', τ')." As a special case we see that a map which is both continuous and open (closed) must send locally connected spaces into locally connected images. The following result is even more special.

19.5 *Let φ be a continuous mapping of a completely compact space (R, τ) onto a T_2-space (R', τ'). If (R, τ) is locally connected, so is (R', τ')* (cf. 19c).

Proof: Let M be an open stable subset of R. Since φ maps onto R', $\varphi\, \complement M = \complement\varphi M$. By 19.2, φ must be a closed mapping. Hence if $\complement M$ is closed, so is $\varphi\, \complement M = \complement\varphi M$ closed, and φM itself is open. Thus φ sends open stable sets onto open sets, and our contention follows from 19.4. ∎

19.6 *Let φ be a continuous mapping of (R, τ) onto (R', τ'). If M is dense in R, then φM is dense in R'. We have* $\mathrm{Sd}\,(R', \tau') \leqslant \mathrm{Sd}\,(R, \tau)$. *If τ' can be induced by a pseudometric, then we also have* $\mathrm{Bd}\,(R', \tau') \leqslant \mathrm{Bd}\,(R, \tau)$.

Proof: If $\bar{M} = R$, then $\overline{\varphi M} \supseteq \varphi \bar{M} = \varphi R = R'$ (17.1). Hence φM is dense in R'. In the special case where M is dense in R and M is of minimal cardinality, we have

$$\mathrm{Sd}\,(R', \tau') \leqq |\,\varphi M\,| \leqq |\,M\,| = \mathrm{Sd}\,(R, \tau).$$

If τ' is induced by a pseudometric, then we see, from 10.11 and 10.7, that

$$\mathrm{Bd}\,(R', \tau') = \mathrm{Sd}\,(R', \tau') \leqq \mathrm{Sd}\,(R, \tau) \leqq \mathrm{Bd}\,(R, \tau). \blacksquare$$

19.7 *If there exists a continuous mapping φ from the completely compact space (R, τ) onto the T_2-space (R', τ'), then we must have* $\mathrm{Bd}\,(R', \tau') \leqslant \mathrm{Bd}\,(R, \tau)$.

Proof: Let $\mathfrak{B}$ be a minimal basis of (R, τ). By 19.2, φ is a closed mapping. Hence

$$\mathfrak{B}' = \{\complement\varphi \complement B : B \in \mathfrak{B}\}$$

is a system of open subsets of (R', τ') with $|\,\mathfrak{B}'\,| \leqslant |\,\mathfrak{B}\,| = \mathrm{Bd}\,(R, \tau)$. Let now M' represent a nonempty open subset of (R', τ'). Then $\varphi^{-1}M'$ is also open, and $\varphi^{-1}M' = \bigcup\{B : B \in \mathfrak{B}_0\}$ for an appropriate subsystem $\mathfrak{B}_0$ of $\mathfrak{B}$, and hence

$$M' = \bigcup\{\varphi B : B \in \mathfrak{B}_0\} \supseteq \bigcup\{\complement\varphi \complement B : B \in \mathfrak{B}_0\} = N'.$$

The complementary filter $\mathfrak{c}$ of the system $\mathfrak{B}_0$ (cf. p. 28) is a closed filter. From $p' \notin N'$ we deduce $\ulcorner p' \urcorner \leqslant \varphi\mathfrak{c}$, hence $\mathfrak{b} = \ulcorner \varphi^{-1}p' \urcorner \wedge \mathfrak{c} \neq \mathfrak{o}$. The filter $\mathfrak{b}$ is also closed and possesses an adherent point q. Consequently $\ulcorner q \urcorner \leqslant \mathfrak{b} \leqslant \ulcorner \varphi^{-1}p' \urcorner$, hence $\varphi q = p'$. Since $\ulcorner q \urcorner \leqslant \mathfrak{b} \leqslant \mathfrak{c}$ we also have $q \in \bigcap\{\complement B : B \in \mathfrak{B}_0\} = \complement\varphi^{-1}\complement M'$, hence $p' = \varphi q \notin M'$. We thus have $N' \supseteq M'$ and finally $M' = N'$. Hence $\mathfrak{B}'$ is even a basis of (R', τ') and therefore $\mathrm{Bd}(R', \tau') \leqslant |\,\mathfrak{B}'\,| \leqslant \mathrm{Bd}(R, \tau)$. $\blacksquare$

19.8 *Let φ and ψ be two continuous mappings of (R, τ) into the T_2-space (R', τ'). If φ and ψ agree on a subset M of R (i.e., $\varphi p = \psi p$ for all $p \in M$), then they must also agree on $\bar{M}$. In the special case where M is dense in R, φ and ψ are everywhere identical.*

Proof: If $p \in \bar{M}$, then $\mathfrak{a} = \tau p \wedge \ulcorner M \urcorner \neq \mathfrak{o}$. Since $\mathfrak{a} \leqslant \ulcorner M \urcorner$, we have $\varphi\mathfrak{a} = \psi\mathfrak{a}$. Since $\mathfrak{a} \leqslant \tau p$, we have $\varphi\mathfrak{a} \leqslant \varphi\tau p \leqslant \tau'\varphi p$ and $\psi\mathfrak{a} \leqslant \psi\tau p \leqslant \tau'\psi p$, hence $\tau'\varphi p \wedge \tau'\psi p \neq \mathfrak{o}$. Since τ' is a T_2-topology, we finally have $\varphi p = \psi p$. $\blacksquare$

Definition 19b: *Let M be a subset of R and φ a mapping from M into R'. We say that a mapping $\psi : R \to R'$ is a* continuation *or* extension *of φ to R*

if ψ agrees with φ on M; i.e., φ is the restriction of ψ to M. If φ and ψ above are continuous mappings from (M, τ_M) and (R, τ), respectively, into (R', τ') then we call ψ a continuous extension *of φ to (R, τ).*

19.9 *A continuous mapping φ from a subspace (M, τ_M) dense in (R, τ) into a T_3-space (R', τ') admits of a continuous extension to (R, τ), if and only if for all $p \in R$ the filter φ Tr (τp) converges in (R', τ').*

Proof: If ψ is a continuous extension of φ, then $\varphi \operatorname{Tr}(\tau p) \leqslant \psi\tau p \leqslant \tau'\psi p$, and $\varphi \operatorname{Tr}(\tau p)$ must be convergent. Conversely, suppose that to every $p \in R \cap \complement M$ we can associate the image $\psi p \in R'$ such that $\varphi \operatorname{Tr}(\tau p) \leqslant \tau'\psi p$. For points $p \in M$ let us also assume that $\psi p = \varphi p$. Then ψ is a extension of φ to R. If U' is a neighborhood belonging to $\tau'\psi p$, then, by the continuity of φ and by $(\mathbf{T}_3)$, there must exist a set $S \in \operatorname{Tr}(\tau p)$ such that $\overline{\varphi S} \subseteq U'$. Furthermore, there exists an open neighborhood $U \in \tau p$ with $U \cap M \subseteq S$. From $q \in U$ it follows that $\tau q \leqslant \ulcorner U \urcorner$, hence $\operatorname{Tr}(\tau q) \leqslant \ulcorner S \urcorner$. Since M is dense in (R, τ), we have $\operatorname{Tr}(\tau q) \neq 0$, and finally

$$\tau'\psi q \wedge \ulcorner \varphi S \urcorner \geqslant \varphi \operatorname{Tr}(\tau q) \wedge \ulcorner \varphi S \urcorner \geqslant \varphi(\operatorname{Tr}(\tau q) \wedge \ulcorner S \urcorner) > 0.$$

Hence $\psi q \in \overline{\varphi S} \subseteq U'$ and, furthermore, $\psi U \subseteq U'$. But this tells us that ψ is continuous. ∎

The convergence criterion of the last theorem can be put into a form more useful for applications.

19.10 *A continuous mapping φ from a subset (M, τ_M) dense in (R, τ) into a T_3-space (R', τ') admits of a continuous extension to (R, τ) if and only if the following conditions are satisfied:*

(a) *If $A \subseteq M$, $B \subseteq M$ and $\overline{\varphi A} \cap \overline{\varphi B} = \emptyset$ then $A \cap B = \emptyset$.*

(b) *For each $p \in R$, the filter φ Tr (τp) possesses an adherent point in (R', τ'). If, in addition τ' is a T_2-topology, then there exists one and only one extension. Whenever (R', τ') is a completely compact T_3-space, we may omit condition* (b) *and the above theorem will still hold.*

Proof: Let us assume that ψ is a continuous extension of φ to (R, τ). If $p \in \bar{A} \cap \bar{B}$, then $\operatorname{Tr}(\tau p) \wedge \ulcorner A \urcorner \neq 0$ and $\operatorname{Tr}(\tau p) \wedge \ulcorner B \urcorner \neq 0$. Since

$$\varphi \operatorname{Tr}(\tau p) \leqslant \psi\tau p \leqslant \tau'\psi p,$$

we must have $\tau'\psi p \wedge \ulcorner \varphi A \urcorner \neq 0$ and $\tau'\psi p \wedge \ulcorner \varphi B \urcorner \neq 0$. Hence $\psi p \in \overline{\varphi A} \cap \overline{\varphi B}$; i.e., (a) is satisfied. Condition (b) follows from the convergence of $\varphi \operatorname{Tr}(\tau p)$ to ψp. To prove the converse, let us suppose that conditions (a) and (b) are satisfied. For every $p \in R$, let p' denote a point adherent to $\varphi \operatorname{Tr}(\tau p)$, and let us suppose that $\varphi \operatorname{Tr}(\tau p)$ does not converge to p'. Then there exists an open neighborhood

$U' \in \tau'p'$ such that $\varphi \operatorname{Tr}(\tau p) \wedge \overline{\complement U'} \neq \mathfrak{o}$. By $(\mathbf{T}_3)$ there also exists a $V' \in \tau'p'$ such that $\bar{V}' \subseteq U'$. If we put $A = \varphi^{-1}\bar{V}'$ and $B = \varphi^{-1}\complement U'$, then $p \in \bar{A} \cap \bar{B}$ and $\overline{\varphi A} \cap \overline{\varphi B} \subseteq \overline{V'} \cap \overline{\complement U'} = \emptyset$, which contradicts (a). Hence every filter $\varphi \operatorname{Tr}(\tau p)$ converges in (R', τ') and our contention follows from 19.9.

The uniqueness statement follows directly from 19.8. In the case where (R, τ) is completely compact, we know that condition (b) is automatically satisfied and it can therefore be omitted from our hypotheses. ∎

The above-proved theorem furnishes a criterion for the existence of a continuous extension of a mapping from a dense subset to the whole space. The next theorem gives a criterion for continuous extension from closed subsets.

19.11 (H. Tietze) *Given the closed subset M of the space (R, τ) and the continuous mapping $\varphi : (M, \tau_M) \to \mathbf{R}$ then there exists a continuous extension of φ to (R, τ) if and only if τ is a T_4-topology.*

Proof: Let the space (R, τ) be such that the above-described extension is always possible. If A and B are two disjoint closed sets; then $M = A \cup B$ is likewise closed. We define a continuous mapping $\varphi : (M, \tau_M) \to \mathbf{R}$ by $\varphi p = 0$ for all $p \in A$ and $\varphi q = 1$ for all $q \in B$. By hypothesis φ admits of a continuous extension ψ to all of (R, τ). The sets

$$U = \{p : p \in R, \psi p < \tfrac{1}{2}\} \qquad \text{and} \qquad V = \{p : p \in R, \psi p > \tfrac{1}{2}\}$$

are then open. We have $A \subseteq U$, $B \subseteq V$, and $U \cap V = \emptyset$. Hence τ is a T_4-topology.

Conversely, let M be a closed subset of the T_4-space (R, τ). Let us suppose that the given continuous map $\varphi\ (M, \tau_M) \to \mathbf{R}$ is such that $\varphi M \subseteq [-1, +1]$. Let $s = \frac{1}{3}$ and $t = \frac{2}{3}$. Then the following inductive prescription defines two sequence of continuous maps (φ_n) and (ψ_n), where $\varphi_n : (M, \tau_M) \to \mathbf{R}$ and $\psi_n : (R, \tau) \to \mathbf{R}$ in such a way that:

(a) $|\varphi_n p| \leqslant t^n$ for all $p \in M$.

(b) $|\psi_n x| \leqslant st^n$ for all $x \in R$.

(c) $|\varphi p - (\psi_0 p + - + \psi_n p)| \leqslant t^{n+1}$ for all $p \in M$.

Suppose we put $\varphi_0 = \varphi$. If φ_n has already been constructed, then the sets

$$A_n = \{p : p \in M, \varphi_n p \leqslant -st^n\} \qquad \text{and} \qquad B_n = \{p : p \in M, \varphi_n p \geqslant +st^n\}$$

are disjoint and closed in M, hence closed in R (for M is closed in R). By 18.6 there exists a continuous mapping $\sigma_n : (R, \tau) \to [0, 1]$ such that $\varphi_n A_n = \{0\}$ and $\sigma_n B_n = \{1\}$. We can then define a continuous map $\varphi_n : (R, \tau) \to \mathbf{R}$ by $\psi_n x = 2st^n(\sigma_n x - \frac{1}{2})$. This map satisfies (b), and we have $\psi_n A_n = \{-st^n\}$ and $\psi_n B_n = \{+st^n\}$. Finally, let us define φ_{n+1} by $\varphi_{n+1} p = \varphi_n p - \psi_n p$ for all $p \in M$.

Then φ_{n+1} is a continuous mapping satisfying (a). Furthermore

$$|\varphi p - (\psi_0 p + \ldots + \psi_n p)| = |\varphi_{n+1} p| \leqslant t^{n+1},$$

hence (c) also holds.

By (b), the series $\Sigma(\psi_n p)$ converges for all $p \in R$. Hence $\psi p = \Sigma(\psi_n p)$ defines a mapping $\psi : R \to \mathbf{R}$ which, by (c) must be an extension of φ to R. The continuity of ψ follows from the fact that by (b) the series $\Sigma(\psi_n p)$ must be uniformly convergent on all R.

If we even have $\varphi M \subseteq]-1, +1[$, the above proof only tells us that ψ is a continuous extension of φ such that $\psi R \subseteq]-1, +1[$. But then M and $N = \psi^{-1}\{-1, +1\}$ are disjoint closed subsets of (R, τ), and hence, by 18.6, there exists a continuous mapping $\chi : (R, \tau) \to [0, 1]$ such that $\chi N = \{0\}$ and $\chi M = \{1\}$. If we put $\psi' p = (\psi p) \cdot \chi(p)$ $(p \in R)$ then ψ' is a continuous extension of φ to (R, τ) and $\psi' R \subseteq]-1, +1[$.

Lastly, if φ is not assumed to satisfy the limitative stipulation $\varphi M \subseteq [-1, +1]$, we still can obtain a continuous extension by proceeding as follows: By $\varphi' p = \varphi p/(1 + |\varphi p|)$ (contraction mapping, cf. 16 I) we can define a continuous mapping $\varphi' : (M, \tau_M) \to]-1, +1[$. By the previous case, φ' admits of a continuous extension $\psi' : (R, \tau) \to]-1, +1[$. Then $\psi p = \psi' p/(1 - |\psi' p|)$ gives us the continuous extension of φ to (R, τ) being sought. ∎

Notes and Exercises

19A The mapping $\varphi : R \to R'$ is said to be a *quasi-open* mapping of (R, τ) into (R', τ') if it possesses the following property: Whenever U is an open subset of (R, τ) which contains a completely compact component of $\varphi^{-1}p'$, we have as a consequence $p' \in \varphi U$.

The mapping φ is said to be *scattered* whenever the inverse image $\varphi^{-1}p'$ is totally disconnected for every $p' \in R'$.

A quasi-open and scattered mapping of (R, τ) onto (R', τ') must be open.

19B *Exercise:* Let φ be a closed mapping of a T_3-space (R, τ) into a completely compact space (R', τ'). Suppose that for every $p' \in R'$, the set $\varphi^{-1}p'$ possesses no point of accumulation. Then φ must be continuous (the sets $\varphi^{-1}p'$ are often called the *fibers* of the mapping φ).

19C A continuous mapping $\varphi : (R, \tau) \to (R', \tau')$ is called *proper* if the inverse image $\varphi^{-1}M'$ of every completely compact subset M' of R' is itself completely compact.

Exercise: Let (R', τ') be a locally compact T_2-space and $\varphi : (R, \tau) \to (R', \tau')$ a proper mapping.

1. Then (R, τ) must also be l.c.c.

2. If M is a closed subset of R and M' a completely compact subset of R', then $\varphi M \cap M'$ is closed.
3. φ is a closed mapping.
4. If φ is a mapping onto R' and (R, τ) is locally connected, then so is (R', τ').

20. Transference of Topologies

Mappings can be used to stipulate topologies: For instance, let φ be a mapping from R onto R'. If there already exists a topology on one of these sets, R or R', we can "transfer" this topology to the other set by means of φ. In this section we shall study just such a transfer. In what follows, φ will always denote a mapping from R onto R', while τ' will denote a topology on R'.

If $\mathfrak{S}'$ stands for the family of all open subsets of R', then the family

$$\mathfrak{S} = \varphi^{-1}\mathfrak{S}' = \{\varphi^{-1}M' : M' \in \mathfrak{S}'\}$$

verifies the hypotheses of 5.7. Hence there exists a unique topology on R, having $\mathfrak{S}$ as its family of open sets. We shall call this topology on R the *topology transferred by* φ^{-1} and denote it by ${}_{\varphi}\tau'$. The fact that in this notation φ is written to the left and below, as a sort of index, should remind us that the transference is being done by φ^{-1}, from R' back to R.

20.1 *Let φ be simultaneously a continuous, open, and closed mapping of $(R, {}_{\varphi}\tau')$ onto (R', τ').*

Then ${}_{\varphi}\tau'$ is the coarsest topology on R with respect to which φ is a continuous mapping onto (R', τ').

For all $p \in R$, we have ${}_{\varphi}\tau'p = \varphi^{-1}\tau'\varphi p$.

A subset M of R is ${}_{\varphi}\tau'$-open (${}_{\varphi}\tau'$-closed) if and only if M is stable (i.e., $M = \varphi^{-1}M$, cf. 19a) *and φM is a τ'-open (τ'-closed) set.*

For all subsets M of $(R, {}_{\varphi}\tau')$ we have $\bar{M} = \varphi^{-1}\overline{\varphi M}$.

Proof: By the very definition of ${}_{\varphi}\tau'$, M is ${}_{\varphi}\tau'$-open if and only if there exists a τ'-open set M' such that $M = \varphi^{-1}M'$. This is equivalent to $M = \varphi^{-1}\varphi M$ and φM being open. The characterization of ${}_{\varphi}\tau'$-closed sets then follows by complementation, using $\complement\varphi^{-1}M' = \varphi^{-1}\complement M'$. Hence φ maps open sets into open sets, closed sets onto closed sets, and it must therefore be a map both open and closed. In addition, φ is continuous, for if M' is open so is $\varphi^{-1}M'$ (17.1). Thus U is an open neighborhood belonging to ${}_{\varphi}\tau'p$ if and only if there exists an open neighborhood $U' \in \tau'\varphi p$ such that $U = \varphi^{-1}U'$. Hence ${}_{\varphi}\tau'p = \varphi^{-1}\tau'\varphi p$. If τ is any topology on R, where $\varphi : (R, \tau) \to (R', \tau')$, then $\varphi\tau p \leqslant \tau'\varphi p$, hence $\tau p \leqslant \varphi^{-1}\tau'\varphi p = {}_{\varphi}\tau'p$. This tells us that ${}_{\varphi}\tau'$ is the coarsest topology on R with respect to which φ is continuous. Since φ is continuous and closed, we have, by 17.1 and 17.8, $\varphi\bar{M} = \overline{\varphi M}$. But since $\bar{M}$ is closed, $\bar{M} = \varphi^{-1}\varphi\bar{M}$ and therefore $\bar{M} = \varphi^{-1}\overline{\varphi M}$. ∎

If R_0 is a third fundamental set and ψ a mapping of R_0 onto R, we see immediately that ${}_{(\varphi\psi)}\tau' = {}_{\psi}({}_{\varphi}\tau')$. If, moreover, a topology τ_0 has been stipulated on R_0, then we can tell when ψ is continuous by the following theorem.

20.2 $\psi : (R_0, \tau_0) \to (R, {}_{\varphi}\tau')$ *is equivalent to* $\varphi\psi : (R_0, \tau_0) \to (R', \tau')$.

Proof: Since φ is continuous, the continuity of ψ implies that of $\varphi\psi$. Conversely, suppose $\varphi\psi$ is continuous. Then for every $p_0 \in R_0$ we have $\varphi\psi\tau_0 p_0 \leqslant \tau'\varphi\psi p_0$ and hence $\psi\tau_0 p_0 \leqslant \varphi^{-1}\tau'\varphi\psi p_0 = {}_{\varphi}\tau'\psi p_0$, which means that ψ is continuous. ∎

A property of the topology τ' *is said to be* invariant under transfer, *or a* transfer-invariant *whenever it is preserved under transfer, i.e., if it holds for* τ' *then it holds for* ${}_{\varphi}\tau'$, *whatever the fundamental set R and the map* φ *from R onto* R' *may be.*

20.3 *The following notions or properties are transfer invariants:* $(\mathbf{T}_3)$, $(\mathbf{T}_{3a})$, $(\mathbf{T}_4)$, $(\mathbf{T}_5)$, "*being probed by a given space,*" *complete compactness, k-compactness, local complete compactness, paracompactness, connectedness, local connectedness.*

Proof: By 20.1, $\bar{M} = \varphi^{-1}\overline{\varphi M}$ for all subsets M of R. Hence $\overline{\varphi^{-1}M'} = \varphi^{-1}\overline{M'}$ for all subsets M' of R' and so $\overline{{}_{\varphi}\tau'p} = \overline{\varphi^{-1}\tau'\varphi p} = \varphi^{-1}\overline{\tau'\varphi p}$. If τ' is a T_3-topology we must have $\overline{\tau'\varphi p} = \tau'\varphi p$, whence $\overline{{}_{\varphi}\tau'p} = \varphi^{-1}\tau'\varphi p = {}_{\varphi}\tau'p$; i.e. ${}_{\varphi}\tau'$ is also a T_3-topology.

If $\varphi\mathfrak{a} \leqslant \tau'\varphi p$, then $\mathfrak{a} \leqslant \varphi^{-1}\tau'\varphi p = {}_{\varphi}\tau'p$. Hence $\{\varphi\}$ is a probing system, $(R, {}_{\varphi}\tau')$ being probed by means of (R', τ'). The invariance of (T_{3a}) and of "being probed" now both follow from 18.10.

Since closed sets of $(R, {}_{\varphi}\tau')$ are stable by 20.1, closed disjoint sets and separated sets must get mapped onto sets of similar type. This fact immediately implies the invariance of $(\mathbf{T}_4)$, $(\mathbf{T}_5)$, connectedness, and local connectedness.

If $\mathrm{Rk}\,(\mathfrak{a}) \leqslant k$, then $\mathrm{Rk}\,(\varphi\mathfrak{a}) \leqslant k$ also, and if $\varphi\mathfrak{a} \leqslant \tau'\varphi p$, then $\mathfrak{a} \leqslant {}_{\varphi}\tau'p$. These results imply the transfer invariance of complete compactness and k-compactness. Moreover, if M' is a completely compact subset of R', then $\varphi^{-1}M'$ is also completely compact under the transfer topology ${}_{\varphi}\tau'$. Since $U' \in \tau'\varphi p$ implies $\varphi^{-1}U' \in {}_{\varphi}\tau'p$, we see that local complete compactness is also transfer invariant.

Finally, let us suppose that (R', τ') is paracompact and that $\mathfrak{H}$ is an open covering of $(R, {}_{\varphi}\tau')$, for which there exists an open and locally finite covering $\mathfrak{K}'$ such that $\mathfrak{K}' \lll \mathfrak{H}'$. The system $\mathfrak{K} = \{\varphi^{-1}K' : K' \in \mathfrak{K}'\}$ is an open covering of R such that $\mathfrak{K} \lll \mathfrak{H}$. If U' is a neighborhood in R' and U' has a nonempty intersection with at most a finite number of sets of $\mathfrak{K}'$, then $\varphi^{-1}U'$ has a nonempty intersection with at most a finite number of sets of $\mathfrak{K}$. Hence $\mathfrak{K}$ is also locally finite and $(R, {}_{\varphi}\tau')$ is paracompact. ∎

The separation axioms ($\mathbf{T}_1$) and ($\mathbf{T}_2$) are, in general, not transfer invariant. They are, however, invariants whenever φ is a 1-1 mapping, and conversely, for φ, being open and continuous, becomes a homeomorphism as soon as it is 1-1. If, nevertheless, there exist two distinct points $p, q \in R$ with $\varphi p = \varphi q$, then $\overline{q} \leqslant {}_{\varphi}\tau' p$, whence ${}_{\varphi}\tau'$ can not be a T_1-topology.

Note that the definition of the transfer topology is still meaningful when φ is not a mapping onto R', but only into R'. In the above theorems, all we need do is replace (R', τ') by the subspace φR, and then all results will still hold true. For questions of continuity, this relativization is, of course, unimportant (17.4).

The following generalization is often useful: Let $\{(R'_\iota, \tau'_\iota) : \iota \in I\}$ be a family of spaces, and for each $\iota \in I$, let φ_ι be a mapping from R into R'_ι. If we put $\Phi = \{\varphi_\iota : \iota \in I\}$ then the topology ${}_{\Phi}\tau'$ on R, defined by ${}_{\Phi}\tau' = \bigwedge \{{}_{\varphi_\iota}\tau' : \iota \in I\}$, is known as the *topology transferred* by Φ^{-1}. For each $p \in R$ we then have

$$ {}_{\Phi}\tau' p = \bigwedge \{\varphi_\iota^{-1}\tau'_\iota \varphi_\iota p : \iota \in I\}. $$

This result ties in with the theory of probings, as follows: If Φ is a family of continuous mappings from (R, τ) into (R', τ'), then Φ is a probing of (R, τ) by means of (R', τ') if and only if $\tau = {}_{\varphi}\tau'$.

20.4 *${}_{\Phi}\tau'$ is the coarsest topology on R with respect to which each φ_ι is a continuous mapping into (R'_ι, τ'_ι).*

Proof: Each φ_ι is continuous because ${}_{\Phi}\tau' \leqslant {}_{\varphi_\iota}\tau'_\iota$. Conversely if

$$ \varphi_\iota : (R, \tau) \to (R'_\iota, \tau'_\iota) $$

then, by 20.1, $\tau \leqslant {}_{\varphi_\iota}\tau'_\iota$ and then $\tau \leqslant \bigwedge {}_{\varphi_\iota}\tau'_\iota = {}_{\Phi}\tau'$. ∎

As in 20.2, we have:

20.5 $\psi : (R_0, \tau_0) \to (R, {}_{\Phi}\tau')$ is equivalent to $\varphi_\iota\psi : (R_0, \tau_0) \to (R'_\iota, \tau'_\iota)$ for all $\iota \in I$.

Proof: Using 17.6, this follows immediately from 20.2. ∎

Notes and Exercises

20A The family of mappings $\Phi = \{\varphi_\iota : \iota \in I : \varphi_\iota : R \to R'_\iota\}$ is said to *separate points*, whenever, given any two distinct points $p, q \in R$, we can find an $\iota \in I$ such that $\varphi_\iota p \neq \varphi_\iota q$.

Exercise: Let $\{(R'_\iota, \tau'_\iota) : \iota \in I\}$ be a system of T_1- or T_2-spaces. Then ${}_{\varphi}\tau'$ is likewise a T_1-, respectively T_2-topology, if and only if Φ separates points.

20B *Exercise:* The notions of rank, basis degree, and separability degree are all transfer invariants: We have

$$ \mathrm{Rk}\,({}_{\varphi}\tau') = \mathrm{Rk}\,(\tau'), \qquad \mathrm{Bd}\,({}_{\varphi}\tau') = \mathrm{Bd}\,(\tau'), \qquad \mathrm{Sd}\,({}_{\varphi}\tau') = \mathrm{Sd}\,(\tau'). $$

20C *Exercise:* Let $\{(R'_\iota, \tau'_\iota) : \iota \in I\}$ be a system or T_3- of T_{3a}-spaces and let $\Phi = \{\varphi_\iota : \iota \in I\}$, where $\varphi_\iota : R \to R'_\iota$. Then $_\varphi\tau'$ is likewise a T_3-, respectively T_{3a}-topology. If all the spaces (R'_ι, τ'_ι) can be probed by the space (R'', τ''), $(R, {}_\varphi\tau')$ can also be probed by (R'', τ'') (cf. 18.14).

21. Quotient Spaces

As in the proceding section, let φ be a mapping from R onto R'. But now let a topology τ be already given on R. We shall show how to transfer this topology to the right, from R to R', using an extremal condition, reminiscent of that condition that characterized $_\varphi\tau'$ as the coarsest topology rendering φ continuous. Let Σ' be the system of all topologies τ' on R', for which φ is a continuous mapping of (R, τ) onto (R', τ'). Then Σ' certainly contains the trivial topology, and by 17.6, we know that τ_φ defined by $\tau_\varphi = \bigwedge \{\tau' : \tau' \in \Sigma'\}$ is the finest topology on R', for which φ is still a continuous mapping from (R, τ) to R'. This topology is known as the *quotient topology* generated by φ (cf. 21a) or *the topology transferred by* φ.

21.1 *A subset M' of R' is τ_φ-open (τ_φ-closed) if and only if $\varphi^{-1}M'$ is τ-open (τ-closed).*

Proof: The system of all subsets M' of R', for which $\varphi^{-1}M'$ is open, satisfies the conditions of 5.7 and thereby defines a topology τ' on R' under which φ is continuous by 17.1. Hence $\tau' \in \Sigma'$ and $\tau_\varphi \leqslant \tau'$. If, on the other hand, M' is a τ_φ-open set, then, since $\varphi : (R, \tau) \to (R', \tau_\varphi)$, we see that $\varphi^{-1}M'$ is also open and M' is therefore τ'-open, which tells us that, conversely, $\tau' \leqslant \tau_\varphi$. The contention on closed sets follows by similar reasoning. ∎

21.2 $\psi : (R', \tau_\varphi) \to (R'', \tau'')$ *is equivalent to* $\psi\varphi : (R, \tau) \to (R'', \tau'')$.

Proof: Since $\varphi : (R, \tau) \to (R', \tau_\varphi)$, the continuity of ψ implies that of $\psi\varphi$. To prove the converse, suppose $\psi\varphi$ is continuous. Then for any point $p \in R$ we have $\psi\varphi\tau p \leqslant \tau''\psi\varphi p$, hence $\varphi\tau p \leqslant \psi^{-1}\tau''\psi\varphi p = {}_\psi\tau''\varphi p$. This implies that $\varphi : (R, \tau) \to (R', {}_\psi\tau'')$ and hence $\tau_\varphi \leqslant {}_\psi\tau''$. Since $\psi : (R', {}_\psi\tau'') \to (R'', \tau'')$, and using 17.6, we certainly have $\psi : (R', \tau_\varphi) \to (R'', \tau'')$. ∎

21.3 *Let ψ be a mapping from R' onto a third fundamental set R''. Then* $\tau_{(\psi\varphi)} = (\tau_\varphi)_\psi$.

Proof: Since $\varphi : (R, \tau) \to (R', \tau_\varphi)$ and $\psi : (R', \tau_\varphi) \to (R'', (\tau_\varphi)_\psi)$, we have $(\psi\varphi : (R, \tau) \to (R'', (\tau_\varphi)_\psi)$, hence $(\tau_\varphi)_\psi \geqslant \tau_{(\psi\varphi)}$. Conversely, if

$$\psi\varphi : (R, \tau) \to (R'', \tau_{(\psi\varphi)}),$$

then, by 21.2, $\psi : (R', \tau_\varphi) \to (R'', \tau_{(\psi\varphi)})$, and therefore $(\tau_\varphi)_\psi \leqslant \tau_{(\psi\varphi)}$. ∎

For transfer in the direction of this section, i.e., from left to right, we have fewer invariants than for transfer in the opposite direction. In full generality, we have only

21.4 *If τ is completely compact, k-compact, connected, or locally connected, then the quotient topology τ_φ is of the same nature.*

Proof: The preservation of complete compactness, of k-compactness, and of connectedness, all follow from the continuity of φ and from 19.1 and 19.3. If M is a τ-open stable set (19A), then by 21.1, φM is a τ_φ-open set. By 19.4, we therefore see that locally connectedness is also preserved. ∎

21.5 *The quotient topology τ_φ is a T_1-topology if and only if, for each point $p' \in R'$ the inverse image $\varphi^{-1}p'$ is a τ-closed set.*

Proof: By 21.2, we see that the above condition is equivalent to the fact that all one-point sets in (R', τ_φ) are closed (8.1). ∎

In general φ is neither an open nor a closed mapping of (R, τ) onto (R', τ'). However, we do have the following result.

21.6 *If φ is both continuous and open, or both continuous and closed, from (R, τ) onto (R', τ'), then $\tau' = \tau_\varphi$.*

Proof: Since $\varphi : (R, \tau) \to (R', \tau')$ we have $\tau' \geqslant \tau_\varphi$. And since

$$\varphi : (R, \tau) \to (R', \tau_\varphi),$$

we have $\varphi\tau p \leqslant \tau_\varphi \varphi p$ for each $p \in R$. If now φ is an open mapping from (R, τ) onto (R', τ'), then $\tau'\varphi p \leqslant \varphi\tau p \leqslant \tau_\varphi\varphi p$ and hence we have the reverse inequality $\tau' \leqslant \tau_\varphi$. If φ is a closed mapping of (R, τ) onto (R', τ'), then, using 17.8, we likewise have $\tau'p = \varphi\varphi^{-1}\tau'p' = \bigvee \{\varphi\tau q : \varphi q = p'\} \leqslant \tau_\varphi p'$. ∎

Every mapping $\varphi : R \to R'$ induces an equivalence relation $\tilde{\varphi}$ where $(p, q) \in \tilde{\varphi}$ if and only if $\varphi p = \varphi q$. Conversely, every equivalence relation can be thought of as induced by an appropriate mapping: Let ρ be the equivalence in question, defined over R, and let κ be that mapping, from R onto the quotient set R/ρ, which associates with each $p \in R$ the equivalence class $\rho\{p\}$ as its image. This mapping κ is called the *canonical mapping* associated with, or "induced by," the equivalence ρ. As a direct consequence of definitions we see that $\rho = \tilde{\kappa}$. If φ is a mapping of R onto R', let κ be the canonical map associated with the equivalence $\tilde{\varphi}$. Then $\kappa\varphi^{-1}$ is a 1-1 mapping of R' onto $R/\tilde{\varphi}$, with inverse map $\varphi\kappa^{-1}$.

Definition 21a: *Let ρ be an equivalence relation defined in R with associated canonical mapping κ. Then the space*

$$(R, \tau)/\rho = (R/\rho, \tau_\kappa)$$

is called the quotient-space *of (R, τ) modulo ρ.*

By the results already established for quotient topologies, we have immediately:

21.7 *The cannonical mapping κ is continuous, from (R, τ) onto $(R, \tau)/\rho$.*

If (R, τ) is completely compact, k-compact, connected, or locally connected, then so is $(R, \tau)/\rho$.

$(R, \tau)/\rho$ is a T_1-space if and only if the equivalence classes generated by ρ in (R, τ) are closed subsets of (R, τ).

Consider the mapping $\psi : R/\rho \to R'$. Then the mappings $\psi : (R, \tau)/\rho \to (R', \tau')$ and $\psi\kappa : (R, \tau) \to (R', \tau')$ are equivalent.

21.8 *Let φ be a mapping from R onto R'. Then*

$$(R', \tau_\varphi) \cong (R, \tau)/\tilde{\varphi}.$$

If κ denotes the cannonical mapping associated with $\tilde{\varphi}$ then $\psi = \kappa\varphi^{-1}$ is a topological mapping which will establish the above homeomorphism.

Proof: ψ is that 1-1 mapping which to every point $p' \in R'$ associates the class $\tilde{\varphi}\{p\}$, where $\varphi p = p'$. We have $\psi\varphi = \kappa$. Since κ, and hence $\psi\varphi$, map (R, τ) onto $(R, \tau)/\tilde{\varphi}$ continuously, we see, by 21.2, that ψ is continuous. Moreover, $\psi^{-1}\kappa = \varphi$. Since $\varphi : (R, \tau) \to (R', \tau_\varphi)$ and since ψ^{-1} is also continuous by 21.7, ψ is also a homeomorphism. ∎

Suppose ρ and σ are two equivalence relations in R with $\rho \subseteq \sigma$. Then the subset

$$\sigma/\rho = \{(\rho\{p\}, \rho\{q\}) : (p, q) \in \sigma\}$$

of $R/\rho \times R/\rho$ is an equivalence relation in R/ρ. It is called the *quotient equivalence* of σ by ρ.

21.9 $$(R, \tau)/\sigma \cong (R, \tau)/\rho/\,\sigma/\rho.$$

Let κ_ρ, κ_σ and κ' be the cannonical mapping associated with ρ, σ, and σ/ρ. Then $\psi = \kappa'\kappa_\rho\kappa_\sigma^{-1}$ is a topological mapping establishing the above homeomorphism.

Proof: ψ is that 1-1 mapping which maps the class $\sigma\{p\}$ onto the class $\sigma/\rho\{\sigma\{p\}\}$ We have $\psi\kappa_\sigma = \kappa'\kappa_\rho$. Since κ_ρ and κ', and therefore $\psi\kappa_\sigma$ are continuous mappings, we see by 21.7 that ψ is also continuous. Moreover, $\psi^{-1}\kappa'\kappa_\rho = \kappa_\sigma$. Since κ_σ is continuous, we likewise see that $\psi^{-1}\kappa'$ is continuous and, finally, that ψ^{-1} is continuous. ∎

We can also ask the converse question: Given a topology τ' on the quotient-set R/ρ, can we put a topology τ on R such that $(R, \tau)/\rho = (R/\rho, \tau')$? The answer is affirmative.

21.10 *Let κ be the cannonical mapping associated with the equivalence relation ρ defined over R. Then*

$$(R, {}_\kappa\tau')/\rho = (R/\rho, \tau').$$

Proof: By 20.1, κ is a continuous and open mapping from $(R, {}_\kappa\tau')$ onto $(R/\rho, \tau')$. Hence, by 21.6, τ' is the quotient-topology $({}_\kappa\tau')_\kappa$ of $(R, {}_\kappa\tau')/\rho$. ∎

21.11 *In a space* (R, τ) *the subset*

$$\rho = \{(p, q) : \tau p = \tau q\}$$

defines an equivalence relation on $R \times R$. *Then the cannonical mapping* κ *from* R *onto* R/ρ *is a continuous, open, and closed mapping from* (R, τ) *onto* $(R, \tau)/\rho$. *The quotient space* $(R, \tau)/\rho$ *is probed by* (R, τ). *If* τ *is a* T_3*-topology* (T_{3a}*-topology*) *then* $(R, \tau)/\rho$ *is regular* (*completely regular*).

Proof: Let M be an open subset of (R, τ). If $q \in \kappa^{-1}\kappa M$ then there exists a $p \in M$ with $\kappa p = \kappa q$, and hence such that $\tau p = \tau q$. This implies $\overline{q} \leqslant \tau p \leqslant \overline{M}$, hence $q \in M$ and $M = \kappa^{-1}\kappa M$. By 20.1 and 21.1, we therefore have $\tau = {}_\kappa(\tau_\kappa)$, and by 20.1 again, κ must be a continuous, open, and closed mapping of (R, τ) onto $(R, \tau)/\rho$.

To each class $a \in R/\rho$ let us associate a unique point $p_a \in R$ such that $\kappa p_a = a$. Then $\psi a = p_a$ defines a mapping $\psi : R/\rho \to R$. Then $\kappa\psi$ is the identity map on R/ρ, hence a continuous mapping of $(R, \tau)/\rho$ onto itself. If in 20.2 we replace the mapping φ by κ and both the spaces (R_0, τ_0), (R', τ') by $(R, \tau)/\rho = (R/\rho, \tau_\kappa)$, hence replacing $(R, {}_\varphi\tau')$ by $(R, {}_\kappa(\tau_\kappa)) = (R, \tau)$, we see that ψ must be a continuous map from $(R, \tau)/\rho$ into (R, τ). For every filter $\mathfrak{a}$ of R/ρ and every class $a \in R/\rho$, $\psi\mathfrak{a} \leqslant \tau\psi a$ implies $\mathfrak{a} = \kappa\psi\mathfrak{a} \leqslant \kappa\tau\psi a = \tau_\kappa\kappa\psi a = \tau_\kappa a$. Hence $\{\psi\}$ is a probing system (18a), $(R, \tau)/\rho$ being probed by means of (R, τ). We see, by 18.2 and 18.10, that if (R, τ) is a T_3- or T_{3a}-space, then so is $(R, \tau)/\rho$.

Let τ be now assumed to be a T_3-topology. Then so is τ_κ. From $\tau_\kappa\kappa p \wedge \tau_\kappa\kappa q \neq \mathfrak{o}$, we have, by 8.6, $\tau_\kappa\kappa p = \tau_\kappa\kappa q$. Since $\tau = {}_\kappa(\tau_\kappa)$, this implies further that $\tau p = \kappa^{-1}\tau_\kappa\kappa p = \kappa^{-1}\tau_\kappa\kappa q = \tau q$, and hence $\kappa p = \kappa q$. Thus τ_κ is not only a T_3-topology but also a T_2-topology and thereby regular. Since (T_{3a}) implies (T_3), we see that if τ is a T_{3a}-topology, then τ_κ must be completely regular. ∎

We terminate with a few remarks on subspaces of quotient spaces: Let M be a subset of R. If ρ is an equivalence over R, with κ its associated canonical mapping, then ρ induces in M the equivalence relation $\rho_M = \rho \cap (M \times M)$, whose associated canonical map we shall denote by κ_M. The mapping $\psi = \kappa_M^{-1}$ is then a 1-1 mapping of M/ρ_M into R/ρ, sending M/ρ_M onto a subset $\tilde{M}$ of R/ρ. If τ is a topology on R, then $\tilde{M}$ above can be viewed as the subspace $(\tilde{M}, \tilde{\tau})$ of $(R, \tau)/\rho$. Since κ is continuous and $\psi\kappa_M = \kappa$, we see, by 21.7, that ψ is a continuous mapping from $(M, \tau_M)/\rho_M$ onto $(\tilde{M}, \tau)$. Nevertheless, ψ is not, in general, a topological mapping. The quotient space of the subspace (M, τ_M) of (R, τ) is thus not in general homeomorphic to the corresponding subspace of $(R, \tau)/\rho$ (cf. 21C).

Notes and Exercises

21A *Exercise:* Let φ be a mapping from R onto R'. For the topologies τ of R and τ' of R' we then have

$$_{\varphi}(\tau_{\varphi}) \geqq \tau \qquad \text{and} \qquad (_{\varphi}\tau')_{\varphi} = \tau'.$$

Show by an example that equality need not hold in the first relation above.

21B In the n-dimensional sphere S^n (cf. 16A), let ρ_n be the equivalence relation which puts in the same class two diametrically opposite points of S^n. In the n-dimensional "solid ball" $K^n = \{x : x \in \mathbf{R}^n, \ \| x \| \leqslant 1\}$ let σ_n be the equivalence relation defined as follows: $(x, y) \in \sigma_n$ if and only if either $x = y$, or else $(x, y) \in S_{n-1}$ and $(x, y) \in \rho_{n-1}$.

Exercise: Projective space P^n (cf. 16B) is homeomorphic to both the quotient spaces, S^n modulo ρ^n and K^n modulo σ_n (of course, S^n and K^n are to bear the topologies induced by those of $\mathbf{R}^{n-1}$ and $\mathbf{R}^n$, respectively).

21C In $\mathbf{R}^2$ let us say that the points (x, y) and (x', y') are equivalent if and only $x = x'$, i.e., they have the same projection upon the x-axis. Call this equivalence ρ and let the canonical mapping associated with ρ be denoted by κ. Let

$$M = \{(x, y) : y = x + 1, \ x \leqq 0\} \cup \{(x, y) : y = x - 1, \ x > 0\},$$

and let τ be the standard topology on $\mathbf{R}^2$.

Exercise: The quotient space $(M, \tau_M)/\rho_M$ is not homeomorphic to the subspace generated in $(\mathbf{R}^2, \tau)/\rho$ by κM.

21D Once the notion of quotient topology has been defined, it is natural to examine two other, allied types of transfer of topologies: Let φ be a mapping from R onto R' and let τ denote the topology on R. Let Σ_0' be the collection of all topologies τ' on R', for which φ is an open mapping from (R, τ) onto (R', τ'). We then see, by 17.7, that $\tau_{\varphi}^0 = \bigvee \{\tau' : \tau' \in \Sigma_0'\}$ is the coarsest topology on R' for which φ is still an open mapping from (R, τ). Likewise, let Σ_a' denote the collection of all topologies τ' on R', for which φ is a closed mapping from (R, τ) onto (R', τ'). By 17.11 we then see that $\tau_{\varphi}^a = \bigvee \{\tau' : \tau' \in \Sigma_a'\}$ is the coarsest topology on R', for which φ is a closed mapping from (R, τ).

Exercise

1. Let $\mathfrak{B}$ be a basis of (R, τ). Then $\varphi\mathfrak{B} = \{\varphi B : B \in \mathfrak{B}\}$ is a subbasis of (R', τ_{φ}^0).

2. Let $\mathfrak{S}$ be the family of all-closed subsets of R. Then a subset M' of R'

is τ_φ^a-closed if and only if M' is expressible as the intersection of sets belonging to $\varphi\mathfrak{S} = \{\varphi S : S \in \mathfrak{S}\}$.

3. $\tau_\varphi^0 \leqslant \tau_\varphi$ and $\tau_\varphi^a \leqslant \tau_\varphi$. With the aid of these relations it is possible to give a new proof of 21.6.

22. Product Spaces

In Section 1 we defined the product set $R_1 \times ... \times R_n$ as the set of all ordered n-tuples $(x_1, ..., x_n)$ where $x_\nu \in R_\nu$. The n-tuple $(x_1, ..., x_n)$ can also be thought of as a mapping ξ from the set of indices $I = \{1, ..., n\}$ into the union of the sets R_ν, $\cup R_\nu$, satisfying the condition that each x_ν appears as image of ν under the mapping ξ, i.e., $\xi\nu = x_\nu$. We can then think of $R_1 \times ... \times R_n$ as a set of mappings—and it is just this way of thinking which will allow us to define the product set of an infinite family of sets.

Definition 22a: *The set of all mappings $\xi : I \to \cup \{R_\iota : \iota \in I\}$ where $\xi_\iota \in R_\iota$ for all $\iota \in I$ is called the* product *of the family of sets $\{R_\iota : \iota \in I\}$ and will be denoted by $\times\{R_\iota : \iota \in I\}$ or by $\Pi\{R_\iota : \iota \in I\}$. The sets R_ι are known as the* coordinate sets *or* factors *of the product.*

For every $\kappa \in I$ we can define a mapping π_κ from $\times R_\iota$ onto the factor R_κ by $\pi_\kappa\xi = \xi_\kappa$. This mapping is called the *projection mapping* at the κth index or κth coordinate. Given any $\xi \in \times R_i$, then $\pi_\iota\xi = x_i$ is a point belong to R_ι, called the *coordinate of index ι of the point ξ*. We have $\{\xi\} = \cap \{\pi_\iota^{-1}x_\iota : \iota \in I\}$. Conversely, given a point x_ι in each of the sets R_ι, we see that $\cap \{\pi_\iota^{-1}x_\iota : \iota \in I\}$ is a one-point subset $\{\xi\}$ of $\times R_\iota$ and $\pi_\iota\xi = x_\iota$. Hence the points $\xi \in \times R_i$ are uniquely determined by their coordinates x_i; and conversely—we can, in fact, embody and summarize this result in the notation $\xi = \times \{x_\iota : \iota \in I\}$.

For every $\iota \in I$ let M_ι be a subset of R_ι. We can then identify $M = \times M_\iota$ with the set $\cap \pi_\iota^{-1}M_\iota$ and hence think of M as a subset of $\times R_\iota$.

22.1 *If the sets M_ι are nonempty, then*

$$\pi_\kappa \times M_\iota = \pi_\kappa(\cap \pi_\iota^{-1}M_\iota) = M_\kappa .$$

But if one of the M_ι is empty, then $\times M_\iota = \varnothing$ and $\pi_\kappa \times M_\iota = \varnothing$.

Proof: Since $\times M_\iota \subseteq \pi_\kappa^{-1}M_\kappa$ we see that $\pi_\kappa \times M_i \subseteq M_\kappa$. If x_κ is a fixed point in M_κ and if for $\iota \neq \kappa$ we choose an arbitrary point $x_\iota \in M_\iota$, then $\xi = \times x_\iota$ is a point in $\times M_i$ and $\pi_\kappa\xi = x_\kappa$. Hence $\pi_\kappa \times M_\iota = M_\kappa$. But if, say, M_{ι_0} is the empty set, then both $\pi_{\iota_0}^{-1}M_{\iota_0}$ and $\times M_\iota$ are also empty, and finally $\pi_\kappa \times M_\iota = \varnothing$. ∎

Let $\mathfrak{a}_\iota$ be a filter in R_ι, for every index ι. The filter $\wedge \{\pi_\iota^{-1}\mathfrak{a}_\iota : \iota \in I\}$ in $\times R_\iota$ will be denoted by $\times \mathfrak{a}_\iota$ and called the *product filter* of the $\mathfrak{a}_i$. Then the family of sets of the form $\pi_{\iota_1}^{-1}A_{\iota_1} \cap \dots \cap \pi_{\iota_n}^{-1}A_{\iota_n}$, where $A_{\iota_\nu} \in \mathfrak{a}_{\iota_\nu}$, constitute a basis of $\times\, A_\iota \in \mathfrak{a}_\iota$. Note, however, a set of the form $\times\, A_\iota$, where $A_\iota \in \mathfrak{a}_\iota$, does not, in general, belong to the product filter $\times\, \mathfrak{a}_\iota$; $\times\, A_\iota \in \times\, \mathfrak{a}_\iota$ if and only if $A_\iota = R_\iota$ for all but a finite number of indices ι. Using 22.1, we immediately have the following result on product filters.

22.2 *If the filters* $\mathfrak{a}_\iota$ *are distinct from* $\mathfrak{o}$, *then*

$$\pi_\kappa \times \mathfrak{a}_\iota = \mathfrak{a}_\kappa .$$

Nevertheless, if only one of the filters $\mathfrak{a}_\iota$ *is the null filter, we have* $\times\, \mathfrak{a}_\iota = \mathfrak{o}$.

If $\mathfrak{a}_\iota \leqslant \mathfrak{b}_\iota$, *then* $\times\, \mathfrak{a}_\iota \leqslant \times\, \mathfrak{b}_\iota$.

Let us now suppose that each of the sets R_ι bears a topology τ_ι. Then, by 20.4, $\tau = \bigwedge_\iota \{_\pi \tau_\iota : \iota \in I\}$ is the coarsest topology on $\times\, R_\iota$ for which each projection π_ι is a continuous mapping onto (R_ι, τ_ι). This topology τ is known as the *product topology* of the τ_ι's and is denoted by $\times\, \tau_\iota$. For an arbitrary point $\xi \in \times\, R_\iota$ we have

$$(\times\, \tau_\iota)\, \xi = \bigwedge \pi_\iota^{-1}\tau_\iota\pi_\iota\xi = \times\, (\tau_\iota\pi_\iota\xi).$$

Hence the sets of the form $\pi_{\iota_1}^{-1}U_{\iota_1} \cap \dots \cap \pi_{\iota_n}^{-1}U_{\iota_n}$, where $U_{\iota_\nu} \in \tau_{\iota_\nu}\pi_{\iota_\nu}\xi$, constitute a basis of $(\times\, \tau_\iota)\, \xi$.

Definition 22b: *The space* $(\times\, R_\iota, \times\, \tau_\iota)$ *is called the* product-space *of the individual spaces* (R_ι, τ_ι). *It is more simply denoted by* $\times\, (R_\iota, \tau_\iota)$.

22.3 *The projection* π_κ *is a continuous and open mapping, from* $\times\, (R_\iota, \tau_\iota)$ *onto* (R_κ, τ_κ).

A filter $\mathfrak{a}$ *of* $\times\, R_\iota$ *converges in* $\times\, (R_\iota, \tau_\iota)$ *to a point* ξ *if and only if each coordinate filter* $\pi_\iota\mathfrak{a}$ *converges in* (R_ι, τ_ι) *to the point* $\pi_\iota\xi$.

For every index ι *let* $\mathfrak{B}_\iota$ *be a subbasis of* (R_ι, τ_ι). *Then*

$$\mathfrak{B} = \{\pi_\iota^{-1}B : B_\iota \in \mathfrak{B}_\iota, \iota \in I\}$$

is a subbasis of $\times\, (R_\iota, \tau_\iota)$.

$\overline{\times\, M_\iota} = \times\, \bar{M}_\iota$. *If* $M_\iota = R_\iota$ *for all but a finite number of* ι, *then* $\widehat{\times\, M}^{\circ} = \times\, \widehat{M_\iota}^{\circ}$.

For arbitrary subsets M *of* $\times\, R_\iota$ *we always have* $\bar{M} \subseteq \times\, \overline{\pi_\iota M}$ *and* $\overset{\circ}{M} \subseteq \times\, \widehat{\pi_\iota M}^{\circ}$.

Proof: By 22.2 we have $\pi_\kappa(\times\, \tau_\iota)\, \xi = \pi_\kappa \times (\tau_\iota\pi_\iota\xi) = \tau_\kappa\pi_\kappa\xi$. Hence π_κ is a continuous and open mapping.

From $\mathfrak{a} \leqslant (\times\, \tau_\iota)\, \xi$ we deduce the inequality $\pi_\kappa\mathfrak{a} \leqslant \pi_\kappa(\times\, \tau_\iota)\, \xi = \tau_\kappa\pi_\kappa\xi$ and from this last inequality we deduce conversely that $\mathfrak{a} \leqslant \times\, \pi_\kappa\mathfrak{a} \leqslant \times\, \tau_\kappa\pi_\kappa\xi$.

Since π_ι is continuous the sets $\pi_\iota^{-1}B_\iota$ are open. The sets of the form $B_\iota^1 \cap \dots \cap B_\iota^n$ with $B_\iota^\nu \in \mathfrak{B}_\iota$ and $\pi_\iota\xi \in B_\iota^\nu$ constitute a basis of $\tau_\iota\pi_\iota\xi$. And since

$$(\times\, \tau_\iota)\, \xi = \bigwedge \pi_\iota^{-1}\tau_\iota\pi_\iota\xi,$$

the sets of the form $\pi_{\iota_1}^{-1}B_{\iota_1} \cap \ldots \cap \pi_{\iota_n}^{-1}B_{\iota_n}$, where $B_{\iota_\nu} \in \mathfrak{B}_{\iota_\nu}$ and $\pi_{\iota_\nu}\xi \in B_{\iota_\nu}$, constitute a basis of $(\times\, \tau_\iota)\,\xi$. Therefore $\mathfrak{B}$ must be a subbasis of $\times\,(R_\iota, \tau_\iota)$.

Using 17.1, 22.1, and the continuity of π_κ we see that

$$\pi_\kappa\, \overline{\times\, M_\iota} \subseteq \overline{\pi_\kappa \times M_\iota} = \overline{M}_\kappa,$$

and then $\overline{\times\, M_\iota} \subseteq \bigcap \pi_\kappa^{-1}\overline{M}_\kappa = \times\, \overline{M}_\kappa$. From $\xi \notin \overline{\times\, M_\iota}$ we deduce the existence of neighborhoods $U_\iota \in \tau_\iota \pi_\iota \xi (U_\iota \neq R_\iota$ for at most a finite number of indices) such that $\emptyset = (\times\, U_\iota) \cap (\times\, M_\iota) = \times\,(U_\iota \cap M_\iota)$. We have $U_\iota \cap M_\iota = \emptyset$, whence $\pi_\iota\xi \notin \overline{M}_\iota$ for some index ι and thus $\xi \notin \times\, \overline{M}_\iota$. This implies $\overline{\times\, M_\iota} = \times\, \overline{M}_\iota$. Since π_κ is open, $\pi_\kappa\overset{\circ}{\widehat{(\times\, M_\iota)}}$ is an open set contained in M_κ. Hence $\pi_\kappa\overset{\circ}{\widehat{(\times\, M_\iota)}} \subseteq \overset{\circ}{\widehat{M}}_\kappa$ and therefore $\overset{\circ}{\widehat{\times\, M_\iota}} \subseteq \bigcap \pi_\kappa^{-1}\overset{\circ}{\widehat{M}}_\kappa = \times\,(\overset{\circ}{M}_\kappa)$. If in addition $M_\iota = R_\iota$ for $\iota \neq \iota_1, \ldots, \iota_n$, then, since π_ι is continuous, we see that

$$\times\, \overset{\circ}{\widehat{M}}_\iota = \pi_{\iota_1}^{-1}\overset{\circ}{\widehat{M}}_{\iota_1} \cap \ldots \cap \pi_{\iota_n}^{-1}\overset{\circ}{\widehat{M}}_{\iota_n}$$

is an open set contained in $\times\, M_\iota$, which implies the reverse inequality $\times\, \overset{\circ}{\widehat{M}}_\iota \subseteq \overset{\circ}{\widehat{(\times\, M_\iota)}}$. Since $M \subseteq \times\, \pi_\iota M$ for any subset M of $\times\, R_\iota$, we have

$$\bar{M} \subseteq \overline{\times\, \pi_\iota M} = \times\, \overline{\pi_\iota M} \qquad \text{and} \qquad \overset{\circ}{M} \subseteq \overset{\circ}{\widehat{\times\, \pi_\iota M}} \subseteq \times\, \overset{\circ}{\widehat{\pi_\iota M}}. \blacksquare$$

Examples

22.I Let τ be the standard topology on the sphere S^1 (i.e., the topology induced by $\mathbf{R}^2$ on S^1). Then $(S^1, \tau) \times (S^1, \tau)$ is homeomorphic to the torus, where the torus bears the usual topology, that induced on it by $\mathbf{R}^3$.

22.II In the product space $(R', \tau') = (R, \tau) \times [-1, +1]$ the relation defined by

$$\rho = \{(\xi, \eta) : \xi = \eta \qquad \text{or} \qquad \pi_2\xi = \pi_2\eta = \pm 1\}$$

is an equivalence. The quotient space $(R', \tau')/\rho$ is known as the *suspension* of (R, τ). The suspension of the sphere is homeomorphic to S^{n+1}.

22.III For every $n \in \mathbf{N}$ let (R_n, τ_n) denote the set $\{0, 1\}$ under the discrete topology. Then in the product space $(W, \tau) = \times\, \{(R_n, \tau_n) : n \in \mathbf{N}\}$, the sets

$$U_n(\xi) = \{\eta : \eta \in W, \pi_\nu\eta = \pi_\nu\xi \quad \text{for} \quad \nu = 1, \ldots, n\} \qquad (n \in \mathbf{N})$$

form a basis of the neighborhoods filter $\tau\xi$. The map

$$\varphi\xi = \sum_{n=1}^{\infty} \frac{2\pi_n\xi}{3^n}$$

is a 1-1 mapping from W onto the Cantor discontinuum (cf. p. 106). Under φ, the basis $\{U_n(\xi) : n \in \mathbf{N},\ \xi \in W\}$ of (W, τ) gets mapped onto the basis $\{J_\nu^n \cap D : n \in \mathbf{N},\ \nu = 1, 2, \ldots, 2^n\}$ of D. By 16.1, φ is therefore a homeomorphism of (W, τ) onto the Cantor discontinuum.

22.4 *For each $\iota \in I$ let M_ι be a subset of R_ι. In addition, let $\tau = \times\ \tau_\iota$. Then for the subset $M = \times\ M_\iota$ of $\times\ R_\iota$ we have $(M, \tau_M) = \times\ (M_\iota, (\tau_\iota)_M)$.*

Proof: The above contention follows immediately from the equation

$$\times\ (U_\iota \cap M_\iota) = (\times\ U_\iota) \cap (\times\ M_\iota). \blacksquare$$

22.5 *Let ξ be a point in $\times\ R_\iota$. Then the subset*

$$T_{\xi,\kappa} = \bigcap \{\pi : \pi_\iota \xi_\iota : \iota \neq \kappa\} = \{\eta : \pi_\iota \eta = \pi_\iota \xi \quad \text{for} \quad \iota \neq \kappa\}$$

generates a subspace of $\times\ (R_\iota, \tau_\iota)$ which is homeomorphic to the coordinate space (R_κ, τ_κ); a topological mapping establishing this homeomorphism is the restriction θ_κ of π_κ onto $T_{\xi,\kappa}$.

Proof: If $\eta_1, \eta_2, \in T_{\xi,\kappa}$ then $\pi_\iota \eta_1 = \pi_\iota \xi = \pi_\iota \eta_2$ for $\iota \neq \kappa$. If in addition $\theta_\kappa \eta_1 = \theta_\kappa \eta_2$, then $\pi_\kappa \eta_1 = \pi_\kappa \eta_2$, which implies $\eta_1 = \eta_2$. Hence θ_κ is a 1-1 mapping of $T_{\xi,\kappa}$ onto R_κ. Since $\pi_\kappa(\times\ \tau_\iota)\ \eta = \tau_\kappa(\times\ \tau_\iota)\ \eta = \tau_\kappa \pi_\kappa \eta$, we see that $\theta_\kappa((\times\ \tau_\iota)\ \eta \wedge \overline{T_{\xi,\kappa}}) = \tau_\kappa \theta_\kappa \eta$ for all $\eta \in T_{\xi,\kappa}$. Therefore θ_κ is continuous and open, and hence a topological mapping. $\blacksquare$

22.6 *The properties* $(\mathbf{T}_1)$, $(\mathbf{T}_2)$, $(\mathbf{T}_3)$, $(\mathbf{T}_{3a})$, "*the capacity to be probed by a space* (R', τ')", *complete compactness, connectedness—these properties all hold in $\times\ (R_\iota, \tau_\iota)$ if and only if they hold in all the coordinate spaces.*

Proof: If $(\mathbf{T}_1)$-$(\mathbf{T}_{3a})$ and the capacity to be probed hold for $\times\ (R_\iota, \tau_\iota)$ then we know they hold for all subspaces. Hence, by 22.5 they hold for all the coordinate spaces. Since π_κ is a continuous mapping from $\times\ (R_\iota, \tau_\iota)$ onto (R_κ, τ_κ), by 19.1 and 19.3 we see that complete compactness and connectedness also hold in the coordinate spaces as soon as they hold in $\times\ (R_\iota, \tau_\iota)$.

To prove the converse; let us suppose that all the coordinate spaces are T_1-spaces. If $\overline{\xi} \leqslant (\times\ \tau_\iota)\ \eta$, then $\overline{\pi_\kappa \xi} \leqslant \pi_\kappa(\times\ \tau_\iota)\ \eta = \tau_\kappa \pi_\kappa \eta$, and therefore $\pi_\kappa \xi = \pi_\kappa \eta$ for all $\kappa \in I$, which in turn implies $\xi = \eta$. If all coordinate spaces satisfy axiom $(\mathbf{T}_2)$, then $\xi \neq \eta$ implies $\pi_\kappa \xi \neq \pi_\kappa \eta$ for some $\kappa \in I$, and thus $(\times\ \tau_\iota)\ \xi \wedge (\times\ \tau_\iota)\ \eta \leqslant \pi_\kappa^{-1} \tau_\kappa \pi_\kappa \xi \wedge \pi_\kappa^{-1} \tau_\kappa \pi_\kappa \eta = \pi_\kappa^{-1}(\tau_\kappa \pi_\kappa \xi \wedge \tau_\kappa \pi_\kappa \eta) = \mathfrak{o}$. If now $(\mathbf{T}_3)$ holds in all the coordinate spaces (R_ι, τ_ι), then, since π_ι is continuous, the filters $\pi_\iota^{-1} \tau_\iota \pi_\iota \xi$ must be closed, and therefore, by 5.4, $(\times\ \tau_\iota)\ \xi = \bigwedge \pi_\iota^{-1} \tau_\iota \pi_\iota \xi$ is also a closed filter.

In addition, let us suppose that for each $\iota \in I$ the space (R_ι, τ_ι) can be probed by (R', τ'), and let Φ_ι be a probing system. Then

$$\Phi = \{\varphi_\iota \pi_\iota : \iota \in I,\ \varphi_\iota \in \Phi_\iota\}$$

is a probing system of $\times (R_\iota, \tau_\iota)$: For if $\varphi_\iota \pi_\iota \mathfrak{a} \leqslant \tau' \varphi_\iota \pi_\iota \xi$ for all $\varphi_\iota \in \Phi_\iota$ and $\iota \in I$, then $\pi_\iota \mathfrak{a} \leqslant \tau_\iota \pi_\iota \xi$ for all $\iota \in I$ and also, by 22.3, $\mathfrak{a} \leqslant (\times \tau_\iota) \xi$. The preservation of $(\mathbf{T}_{3a})$ is then a special case.

If the spaces (R_ι, τ_ι) are completely compact, then the coordinate filters $\pi_\iota \mathfrak{v}$ of any ultrafilter $\mathfrak{v}$ of $\times R_\iota$ must converge to points $x_\iota \in R_\iota$. By 22.3, this means that τ is convergent to $\times x_\iota$; i.e., $\times (R_\iota, \tau_\iota)$ is also completely compact.

Finally, let us suppose that all the spaces (R_ι, τ_ι) are connected, and let K be a component of $\times (R_\iota, \tau_\iota)$. Since, by 14.8, K must be closed if $\xi \notin K$ then there exists a neighborhood $U \in (\times \tau_\iota) \xi$ such that $U \cap K = \emptyset$. We can therefore assume $U = \pi_{\iota_1}^{-1} U_{\iota_1} \cap \dots \cap \pi_{\iota_n}^{-1} U_{\iota_n}$, with n minimal. Hence there exists a point $\eta \in K \cap \pi_{\iota_1}^{-1} U_{\iota_1} \cap \dots \cap \pi_{\iota_{n-1}}^{-1} U_{\iota_{n-1}}$. This means that the subspace $T_{n,\iota_{n-1}}$ which is homeomorphic to $(R_{\iota_{n-1}}, \tau_{\iota_{n-1}})$ (22.5) must be connected, and since $\eta \in K \cap T_{\eta,\iota_{n-1}}$ we have $T_{n,\iota_{n-1}} \subseteq K$. On the other hand we have $T_{\eta,\iota_{n-1}} \cap U \neq \emptyset$, contradicting $K \cap U = \emptyset$. Hence $\xi \notin K$ can not hold for any point of $\times R_\iota$. Consequently $\times R_\iota = K$ and $\times (R_\iota, \tau_\iota)$ must be connected. ∎

22.7 *$\times (R_\iota, \tau_\iota)$ is locally completely compact (locally connected) if and only if all its coordinate spaces are locally completely compact (locally connected) and all but a finite number of them are completely compact (connected).*

Proof: Let us first suppose that the spaces (R_ι, τ_ι) are l.c.c. and for $\iota \neq \iota_1, \dots, \iota_n$ suppose they are even completely compact. Then, to the point ξ we can associate a completely compact neighborhood $U_{\iota_\nu} \in \tau_{\iota_\nu} \pi_{\iota_\nu} \xi$ and this for $\nu = 1, \dots, n$. If we further set $U_\iota = R_\iota$ for $\iota \neq \iota_1, \dots, \iota_n$, then $U = \times U_\iota$ is a neighborhood of ξ, completely compact by 22.6. Conversely, let $\times (R_\iota, \tau_\iota)$ be locally completely compact, and let U be a completely compact neighborhood of ξ. Then $\pi_\iota U = R_\iota$ for all but a finite number of indices—let us denote these indices where $\pi_\iota U \neq R_\iota$ by $\iota_1, \dots, \iota_n$. Since π_ι is continuous, we see by 19.1 that (R_ι, τ_ι) is completely compact for $\iota \neq \iota_1, \dots, \iota_n$. For the very same reason, $\pi_{\iota_\nu} U$ is a completely compact neighborhood of $\pi_{\iota_\nu} \xi$ for $\nu = 1, \dots, n$. But ξ was arbitrary, and therefore $(R_{\iota_\nu}, \tau_{\iota_\nu})$ must be locally completely compact.

Now let us suppose that the spaces (R_ι, τ_ι) are locally connected for all ι and connected for $\iota \neq \iota_1, \dots, \iota_n$. If U is a neighborhood of ξ, we again have $\pi_\iota U = R_\iota$ for $\iota \neq \iota_1, \dots, \iota_m$ with $m \geqslant n$. For $\mu = 1, \dots, m$ let V_{ι_μ} be a connected neighborhood of $\pi_{\iota_\mu} \xi$ with $V_{\iota_\mu} \subseteq \pi_{\iota_\mu} U$. For all other indices let us suppose $V_\iota = R_\iota$. Then, by 22.6, $V = \times V_\iota$ is a connected neighborhood of ξ contained in U. Conversely, let $\times (R_\iota, \tau_\iota)$ be locally connected. Since

$$\pi_\kappa(\times \tau_\iota) \xi = \tau_\kappa \pi_\kappa \xi$$

we see that $\tau_\kappa \pi_\kappa \xi$ possesses a basis consisting of connected sets, and (R_κ, τ_κ) is thereby locally connected. If U is a connected open subset of the product space, then $\pi_\iota U = R_\iota$ for all but a finite number of indices ι. Hence all but a finite number of the spaces (R_ι, τ_ι) are also connected. ∎

The product of two paracompact spaces need not be paracompact (cf. 22C). On the contrary, the product of a paracompact and a completely compact space is always paracompact. We even have the more general result.

22.8 *Let (R_1, τ_1) be a paracompact T_3-space and let (R_2, τ_2) be countable at infinity* (12b). *Then $(R_1, \tau_1) \times (R_2, \tau_2)$ is paracompact. If (R_2, τ_2) is completely compact we need not even assume (R_1, τ_1) to be a T_3-space.*

Proof: R_2 is the union of a countable number of completely compact sets $M_n (n \in \mathbf{N})$. Let $\mathfrak{H}$ be an open covering of $(R_1, \tau_1) \times (R_2, \tau_2)$. With each $\xi \in R_1 \times R_2$, we can associate neighborhoods $U_\xi \in \tau_1 \pi_1 \xi$ and $V_\xi \in \tau_2 \pi_2 \xi$ such that $U_\xi \times V_\xi$ is contained in some suitable $H \in \mathfrak{H}$. For a fixed $p \in R_1$, the family $\{V_\xi : \pi_1 \xi = p, \pi_2 \xi \in M_n\}$ is an open covering of M_n. Hence there exists a finite subset E_p^n of $\pi_1^{-1} p \cap \pi_2^{-1} M_n$, for which the family $\{V_\xi : \xi \in E_p^n\}$ is also a covering of M_n. If we put $U_p^n = \cap \{U_\xi : \xi \in E_p^n\}$, then $\mathfrak{U}_n = \{U_p^n : p \in R_1\}$ is an open covering of R, for which we can find an open and locally finite covering $\mathfrak{K}_n$ of R_1 such that $\mathfrak{K}_n \ll \mathfrak{U}_n$. Finally, to each $K \in \mathfrak{K}$ we can associate a unique point $p_K \in R_1$ such that $K \subseteq U_{p_K}^n$. Then $\mathfrak{L}_n = \{K \times V_\xi : K \in \mathfrak{K}_n, \xi \in E_{p_K}^n\}$ is an open covering of $R_1 \times M_n$ with $\mathfrak{L}_n \ll \mathfrak{H}$ and hence $\mathfrak{L} = \cup \{\mathfrak{L}_n : n \in \mathbf{N}\}$ is an open covering of $R_1 \times R_2$ such that $\mathfrak{L} \ll \mathfrak{H}$. Furthermore, the families $\mathfrak{L}_n$ are all locally finite: Given any $\eta \in R_1 \times R_2$ we can certainly find a neighborhood $U \in \tau_1 \pi_1 \eta$ which has a nonempty intersection with only a finite number of $K \in \mathfrak{K}_n$. Then $\pi_1^{-1} U$ is a neighborhood of η. And since the sets $E_{p_K}^n$ are finite, the set $\pi_1^{-1} U$ possesses a nonempty intersection with only a finite number of sets $K \times V_\xi$ belonging to $\mathfrak{L}_n$. Hence every open covering of $R_1 \times R_2$ possesses an open and $\aleph_0$-locally finite refinement. Our contention then follows from 13.11.

If R_2 is also completely compact, then all the M_n can be taken equal to R_2, and then the above proof furnishes us with an open and locally finite refinement of $\mathfrak{H}$, without our having to make use of 13.11. We can therefore dispense with $(\mathbf{T_3})$ in this case. ∎

For each $\iota \in I$ suppose $\varphi_\iota : R' \to R_\iota$ and $\psi_\iota : R_\iota \to R_\iota''$. Then we can define maps $\times\, \varphi_\iota : R' \to \times\, R_\iota$ and $\times\, \psi_\iota : \times\, R_\iota \to \times\, R_\iota''$ by

$$(\times\, \varphi_\iota)\, p' = \times\, (\varphi_\iota p') \qquad \text{and} \qquad (\times\, \psi_\iota)\, \xi = \times\, (\psi_\iota \pi_\iota \xi).$$

We then have

$$\pi_\kappa \times \varphi_\iota = \varphi_\kappa \qquad \text{and} \qquad \pi_\kappa'' \times \psi_\iota = \psi_\kappa \pi_\kappa ,$$

where π_κ'' denotes the κ-coordinate projection mapping from $\times\, R_\iota''$. If we are given the mapping $\varphi : R' \to \times\, R_\iota$, then $\pi_\iota \varphi : R' \to R_\iota$ and $\varphi = \times\, (\pi_\iota \varphi)$.

For subsets M' of R', or subsets M of $\times\ R_\iota$, we see immediately that

$$(\times\ \varphi_\iota)\ M' \subseteq \times\ (\varphi_\iota M') \qquad \text{and} \qquad (\times\ \psi_\iota)\ M \subseteq \times\ (\psi_\iota \pi_\iota M),$$

respectively. This implies that for filters $\mathfrak{a}'$ of R', or $\mathfrak{a}$ of $\times\ R_\iota$, we have, respectively,

$$(\times\ \varphi_\iota)\ \mathfrak{a}' \leqq \times\ (\varphi_\iota \mathfrak{a}') \qquad \text{and} \qquad (\times\ \psi_\iota)\ \mathfrak{a} \leqq \times\ (\psi_\iota \pi_\iota \mathfrak{a}).$$

22.9 $\quad \times\ \varphi_\iota : (R', \tau') \to \times\ (R_\iota, \tau_\iota)$ and $\times\ \psi_\iota : \times\ (R_\iota, \tau_\iota) \to \times\ (R''_\iota, \tau''_\iota)$

are, respectively, equivalent to

$$\varphi_\iota : (R', \tau') \to (R_\iota, \tau_\iota) \qquad \text{and} \qquad \psi_\iota : (R_\iota, \tau_\iota) \to (R''_\iota, \tau''_\iota),$$

for all $\iota \in I$.

Proof: Since π_κ is continuous, we see that the continuity of $\times\ \varphi_\iota$ implies that of $\varphi_\kappa = \pi_\kappa \times \varphi_\iota$. Conversely, the continuity of all φ_ι implies

$$(\times\ \varphi_\iota)\ \tau' p' \leqq \times\ (\varphi_\iota \tau' p') \leqq \times\ (\tau_\iota \varphi_\iota p') = (\times\ \tau_\iota)\ (\times\ \varphi_\iota)\ p',$$

and hence $\times\ \varphi_\iota$ is continuous also. Likewise the continuity of all ψ_ι implies

$$\begin{aligned}(\times\ \psi_\iota)\ (\times\ \tau_\iota)\ \xi &\leqq \times\ (\psi_\kappa \pi_\kappa (\times\ \tau_\iota)\ \xi) = \times\ (\psi_\kappa \tau_\kappa \pi_\kappa \xi) \\ &\leqq \times\ (\tau''_\kappa \psi_\kappa \pi_\kappa \xi) = (\times\ \tau''_\kappa)\ (\times\ \psi_\kappa)\ \xi.\end{aligned}$$

Conversely, suppose $\times\ \psi_\iota$ is continuous. If then p_κ belongs to R_κ, let ξ denote a point of $\times\ R_\iota$ such that $\pi_\kappa \xi = p_\kappa$. Then we have

$$\begin{aligned}\psi_\kappa \tau_\kappa p_\kappa = \psi_\kappa \pi_\kappa (\times\ \tau_\iota)\ \xi &= \pi''_\kappa (\times\ \psi_\iota)\ (\times\ \tau_\iota)\ \xi \leqq \pi''_\kappa (\times\ \tau''_\iota)\ (\times\ \psi_\iota)\ \xi \\ &= \tau''_\kappa \psi_\kappa \pi_\kappa \xi = \tau''_\kappa \psi_\kappa p_\kappa . \blacksquare\end{aligned}$$

22.10 *For each* $\iota \in I$ *let* τ_ι *be a topology on R and let* φ_ι *be the identity map on R. Then* $\varphi = \times\ \varphi_\iota$ *is a homeomorphism of* $(R, \bigwedge \tau_\iota)$ *onto the subspace of* $\times\ (R, \tau_\iota)$ *generated by* φR.

Proof: Evidently, φ is a 1-1 mapping. Since $\bigwedge \tau_\iota \leqslant \tau_\kappa (\kappa \in I)$ we have $\varphi_\kappa : (R, \bigwedge \tau_\iota) \to (R, \tau_\kappa)$, whence, by 22.9, $\varphi : (R, \bigwedge \tau_\iota) \to \times\ (R, \tau_\iota)$. Every neighborhood $U \in (\bigwedge \tau_\iota)\ p$ is of the form $U = U_{\iota_1} \cap \dots \cap U_{\iota_n}$ where $U_{\iota_\nu} \in \tau_{\iota_\nu} p$. We have $\varphi U = \varphi R \cap \bigcap \pi_{\iota_\nu}^{-1} U_{\iota_\nu}^{-1} \in \overline{\varphi R} \wedge (\times \tau_\iota)\ p$. Hence $\varphi(\bigwedge \tau_\iota)\ p \geqslant \overline{\varphi R} \wedge (\times \tau_\iota) p$, which implies that φ is also an open mapping and consequently a homeomorphism. $\blacksquare$

Notes and Exercises

22A The product of a countable number of copies of the interval [0, 1] is a space homeomorphic to the unit cube of Hilbert space (cf. 6D). The topological mapping $\xi \to (\pi_n \xi / n)_{n \in \mathbf{N}}$ will establish this homeomorphism.

22B Let $\{I_\kappa : \kappa \in K\}$ be a collection of indexing sets. Let us also suppose that

$$(R, \tau) = \times \{(R_\iota, \tau_\iota) : \iota \in \cup I_\kappa\},$$
$$(R'_\kappa, \tau'_\kappa) = \times \{(R_\iota, \tau_\iota) : \iota \in I_\kappa\},$$
$$(R'', \tau'') = \times \{(R'_\kappa, \tau'_\kappa) : \kappa \in K\}$$

and let π_ι, π'_ι, and π''_ι be the projection-mappings belonging to these product spaces. Let $\varphi_\kappa = \times \{\pi_\iota : \iota \in I_\kappa\}$ and $\varphi = \times \{\varphi_\kappa : \kappa \in K\}$.

Exercise: φ is a homeomorphism from (R, τ) onto (R'', τ''). The following associativity result is valid

$$\underset{\kappa \in K}{\times} (\underset{\iota \in I_\kappa}{\times} (R_\iota, \tau_\iota)) \cong \times \{(R_\iota, \tau_\iota) : \iota \in \bigcup_{\kappa \in K} I_\kappa\}$$

22C Let τ be the left-order topology on the set of real number $\mathbf{R}$ (cf. 6A; the half open intervals $[a, b[$ form a basis of τ).

Exercise

1. $(\mathbf{R}, \tau)$ is a regular, Lindelof, paracompact, normal space.
2. In the product space $(\mathbf{R}, \tau) \times (\mathbf{R}, \tau)$ the sets

$$A = \{\xi : \pi_1 \xi = -\pi_2 \xi, \quad \pi_1 \xi \text{ irrational}\}$$

and

$$B = \{\xi : \pi_1 \xi = -\pi_2 \xi, \quad \pi_1 \xi \text{ rational}\}$$

are disjoint and closed.

3. But A and B do not possess any disjoint neighborhoods in the product space (proof similar to that for 8D).

Hence the product space $(\mathbf{R}, \tau) \times (\mathbf{R}, \tau)$ is not normal. But by 22.6 this space is regular—and therefore, by 13.3, it cannot be paracompact. The product of two normal, or two paracompact, spaces need not be of the same nature [24, p. 133].

23. Spaces of Mappings

In this section (R, τ) and (R', τ') will denote two fixed spaces and $\hat{A} = \hat{A}(R, R')$ will denote the set of all mappings from R into R'. For each $p \in R$ we can also define a mapping $\epsilon_p : \hat{A} \to R'$ by $\epsilon_p \varphi = \varphi p$. In what follows we shall examine certain topologies on $\hat{A}$, all of which will render the mapping ϵ_p continuous. For these questions, the topological structure of R will, at first, play no role, so that for the moment we may think of R merely as a set.

For each $p \in R$ let $R'_p = R'$ and $\tau'_p = \tau'$. Then the product set

$$R'^R = \times \{R'_p : p \in R\}$$

is composed exactly of all the mappings from R into R'; i.e., $\hat{A} = R'^R$. Moreover, the "pth-coordinate" projection mapping π_p from $\hat{A}$ onto R' is identical to the above defined mapping ϵ_p. Hence the product topology $\hat{\tau} = \times \{\tau'_p : p \in R\}$ is the coarsest topology on $\hat{A}$ under which all the ϵ_p are continuous mapping onto (R', τ'). It is called the *topology of pointwise convergence* (cf. 23.1). For a fixed mapping $\varphi \in \hat{A}$ the sets of the form

$$\hat{U} = \epsilon_{p_1}^{-1} U'_1 \cap \dots \cap \epsilon_{p_n}^{-1} U'_n$$

where

$$p_\nu \in R \qquad \text{and} \qquad U'_\nu \in \tau'\varphi p_\nu \qquad \text{for} \qquad \nu = 1, \dots, n$$

constitute a basis of the neighborhoods filter $\hat{\tau}\varphi$. The set $\hat{U}$ is composed of all $\psi \in \hat{A}$ such that $\psi p_\nu \in U'_\nu$ for $\nu = 1, \dots, n$. We have

$$\hat{\tau}\varphi = \bigwedge \{\epsilon_p^{-1} \tau' \varphi p : p \in R\} \qquad \text{and} \qquad \epsilon_p \hat{\tau}\varphi = \tau' \epsilon_p \varphi = \tau' \varphi p.$$

We shall say that the filter $\hat{\mathfrak{a}}$ of $\hat{A}$ converges at the point $p \in R$ whenever $\epsilon_p \hat{\mathfrak{a}}$ is a convergent filter in (R', τ'). Applying 22.3 to this special case we have:

23.1 *A filter $\hat{\mathfrak{a}}$ of $\hat{A}$ converges in $(\hat{A}, \hat{\tau})$ if and only if $\hat{\mathfrak{a}}$ converges at every point $p \in R$ (pointwise convergence). If $\epsilon_p \hat{\mathfrak{a}} \leqslant \tau' p'$, then $\hat{\mathfrak{a}}$ converges to the mapping $\varphi \in \hat{A}$ defined by $\varphi p = p'$.*

The mapping ϵ_p is continuous and open, from $(\hat{A}, \hat{\tau})$ onto (R', τ').

If $\mathfrak{B}'$ is a subbasis of (R', τ'), then $\mathfrak{B} = \{\epsilon_p^{-1} B' : B' \in \mathfrak{B}', p \in R\}$ is a subbasis of $(\hat{A}, \hat{\tau})$.

If M' is a subset of R', we can think of the set $\hat{M}$ of all mappings from R into M' as a subset of $\hat{A}$; namely, $\hat{M}$ is equivalent to the set of all $\varphi \in \hat{A}$ such that $\varphi R \subseteq M'$. There are two possible natural topologies on $\hat{M}$: on the one hand we have $\hat{\tau}_0$, the topology of pointwise converge in $\hat{M}$, and on the other hand, we have $\hat{\tau}_{\hat{M}}$, the relative topology in $\hat{M}$ induced by $\hat{\tau}$. But 22.4 implies that $\hat{\tau}_0 = \hat{\tau}_{\hat{M}}$. If we apply 22.6 to the special case at hand, we also have:

23.2 *The notions or properties* (T_1), (T_2), (T_3), (T_{3a}), *"being probed by a space* (R'', τ'')," *complete compactness, and connectedness hold in* $(\hat{A}, \hat{\tau})$ *if and only if they hold in* (R', τ').

In addition to the above, note also that $(\hat{A}, \hat{\tau})$ can be probed by (R', τ'): For pointwise convergence means nothing other than the fact that $\{\epsilon_p : p \in R\}$ is a probing system. By 22.7 $(\hat{A}, \hat{\tau})$ is l.c.c. if and only if (R', τ'), and hence also $(\hat{A}, \hat{\tau})$ itself, is completely compact. Thus $(\hat{A}, \hat{\tau})$ is l.c.c. if and only if it is completely compact. Another consequence of 22.7 is the following result.

23.3 $(\hat{A}, \hat{\tau})$ *is locally connected if and only if it is both connected and locally connected.*

In the definition of pointwise convergence only the topology of R' plays a role. In the following more general construction the situation is different, the topology on R playing a role also.

Given any two subsets, M of R and M' of R', we can define a subset of $\hat{A}$ which we shall denote by $\langle M, M' \rangle$ and define by

$$\langle M, M' \rangle = \{\varphi : \varphi \in \hat{A}, \varphi M \subseteq M'\}.$$

Let $\mathfrak{S}$ be a non-empty collection of subsets of R and $\mathfrak{U}'$ the system of all open sets of (R', τ'). Then

$$\mathfrak{B}_{\mathfrak{S}} = \{\langle S, U' \rangle : S \in \mathfrak{S}, U' \in \mathfrak{U}'\}$$

is a covering of $\hat{A}$ and thereby a subbasis of a uniquely determined topology $\hat{\tau}_{\mathfrak{S}}$ of $\hat{A}$. This new topology will, in general, not satisfy the condition that all mappings ϵ_p be continuous. For we know that $\hat{\tau}$, the topology of pointwise convergence, is the coarsest topology satisfying the above condition—hence all ϵ_p are continuous if and only if $\hat{\tau}_{\mathfrak{S}} \leqslant \hat{\tau}$.

23.4 *If* $\mathfrak{S}$ *contains all one-points subsets of* R, *then* $\hat{\tau}_{\mathfrak{S}} \leqslant \hat{\tau}$.

Proof: By 23.1, the system $\{\epsilon_p^{-1} : p \in R, U' \in \mathfrak{U}'\}$ is a subbasis of $(\hat{A}, \hat{\tau})$. Furthermore, $\langle \{p\}, U' \rangle = \epsilon_p^{-1} U'$. Hence the subbasis $\mathfrak{B}_{\mathfrak{S}}$ of $\hat{\tau}_{\mathfrak{S}}$ contains a subbasis of $\hat{\tau}$, and all τ-open sets are also $\hat{\tau}_{\mathfrak{S}}$-open. By 7.1, we therefore have $\hat{\tau}_{\mathfrak{S}} \leqslant \hat{\tau}$. ∎

23.5 *For the system* $\mathfrak{E}$ *of all finite subsets of* R *we have* $\hat{\tau}_{\mathfrak{E}} = \hat{\tau}$.

Proof: By 23.4 we have $\hat{\tau}_{\mathfrak{E}} \leqslant \hat{\tau}$. For a finite subset $E = \{p_1, \ldots, p_n\}$ of R we also have

$$\langle E, U' \rangle = \epsilon_{p_1}^{-1} U' \cap \ldots \cap \epsilon_{p_n}^{-1} U'.$$

Hence every set belonging to $\mathfrak{B}_{\mathfrak{E}}$ is $\hat{\tau}$-open, which implies the reverse inclusion $\hat{\tau} \leqslant \hat{\tau}_{\mathfrak{E}}$. ∎

We can also use topological notions to characterize the sets belonging to the collection $\mathfrak{S}$. In the definition of $\hat{\tau}_{\mathfrak{S}}$ we used both the topologies on R' and R. We obtain an important special case by focusing our attention on the system $\mathfrak{K}$ of all completely compact subsets of (R, τ). The resulting topology $\hat{\tau}_{\mathfrak{K}}$ is then the special case of $\hat{\tau}_{\mathfrak{S}}$ obtained by taking $\mathfrak{S} = \mathfrak{K}$. It is called the *topology of compact convergence.*

23.6 *All the mappings ϵ_p are continuous mappings from $(\hat{A}, \hat{\tau}_{\mathfrak{K}})$ onto (R', τ'). If τ' is a T_1- or T_2-topology, then so is $\hat{\tau}_{\mathfrak{K}}$.*

Proof: Every one-point set is completely compact. By 23.4 we therefore have $\hat{\tau}_{\mathfrak{K}} \leqslant \hat{\tau}$, which implies our first contention. By 23.2, if τ' is a T_1-topology so is $\hat{\tau}$, and since $\hat{\tau}_{\mathfrak{K}} \leqslant \hat{\tau}$, by 23.2 again, if τ' is a T_2-topology so is $\hat{\tau}_{\mathfrak{K}}$. ∎

The separation axiom $(\mathbf{T_3})$ is generally not preserved under the conditions of the above theorem. If, however, we limit ourselves to the subspace of $(\hat{A}, \hat{\tau}_{\mathfrak{K}})$ composed of the set $\hat{C}$ of all continuous mappings of (R, τ) into (R', τ'), then we have the following result.

23.7 *If (R, τ) is a T_3-space, so is the subspace $\hat{C}$ of $(\hat{A}, \hat{\tau}_{\mathfrak{K}})$ composed of all continuous functions.*

Proof: Let K be a completely compact subset of (R, τ), U' an open subset of (R', τ'), and φ a mapping belonging to $\hat{C}$ such that $\varphi \in \langle K, U' \rangle$ and hence $\varphi K \subseteq U'$. Given any $p' \in \varphi K$ we can find an open neighborhood $V'_{p'} \in \tau' p'$ such that $\bar{V}'_{p'} \subseteq U'$. Since φK is completely compact by 19.1, φK can be covered by a finite number of the above neighborhoods $V'_{p'}$. For the union V' of these neighborhoods we have $\varphi K \subseteq V'$ and $\bar{V}' \subseteq U'$. Thus $\varphi \in \langle K, V' \rangle$, and $\langle K, V' \rangle \in \hat{\tau}_{\mathfrak{K}} \varphi$. If $\psi \in \hat{A}$ and $\psi \notin \langle K, \bar{V}' \rangle$ then there exists a $p \in K$ such that $\psi p \notin \bar{V}'$. Hence $\epsilon_p^{-1} \complement \bar{V}'$ is a $\hat{\tau}$-neighborhood, and all the more so a $\hat{\tau}_{\mathfrak{K}}$-neighborhood of ψ, and it is disjoint from $\langle K, V' \rangle$. This implies

$$h_{\hat{\tau}_{\mathfrak{K}}} \langle K, V' \rangle \subseteq \langle K, \bar{V}' \rangle \subseteq \langle K, U' \rangle.$$

Any $\hat{\tau}_{\mathfrak{K}}$-neighborhood of φ contains a neighborhood of the form

$$\hat{U} = \langle K_1, U_1' \rangle \cap \dots \cap \langle K_n, U_n' \rangle \qquad \text{with} \qquad \varphi \in \hat{U}.$$

For the neighborhood $\hat{V}$ constructed by the above procedure we have

$$\hat{V} = \langle K_1, V_1' \rangle \cap \dots \cap \langle K_n, V_n' \rangle$$

and

$$h_{\hat{\tau}_{\mathfrak{K}}}(\hat{V}) \subseteq h_{\hat{\tau}_{\mathfrak{K}}} \langle K_1, V_1' \rangle \cap \dots \cap h_{\hat{\tau}_{\mathfrak{K}}} \langle K_n, V_n' \rangle \subseteq \hat{U}.$$

Hence the neighborhoods filter $\hat{\tau}_{\mathfrak{K}} \varphi$ possesses a basis made up of closed sets. ∎

Notes and Exercises

23A Let the mappings φ and $\varphi_n (n \in \mathbf{N})$ of $\mathbf{R}$ into itself be defined as follows:

$$\varphi x = \begin{cases} \sin \frac{\pi}{x} & \text{for} \quad x \neq 0 \\ 0 & \phantom{\text{for}} \quad x = 0 \end{cases} \quad \text{and} \quad \varphi_n x = \begin{cases} \sin \frac{\pi}{x} & \text{for} \quad |x| \geqslant \frac{1}{n} \\ 0 & \phantom{\text{for}} \quad |x| < \frac{1}{n}. \end{cases}$$

Exercise: Under the topology of compact convergence the Fréchet filter of the sequence $(\varphi_n)_{n \in \mathbf{N}}$ converges to the function φ.

23B Let M be a subset of R. The topology on $\hat{A}$ defined by

$$\hat{\tau}_M = \bigwedge \{ {}_{\epsilon_p}\tau' : p \in M \}$$

is called the *topology of pointwise convergence on* M.

Exercise: In what sense do the theorems 23.1-23.3 remain valid for the topology of pointwise convergence?

23C *Exercise:* The basis degree of the topology $\hat{\tau}$ of pointwise convergence (cf. 18.3) satisfies

$$\mathrm{Bd}\,(\hat{A}, \hat{\tau}) \leqslant |R| \cdot \mathrm{Bd}\,(R', \tau').$$

24. Sums of Spaces

Let $\{R_\iota : \iota \in I\}$ be a decomposition of the fundamental set R; i.e., let $R = \cup R_\iota$, $R_\iota \neq \emptyset$, and $R_{\iota_1} \cap R_{\iota_2} \neq \emptyset$ for $\iota_1 \neq \iota_2$. In this situation we also say that R is the *sum* of the sets R_ι and, in order to emphasize this "sum idea," we write $R = \Sigma \{R_\iota : \iota \in I\}$.

Definition 24a: *A space* (R, τ) *is said to be the* sum (or sum space) *of the spaces* (R_ι, τ_ι), $\iota \in I$*—we symbolize this by* $(R, \tau) = \Sigma \{(R_\iota, \tau_\iota) : \iota \in I\}$*—whenever the following conditions are satisfied:*

(1) $R = \Sigma \{R_\iota : \iota \in I\}$.

(2) *For each* $\iota \in I$, R_ι *is an open subset of* (R, τ).

(3) *For each* $\iota \in I$, τ_ι *is the topology induced by* τ *on* R_ι, *i.e.,* $\tau_\iota = \tau_{R_\iota}$.

Since $R_\iota = \complement \cup \{R_\kappa : \kappa \neq \iota\}$, using (1) and (2) we see that the sets R_ι are also closed. Hence a sum of spaces is never connected, unless perhaps when the set of indices reduces to one element. If the space (R, τ) has only a finite number of components or else is locally connected, then the sum space is the subspace generated by its components (14.8, 15.1).

Every point $p \in R_\kappa (\kappa \in I)$ is also a point of the sum set $\Sigma\{R_\iota : \iota \in I\}$. In order to differentiate between these two ways of looking at p, we shall use p' to mean p looked at as a point of ΣR_ι. Then $\alpha_\kappa p = p'$ defines, for each $\kappa \in I$, a mapping $\alpha_\kappa : R_\kappa \to \Sigma R_\iota$.

24.1 *The mapping* α_κ *is a continuous, open, and closed* 1-1 *mapping from* (R_κ, τ_κ) *into the sum space* $(R, \tau) = \Sigma (R_\iota, \tau_\iota)$, *and* τ *is the finest topology on* R *for which all the* α_κ *are continuous mappings from* (R_κ, τ_κ).

Proof: For each $p_\kappa \in R_\kappa$ we have, using property (3) of sum spaces,

$$\alpha_\kappa \tau_\kappa p_\kappa = \tau p_\kappa \wedge \overset{\sqcap}{R_\kappa}.$$

But by (2) R_κ is a τ-open set, hence $\tau p_\kappa \leqslant \overset{\sqcap}{R_\kappa}$ and $\alpha_\kappa \tau_\kappa p_\kappa = \tau p_\kappa = \tau \alpha_\kappa p_\kappa$. Our contentions now follow immediately. ∎

In 24a we assumed that the spaces (R_ι, τ_ι) were actually subspaces of the sum space (R, τ). Such an assumption is not strictly necessary for defining a sum space. Consider an arbitrary family $\{(R_\iota, \tau_\iota) : \iota \in I\}$ of topological spaces. Without loss of generality, we may assume that the R_ι are pairwise disjoint—for if the R_ι were not disjoint we could associate with each $p_\iota \in R_\iota$ the couple (p_ι, ι) and then the sets $R'_\iota = \{(p_\iota, \iota) : p_\iota \in R_\iota\}$ would be pairwise disjoint. If now we put $R = \cup R_\iota$, then $R = \Sigma R_\iota$ and to each $p \in R$ we can uniquely associate an index $\iota(p)$ such that $p \in R_{\iota(p)}$. We can define a topology τ on R by $\tau p = \alpha_{\iota(p)} \tau_{\iota(p)} p$ and then we have $(R, \tau) = \Sigma (R_\iota, \tau)$. In this sense we can call τ the sum of the topologies τ_ι and designate it by $\Sigma \tau_\iota$.

If M is a subset of ΣR_ι, then the subset $M_\iota = \alpha_\iota^{-1} M = M \cap R_\iota$ is called the ιth-*summand* or *summand of index* ι of M. We manifestly have $M = \Sigma M_\iota$.

24.2 M *is open (closed) in* $\Sigma (R_\iota, \tau_\iota)$ *if and only if each summand* M_ι *is open (closed) in* (R_ι, τ_ι). *We have* $\bar{M} = \Sigma \bar{M}_\iota$ *and* $\overset{\circ}{M} = \Sigma \overset{\circ}{M}_\iota$.[4]

Proof: Since α_ι is continuous (24.1), if M is open or closed so is $M_\iota = \alpha_\iota^{-1} M$. Conversely, if M_κ is open (closed) in (R_κ, τ_κ) then M_κ is also open (closed) in $\Sigma (R_\iota, \tau_\iota)$ (9.3)—this because R_κ is open (closed) in $\Sigma (R_\iota, \tau_\iota)$. The openness of all the summands M_ι therefore implies the openness of $M = \Sigma M_\iota = \cup M_\iota$. For each $p \in R_\kappa$ we see that R_κ is a neighborhood belonging to $(\Sigma \tau_\iota) p$ and possessing a nonempty intersection with M_ι for $\iota \neq \kappa$. Hence $\{M_\iota : \iota \in I\}$ is

[4] $\bar{M}_\iota$, $\overset{\circ}{\widehat{M_\iota}}$ represent the closure and the interior, respectively, of M_ι in (R_ι, τ_ι).

a discrete system of sets in $\Sigma(R_\iota, \tau_\iota)$, and since the M_ι are closed, 13.1 tells us that M is also closed. The equations for the closures (hulls) and interiors (kernels) then follow immediately. ∎

For each $\iota \in I$ let φ_ι be a mapping from R_ι into R'. Then $\varphi p = \varphi_{\iota(p)} p$ defines a mapping $\varphi : \Sigma R_\iota \to R'$, called the *sum map* of the mappings φ_ι and designated by $\Sigma \varphi_\iota$. Conversely, if the mapping $\varphi : \Sigma R_\iota \to R'$ is given, then we also have $\varphi = \Sigma(\varphi \alpha_\iota)$.

24.3 *$\Sigma \varphi_\iota$ is a continuous (open) mapping from $\Sigma(R_\iota, \tau_\iota)$ into (R', τ') if and only if each φ_ι is a continuous (open) mapping from (R_ι, τ_ι) into (R', τ').*

If $\varphi : \Sigma R_\iota \to R'$ is a closed mapping from $\Sigma(R_\iota, \tau_\iota)$ into (R', τ'), then each summand $\varphi\alpha_\iota$ is a closed mapping from (R_ι, τ_ι) into (R', τ').

Proof: The first contention follows from $(\Sigma \varphi_\iota)(\Sigma \tau_\iota)p = \varphi_{\iota(p)}\tau_{\iota(p)}p$. By 24.1, α_ι is a closed mapping, and therefore, if φ is a closed mapping so is $\varphi\alpha_\iota$. ∎

Note that even though all the φ_ι are closed, their sum $\Sigma \varphi_\iota$ need not generally be closed (cf. 24A).

24.4 *If the sum space $\Sigma(R_\iota, \tau_\iota)$ can be probed by (R', τ') then so can all the summands (R_ι, τ_ι). Conversely, suppose the spaces (R_ι, τ_ι) can be probed by the space (R', τ'). Then if in R' there exist two points p', q' such that $\overline{q'} \wedge \tau' p' = \mathfrak{o}$, we can assert that $\Sigma(R_\iota, \tau_\iota)$ can also be probed by (R', τ').* (cf 24B)

Proof: (cf. 18.1) Let $\mathfrak{v}$ be an ultrafilter of R_κ, and for $p \in R_\kappa$ let us suppose $\mathfrak{v} \wedge \tau_\kappa p = \mathfrak{o}$. For the ultrafilter $\alpha_\kappa \mathfrak{v}$ of ΣR_ι we then likewise have

$$\alpha_\kappa \mathfrak{v} \wedge (\Sigma \tau_\iota) p = \alpha_\kappa \mathfrak{v} \wedge \alpha_\kappa \tau_\kappa p = \mathfrak{o}.$$

Hence if $(\Sigma R_\iota, \tau_\iota)$ can be probed by (R', τ'), there exists a continuous mapping $\varphi : \Sigma(R_\iota, \tau_\iota) \to (R', \tau')$ such that $\varphi\alpha_\kappa \mathfrak{v} \wedge \tau' \varphi \alpha_\kappa p = \varphi \alpha_\kappa \mathfrak{v} \wedge \tau' \varphi p = \mathfrak{o}$. Since $\varphi\alpha_\kappa : (R_\kappa, \tau_\kappa) \to (R', \tau')$ we see that (R_κ, τ_κ) can also be probed by (R', τ').

Conversely, suppose the spaces (R_ι, τ_ι) can be probed by (R', τ'). Let $p \in R_\kappa$ and let $\mathfrak{v}$ now be an ultrafilter of ΣR_ι such that $\mathfrak{v} \wedge (\Sigma \tau_\iota) p = \mathfrak{o}$. First, if $\mathfrak{v} \wedge \overline{R_\kappa} \neq \mathfrak{o}$, then $\mathfrak{v} \leqslant \overline{R_\kappa}$ and also $\mathfrak{v} \wedge \tau_\kappa p = \mathfrak{o}$. Hence there exists a continuous mapping $\varphi_\kappa : (R_\kappa, \tau_\kappa) \to (R', \tau')$ such that $\varphi_\kappa \mathfrak{v} \wedge \tau' \varphi_\kappa p = \mathfrak{o}$. If we now define, for $\iota \neq \kappa$, the mapping $\varphi_\iota : R_\iota \to R'$ by $\varphi_\iota R_\iota = \{\varphi_\kappa p\}$, then $\varphi = \Sigma \varphi_\iota$ is, by 24.3, a continuous mapping of $\Sigma(R_\iota, \tau_\iota)$ into (R', τ'). We have

$$\varphi \mathfrak{v} \wedge \tau' \varphi p = \varphi_\kappa \mathfrak{v} \wedge \tau' \varphi_\kappa p = \mathfrak{o}.$$

But if $\mathfrak{v} \wedge \overline{R_\kappa} = \mathfrak{o}$, let $\varphi_\kappa : R_\kappa \to R'$ be defined by $\varphi_\kappa R_\kappa = \{p'\}$ and let $\varphi_\iota : R_\iota \to R'$ for $\iota \neq \kappa$ be defined by $\varphi_\iota R_\iota = \{q'\}$ (cf. assumptions on (R', τ')). Then $\varphi = \Sigma \varphi_\iota$ is again a continuous mapping from $\Sigma(R_\iota, \tau_\iota)$ into (R', τ') and we have $\varphi \mathfrak{v} \wedge \tau' \varphi p = \overline{q'} \wedge \tau' p' = \mathfrak{o}$. Hence $\Sigma(R_\iota, \tau_\iota)$ can be probed by (R', τ'). ∎

24.5 *The properties* $(\mathbf{T}_1)$-$(\mathbf{T}_5)$, $(\mathbf{T}_{3a})$ *included, and complete compactness, paracompactness, local connectedness hold in* $(R, \tau) = \Sigma(R_\iota, \tau_\iota)$ *if and only if they hold in all the spaces* (R_ι, τ_ι).

Proof: The assertion on axiom $(\mathbf{T}_{3a})$ is already covered in 24.4. Since $\tau p = \alpha_\iota \tau_\iota p$, every basis of $\tau_\iota p$ is also a basis of τp; and conversely, every basis of τp contains a basis of $\tau_\iota p$. This now implies the contention on local connectedness.

As soon as (R, τ) is a T_1-, T_2-, or T_3-space (9.4), a T_5-space (9.6), locally completely compact (12.5), or paracompact (13.19)—as the case may be—so is each of its closed subspaces (R_ι, τ_ι).

Conversely, let us suppose that each of the (R_ι, τ_ι) is a T_1-space. If p and q are two distinct points of R and if $p, q \in R_\iota$, then $\overline{p} \wedge \tau q = \alpha_\iota(\overline{p} \wedge \tau_\iota q) = \mathfrak{o}$. If, however, $p \in R_\iota$ and $q \in R_\kappa$ where $\iota \neq \kappa$, then R_ι and R_κ are even disjoint neighborhoods of p and q in (R, τ); i.e., (R, τ) is also a T_1-space. We obtain a similar conclusion by similar reasoning in the case of axiom $(\mathbf{T}_2)$.

Since α_ι is a closed mapping (24.1), if $\tau_\iota p$ is a closed filter so is $\tau p = \alpha_\iota \tau_\iota p$. Hence if all (R_ι, τ_ι) are (T_3)-spaces, so is (R, τ).

Continuing, let (R_ι, τ_ι) be a T_4-space for all $\iota \in I$. If A and B denote disjoint closed subsets of (R, τ), then the summands $A_\iota = A \cap R_\iota$ and $B_\iota = B \cap R_\iota$ are, by 24.2, closed in (R_ι, τ_ι) and they possess disjoint neighborhoods U_ι and V_ι, where U_ι and V_ι are open in R_ι. The sets $U = \Sigma U_\iota$ and $V = \Sigma V_\iota$ are therefore disjoint open neighborhoods of A and B in (R, τ); i.e., (R, τ) is also a T_4-space. We obtain a similar conclusion by similar reasoning in the case of axiom $(\mathbf{T}_5)$.

Every completely compact neighborhood $U \in \tau_\iota p$ is also a completely compact neighborhood of p in (R, τ). If all the spaces (R_ι, τ_ι) are locally completely compact so is (R, τ).

Finally let the spaces (R_ι, τ_ι) be paracompact. If $\mathfrak{H}$ is an open covering of (R, τ), then $\mathfrak{H}_\iota = \{H \cap R_\iota : H \in \mathfrak{H}\}$ is an open covering of (R_ι, τ_ι) for which we can find an open and locally finite covering $\mathfrak{R}_\iota$ such that $\mathfrak{R}_\iota \ll \mathfrak{H}_\iota$. By 24.2 the sets belonging to $\mathfrak{R}_\iota$ are open in (R, τ). Hence $\mathfrak{R} = \cup \mathfrak{R}_\iota$ is an open covering of (R, τ) with $\mathfrak{R} \ll \mathfrak{H}$. Given any point $p \in R_\iota$ we can find a neighborhood $U \in \tau_\iota p$ which possesses a nonempty intersection with only a finite number of sets belonging to $\mathfrak{R}_\iota$. But U is also a neighborhood belonging to τp and U is disjoint for all sets belonging to $\mathfrak{R}_\kappa$ when $\kappa \neq \iota$. Hence U possesses a nonempty intersection with only a finite number of sets of $\mathfrak{R}$. Therefore $\mathfrak{R}$ is locally finite and (R, τ) paracompact. ∎

24.6 *The sum space* $(R, \tau) = \Sigma(R_\iota, \tau_\iota)$ *is completely compact (k-compact) if and only if the indexing set* I *is finite and all the spaces* (R_ι, τ_ι) *are completely compact (k-compact).*

Proof: If the indexing set I is finite and all the subspaces (R_ι, τ_ι) are completely compact (k-compact), then so is their union (R, τ) (11.8). Conversely, if (R, τ) is completely compact (k-compact) then so are the closed subspaces (R_ι, τ_ι) (11.6). If, on the other hand, $\mathbf{N} \subseteq I$ and $A_n = \cup \{R_\nu : \nu \in \mathbf{N}, \nu \geqslant n\}$, then $(A_n)_{n \in \mathbf{N}}$ is a monotonically decreasing sequence of nonempty closed sets with empty intersection. By 11.20, if I is infinite we see that the sum space (R, τ) cannot ever be $\aleph_0$-compact. ∎

Using the concept of the sum of spaces we can give an interesting characterization of a space which is simultaneously paracompact and locally completely compact.

24.7 *A T_3-space (R, τ) is simultaneously paracompact and locally completely compact if and only if it is expressible as the sum of spaces which are both locally completely compact and countable at infinity.*

Proof: Let (R, τ) be locally completely compact and paracompact. For each $p \in R$ let U_p be an open neighborhood belonging to τp with completely compact closure (12.1). Given the open covering $\mathfrak{U} = \{U_p : p \in R\}$ we can find a locally finite open covering $\mathfrak{H}$ of R such that $\mathfrak{H} \ll \mathfrak{U}$. If $H \in \mathfrak{H}$ then $H \subseteq U_p$ for some $p \in R$, $\bar{H} \subseteq \bar{U}_p$, and hence $\bar{H}$ is completely compact (11.6). For each point $q \in \bar{H}$ we can also find an open neighborhood $V_q \in \tau q$, which possesses a nonempty intersection with only a finite number of sets of $\mathfrak{H}$. But since $\bar{H}$ is completely compact it can be covered by a finite number of the neighborhoods V_q and hence it also possesses a nonempty intersection with at most a finite number of sets belonging to $\mathfrak{H}$.

In R we can define the following equivalence relation: $(p, q) \in \rho$ if and only if there exist a finite number of sets $H_0, ..., H_n \in \mathfrak{H}$ such that $p \in H_0$, $q \in H_n$ and $H_{\nu-1} \cap H_\nu \neq \emptyset$ for $\nu = 1, ..., n$. Since the equivalence classes are unions of sets belonging to $\mathfrak{H}$ they must be open subsets of R. Hence (R, τ) is the sum space of the subspaces generated by the equivalence classes. By 24.5 these subspaces are locally completely compact. Every equivalence class R_ι, being a summand of R, must be closed. Hence $H \subseteq R_\iota$ implies $\bar{H} \subseteq R_\iota$. For a fixed point $p \in R_\iota$ let now $K_0 = \{p\}$ and $K_{n+1} = \cup \{\bar{H} : H \in \mathfrak{H}, H \cap \bar{K}_n \neq \emptyset\}$. Then $R_\iota = \cup \{K_n : n \in \mathbf{N}\}$. Moreover, each K_n is, by the first part of the proof at hand, the union of only a finite number of sets $\bar{H}$ and hence completely compact (11.8). This proves that the subspace generated by R_ι is countable at infinity.

To prove the converse, suppose $(R, \tau) = \Sigma (R_\iota, \tau_\iota)$, where the (R_ι, τ_ι) are locally completely compact and countable at infinity. If (R, τ) is a T_3-space, so is (R_ι, τ_ι) and hence, by 13.12, the (R_ι, τ_ι) must be paracompact. (R, τ) is therefore locally completely compact and paracompact, by 24.5. ∎

Notes and Exercises

24A Let $(R', \tau') = \mathbf{R}$ and for each $n \in \mathbf{N}$, let $(R_n, \tau_n) = \mathbf{R}$ also. Let φ_n denote the identity map on $\mathbf{R}$, for all $n \in \mathbf{N}$. Then the φ_n are all closed mappings. The subset $T_n = [1/n, n/n + 1]$ of R_n is closed. Hence by 24.2, the set $T = \Sigma T_n$ is a closed set in $\Sigma(R_n, \tau_n)$. But the image of τ under the sum mapping $\Sigma \varphi_n$ is the open interval $]0, 1[$. Hence, although the φ_n are closed mappings, their sum is not a closed mapping.

24B Let R_1, R_2, and R' be one-point sets, bearing the discrete topology. Then R_1 and R_2 can be probed by R'. But the sum-space of R_1 and R_2 cannot be probed by R'.

Exercise: For each $\iota \in I$ suppose (R_ι, τ_ι) can be probed by (R'_ι, τ'_ι). Then $\Sigma(R_\iota, \tau_\iota)$ can be probed by $\Sigma(R'_\iota, \tau'_\iota)$.

24C *Exercise:* For each $\iota \in I$ let ρ_ι be an equivalence relation on R_ι. Then

$$\rho = \{(p, q) : \iota(p) = \iota(q), (p, q) \in \rho_{\iota(p)}\}$$

defines an equivalence relation on ΣR_ι. We have

$$\Sigma(R_\iota, \tau_\iota)/\rho_\iota \cong (\Sigma(R_\iota, \tau_\iota))/\rho.$$

V

Extensions and Characterization of Topological Spaces

In order to obtain "intuitive insight" into special classes of topological spaces we can proceed in several ways, only a few of which will be pursued in this chapter. For instance, we can seek to describe important topological spaces by means of enough of their properties to completely characterize them, up to homeomorphism. An even more general procedure is to start from some simple, standard space and seek to characterize all those spaces which can be obtained as a continuous image of this standard space. A third procedure is to try to establish so-called embedding theorems—this method characterizes new spaces as being homeomorphic to subspaces of a well-known, standard space. Intimately allied to the problem of embedding is that of extension, which seeks to determine when a given space can be extended to a larger one, containing the given one as subspace. We shall study this problem first.

25. Extensions of Topological Spaces

Let (R, τ) be a subspace of the space (R', τ'). By Section 9, for every $p \in R$, we have $\tau p = \mathrm{Tr}\,(\tau' p')$, and $T^* = \{\mathrm{Tr}\,(\tau' p') : p' \in R'\}$ is a collection of open filters of (R, τ). This set T^* of the trace filters of all the neighborhoods filters of the space (R', τ'), is called the *trace system* of the space (R', τ') in the subspace (R, τ). Among all spaces having a given space (R, τ) as subspace, we shall only take into consideration those in which R is dense, i.e., those containing (R, τ) as a dense subspace.

Definition 25a: *A space* (R', τ') *is said to be an* extension *of the space* (R, τ) *whenever* (R, τ) *is a dense subspace of* (R', τ').

By our previous remarks, we can surely associate with every extension a unique trace system. In general, however, an extension is not characterized by knowing its trace system: Distinct extensions can have the same trace system. In fact, the neighborhoods filters of different points of an extension can possess the same trace filter.

If (R', τ') is an extension of (R, τ), then the trace filters $\mathrm{Tr}\,(\tau' p')$ of all points $p' \in R'$ are different from the null filter, because R is dense in (R', τ'). Hence the trace system T^* of the extension satisfies the following conditions:

(a) T^* is a set of open filters which does not contain the nullfilter.

(b) T^* contains all neighborhoods filters τp for points $p \in R$.

The goal of the following construction is this important result: If, conversely, T^* is a set of filters of the space (R, τ) satisfying conditions (a) and (b), then there exists at least one extension T^* of (R, τ) having T^* as its trace system. Among all possible extensions there exists one which we can distinguish from all others by its characteristic properties.

To justify this result, let T^* now denote a set of filters of (R, τ) satisfying conditions (a) and (b). Let the set $\hat{R}$ contain, on the one hand, all the points of R and, on the other hand, all those filters $\mathfrak{t} \in T^*$ which are distinct from all neighborhoods filters $\tau p (p \in R)$. Then $R \subseteq \hat{R}$ and every point $\hat{p} \in \hat{R} \cap \complement R$ can be represented by a filter belonging to T^*. We can define a mapping σ of $\hat{R}$ onto T^* as follows: If $p \in R$ let $\sigma p = \tau p$ while if $\hat{p} \in \hat{R} \cap \complement R$ let $\sigma \hat{p}$ be the filter in T^* which represents $\hat{p}$. This mapping σ is called the *trace mapping*. The restriction of σ to $\hat{R} \cap \complement R$ is a 1-1 mapping. If different points of R have the same neighborhoods filter under the topology τ, σ itself is not a 1-1 mapping. But if (R, τ) is a T_1-space, then σ is 1-1 also.

For subsets M of R let

$$s(M) = \{\hat{p} : \hat{p} \in \hat{R}, \sigma \hat{p} \leqslant \overset{\frown}{M}\}.$$

If M is an open subset of (R, τ), then $s(M) \cap R = \{p : p \in R, \tau p \leqslant M\} = \overset{\frown}{M}$. In addition, we see immediately, that $s(R) = \hat{R}$ and $s(M_1) \cap s(M_2) = s(M_1 \cap M_2)$. If, therefore, $\mathfrak{U}$ denotes the system of all open subsets of (R, τ), then

$$\hat{\mathfrak{B}} = \{s(U) : U \in \mathfrak{U}\}$$

contains the set $\hat{R}$ and if $\hat{\mathfrak{B}}$ contains a finite number of sets it also contains their intersection. By 5.8 there must therefore exist a unique topology $\hat{\tau}$ on $\hat{R}$ having $\mathfrak{B}$ as basis. For each $\hat{p} \in \hat{R}$ the system $\{s(U) : U \in \mathfrak{U}, \sigma \hat{p} \leqslant \overset{\frown}{U}\}$ is a basis of the neighborhoods filter $\hat{\tau}\hat{p}$. Since T^* is composed of open filters, we have

$$\mathrm{Tr}\,(\hat{\tau}\hat{p}) = \bigwedge \{\overset{\frown}{U} : U \in \mathfrak{U}, \sigma \hat{p} \leqslant \overset{\frown}{U}\} = \sigma \hat{p}.$$

Since $\sigma p \neq \mathfrak{o}$, this equation tells us, first, that R is dense in $(\hat{R}, \hat{\tau})$. Secondly, for points $p \in R$ it implies that $\mathrm{Tr}\,(\hat{\tau}p) = \sigma p = \tau p$. Hence (R, τ) is a subspace of $(\hat{R}, \hat{\tau})$ and $(\hat{R}, \hat{\tau})$ is an extension of (R, τ). Lastly, this equation shows that T^* is the trace system of this extension.

The above-constructed extension $(\hat{R}, \hat{\tau})$ is called the T^*-extension of (R, τ). Its existence implies that the trace systems of extensions of (R, τ) are characterized by properties (a) and (b) above. From now on, in this section, T^* always will denote such a trace system.

25.1 *The T^*-extension $(\hat{R}, \hat{\tau})$ of (R, τ) possesses the following property:*

(*) *If (R', τ') is another extension of (R, τ), also having T^* as its trace system, then there exists a continuous mapping φ from (R', τ') onto $(\hat{R}, \hat{\tau})$ such that the restriction of φ to R is the identity map. The mapping φ also satisfies the* trace condition $\mathrm{Tr}\,(\tau' p') = \mathrm{Tr}\,(\hat{\tau}\varphi p')$.

The space $(\hat{R}, \hat{\tau})$ is uniquely determined, up to homeomorphism, by the above condition: Given any other extension $(\tilde{R}, \tilde{\tau})$ of (R, τ) with the trace system T^* and satisfying condition (*), then there exists a homeomorphism of $(\tilde{R}, \tilde{\tau})$ onto $(\hat{R}, \hat{\tau})$ which reduces to the identity on R.

$\hat{\tau}$ is the coarsest topology on $\hat{R}$ having the given trace system T^*.

Proof: Suppose that $p' \in R'$ satisfies the condition $\mathrm{Tr}\,(\tau' p') \neq \tau p$ for all $p \in R$. Then there exists only one $\hat{p} \in \hat{R} \cap \complement R$ such that $\mathrm{Tr}\,(\hat{\tau}\hat{p}) = \mathrm{Tr}\,(\tau' p')$. Let us define the mapping φ by $\varphi p' = \hat{p}$, while otherwise if $p' \in R'$ does not satisfy the above condition, let $\varphi p'$ be a point of R such that $\mathrm{Tr}\,(\tau' p') = \tau\varphi p'$. If $p \in R$, let $\varphi p = p$. This mapping φ from R' onto $\hat{R}$ manifestly satisfies the trace condition and reduces to the identity mapping on R. If U' is an open neighborhood belonging to $\tau' p'$, then $q' \in U'$ implies first that

$$\mathrm{Tr}\,(\hat{\tau}\varphi q') = \mathrm{Tr}\,(\tau' q') \leqslant \overline{U' \cap R}$$

and then that $\varphi q' \in s(U' \cap R)$, hence $\varphi U' \subseteq s(U' \cap R)$. This implies further that $\varphi\tau' p \leqslant \hat{\tau}\varphi p'$, i.e., φ must be continuous. If, in particular, $R' = \hat{R}$, then φ is the identity mapping and we have $\tau' \leqslant \hat{\tau}$. Hence $\hat{\tau}$ is the coarsest topology on $\hat{R}$ having T^* as trace system.

By condition (*), we see that there must exist continuous mappings φ from $(\hat{R}, \hat{\tau})$ onto $(\tilde{R}, \tilde{\tau})$ and ψ from $(\tilde{R}, \tilde{\tau})$ onto $(\hat{R}, \hat{\tau})$, both of which satisfy the trace condition and both of which reduce to the identity on R. Their product $\psi\varphi$ is a mapping of $\hat{R}$ onto itself, which again reduces to the identity on R. We have $\sigma\psi\varphi\hat{p} = \mathrm{Tr}\,(\hat{\tau}\psi\varphi\hat{p}) = \mathrm{Tr}\,(\hat{\tau}\varphi\hat{p}) = \mathrm{Tr}\,(\hat{\tau}\hat{p}) = \sigma\hat{p}$ for all $\hat{p} \in \hat{R}$. But since the restriction of σ to $\hat{R} \cap \complement R$ is 1-1, the above equality implies that $\psi\varphi\hat{p} = \hat{p}$ also for all $\hat{p} \in \hat{R} \cap \complement R$. Hence $\psi\varphi$ is the identity in $\hat{R}$; ψ is a 1-1 mapping of $\tilde{R}$ onto $\hat{R}$ with φ as its inverse. Since φ and ψ are continuous, $\psi\varphi$ is a homeomorphism. ∎

Using 19.9, we also have:

25.2 *Let φ be a continuous mapping from (R, τ) into a T_3-space (R', τ'). A necessary and sufficient condition for φ to have a continuous extension onto the T^*-extension $(\hat{R}, \hat{\tau})$ is that all the image filters $\varphi\mathfrak{t}$, for all $\mathfrak{t} \in T^*$, converge in (R', τ').*

Let M be a nonempty open subset of an extension (R', τ') of (R, τ). Since R is dense in (R', τ'), the open subset $M = M' \cap R$ of (R, τ) is also nonempty. If therefore $\{M', N'\}$ is a decomposition of (R', τ') into two open sets, then the sets $M = M' \cap R$ and $N = N' \cap R$ form a decomposition of (R, τ) into open sets. Hence if (R, τ) is connected, so is every one of its extensions (for the question of connectedness of T^*-extensions, cf. 25A).

25.3 *A necessary and sufficient condition for the T^*-extension $(\hat{R}, \hat{\tau})$ of (R, τ) to be a T_1-space is that (R, τ) be a T_1-space and that any two filters of T^* be noncomparable; while a necessary and sufficient condition for $(\hat{R}, \hat{\tau})$ to be a T_2-space is that (R, τ) be a T_2-space and that for any two distinct filters $\mathfrak{t}_1, \mathfrak{t}_2 \in T^*$ we always have $\mathfrak{t}_1 \wedge \mathfrak{t}_2 = \mathfrak{o}$.*

Proof: The condition $\overline{\hat{p}} \leqslant \hat{\tau}\hat{q}$ is equivalent to $\sigma\hat{p} \leqslant \mathrm{Tr}\,(\hat{\tau}\hat{q}) = \sigma\hat{q}$. Consequently, if $(\hat{R}, \hat{\tau})$ is a T_1-space, then any two filters of T^* are noncomparable, while (R, τ), being a subspace, is also a T_1-space. Conversely, if no two filters of T^* are comparable, then $\overline{\hat{p}} \leqslant \hat{\tau}\hat{q}$ implies $\sigma\hat{p} = \sigma\hat{q}$. If (R, τ) is a T_1-space, then σ is a 1-1 mapping and thus $\hat{p} = \hat{q}$.

The condition $\hat{\tau}\hat{p} \wedge \hat{\tau}\hat{q} = \mathfrak{o}$ implies $\sigma\hat{p} \wedge \sigma\hat{q} = \mathrm{Tr}\,(\hat{\tau}\hat{p}) \wedge \mathrm{Tr}\,(\hat{\tau}\hat{q}) = \mathfrak{o}$. Conversely, if $\sigma\hat{p} \wedge \sigma\hat{q} = \mathfrak{o}$, there must exist open sets $U \in \sigma\hat{p}$ and $V \in \sigma\hat{q}$ such that $U \wedge V = \varnothing$. Consequently $s(U) \cap s(V) = \varnothing$ and since $s(U) \in \hat{\tau}\hat{p}$ and $s(V) \in \hat{\tau}\hat{q}$ we see that $\hat{\tau}\hat{p} \wedge \hat{\tau}\hat{q} = \mathfrak{o}$. From the equivalence of the conditions $\hat{\tau}\hat{p} \wedge \hat{\tau}\hat{q} = \mathfrak{o}$ and $\sigma\hat{p} \wedge \sigma\hat{q} = \mathfrak{o}$, our second contention follows by reasoning similar to that used to establish the first contention. ∎

Definition 25b: *A system $\mathfrak{A}$ of open subsets of (R, τ) is said to be a T^*-*selection system *if, given any $t \in T^*$ we can always choose an $A \in \mathfrak{A}$ such that $A \in t$.*

25.4 *The T^*-extension $(\hat{R}, \hat{\tau})$ of (R, τ) is completely compact if and only if every T^*-selection system contains a finite T^*-selection system.*

Proof: $\mathfrak{A}$ is a T^*-selection system if and only if $\hat{\mathfrak{A}} = \{s(A) : A \in \mathfrak{A}\}$ is an open covering of $(\hat{R}, \hat{\tau})$. Since $s(A) \cap R = A$, distinct sets of $\mathfrak{A}$ must correspond to distinct sets of $\hat{\mathfrak{A}}$. Hence the condition of the theorem is equivalent to the condition that every covering of $\hat{R}$ contained in the basis $\hat{\mathfrak{B}} = \{s(U) : U \in \mathfrak{U}\}$ possesses a finite subcover of $\hat{R}$. The assertion of the theorem then follows from 11.12. ∎

By a *compactification* of (R, τ) we mean a completely compact extension of this space. In the next two sections we shall examine two special compactifications, both important for the applications. We shall finish this section by showing that every space can be compactified by the adjunction of one new point. Since a completely compact space is already a compactification of itself, we need only examine those spaces which are not completely compact.

In the noncompletely compact space (R, τ) let $\mathfrak{C}$ denote the system of all subsets whose complement is completely compact and closed. $\mathfrak{C}$ contains R, but does not contain the empty set. By 11.8, $\mathfrak{C}$ is a filter basis which generates a filter $\mathfrak{c} \neq \mathfrak{o}$. By $T^* = \{\tau p : p \in R\} \cup \{\mathfrak{c}\}$, we can define a trace system of (R, τ), and the T^*-extension $(\hat{R}, \hat{\tau})$ contains, besides the points of R, only the improper point represented by $\mathfrak{c}$. In addition, let $\mathfrak{A}$ be a T^*-selection system. Then there exists an $A_0 \in \mathfrak{A}$ such that $A_0 \in \mathfrak{C}$. The set $\complement A_0$ is completely compact and can therefore be covered by the finite number of sets $A_1, ..., A_n \in \mathfrak{A}$. Hence $\{A_0, ..., A_n\}$ is also a T^*-selection system and $(\hat{R}, \hat{\tau})$ is completely compact by 25.4.

The above construction is called the *one-point compactification* or the *Alexandroff compactification* of (R, τ). Every ultrafilter $\mathfrak{v}$ which does not converge in (R, τ) converges in $(\hat{R}, \hat{\tau})$, however, to the improper point $\hat{u}$ *represented by* $\mathfrak{c}$, i.e., $\mathfrak{v} \leqslant \mathrm{Tr}\,(\hat{\tau}\hat{u})$.

25.5 *The Alexandroff compactification of a noncompletely compact space* (R, τ) *is a* T_1*-space if and only if* (R, τ) *is a* T_1*-space, while it is a* T_2*-space if and only if* (R, τ) *is a locally completely compact* T_2*-space.*

Proof: Since (R, τ) possesses at least one nonconvergent ultrafilter $\mathfrak{v}$, the relation $\mathrm{Tr}\,(\hat{\tau}\hat{u}) \leqslant \tau p$ can never hold between the improper point $\hat{u}$ and an arbitrary point $p \in R$—for otherwise, since $\mathfrak{v} \leqslant \mathrm{Tr}\,(\hat{\tau}\hat{u})$, $\hat{\mathfrak{v}}$ would have to be convergent. If (R, τ) is a T_1-space, then every one-point set $\{p\}$ is completely compact and closed. Consequently $\complement p \in \mathrm{Tr}\,(\hat{\tau}\hat{u})$ which implies that $\tau p \leqslant \mathrm{Tr}\,(\hat{\tau}\hat{u})$ can never occur. Hence our first contention follows from 25.3. Let us now suppose that (R, τ) is a locally completely compact T_2-space. If U is a completely compact neighborhood belonging to τp then, by 11.7, U is closed and thus $\complement U \in \mathrm{Tr}\,(\hat{\tau}\hat{u})$ and $\tau p \wedge \mathrm{Tr}\,(\hat{\tau}\hat{u}) = \mathfrak{o}$. By 25.3 again, the Alexandroff compactification of (R, τ) must be a T_2-space. Conversely, since $R = \complement\hat{u}$, and (R, τ) is an open subspace, we see by 12.5 that (R, τ) must also be a locally completely compact T_2-space. ∎

By the above theorem we can say that a (noncompletely compact) T_2-space is locally completely compact if and only if, under the one-point compactification, it becomes a completely compact T_2-space. A space is countable at infinity if and only if under its Alexandroff compactification the neighborhoods filter of the point at infinity possesses a countable basis (cf. p. 90). The best-known example of a one-point compactification occurs in the theory of functions of a complex variable where, as we know, the complex plane is enlarged by the addition of a point at infinity to form the Riemann sphere of numbers: the Alexandroff compactification of the plane $\mathbf{R}^2$ is homeomorphic to the sphere S^2: (cf. 16A).

Notes and Exercises

25A *Exercise:* The T^*-extension of a space is connected if and only if, given any decomposition $\{T_1^*, T_2^*\}$ of T^* the condition

$$\bigvee \{\mathfrak{t} : \mathfrak{t} \in T_1^*\} \wedge \bigvee \{\mathfrak{t} : \mathfrak{t} \in T_2^*\} \neq \mathfrak{o}$$

is always satisfied.

25B *Exercise:* The T_1^*-extension of (R, τ) is a subspace of the T_2^*-extension of (R, τ) if and only if $T_1^* \subseteq T_2^*$.

25C By 5.4 we see that the set of all open, nonnull filters of a space (R, τ) contains minimal elements, which we call ***ultra-open filters***. Let the set T^* be composed of all neighborhoods filters τp of points $p \in R$ and of all ultra-open filters $\mathfrak{b}$ such that $\mathfrak{b} \wedge \tau p = \mathfrak{o}$ for all $p \in R$.

Exercise: T^* is a trace system. If (R, τ) is a T_1- or a T_2-space then so is the T^*-extension $(\hat{R}, \hat{\tau})$ of (R, τ). If $(\hat{R}, \hat{\tau})$ is a subspace of $(\tilde{R}, \tilde{\tau})$, then $\hat{R}$ is closed in $\tilde{R}$ whenever $\tilde{\tau}$ is a T_2-topology. In general, a T_2-space (R, τ), which is a closed subspace of any other T_2-space containing (R, τ) as a subspace, is known as a *Hausdorff-closed* space.

25D *Exercise:* The Alexandroff compactification of a locally completely compact (but not completely compact) T_2-space is, up to homeomorphism, the only extension of this space to a completely compact T_2-space that can be effected by the adjunction of only one point.

25E *Exercise:* Let $(\hat{R}, \hat{\tau})$ and $(\hat{R}', \hat{\tau}')$ be two Alexandroff compactifications of the locally completely compact (but not completely compact) T_2-spaces (R, τ) and (R', τ'). Then a continuous mapping $\varphi : (R, \tau) \to (R', \tau')$ is proper (cf. 19C) if and only if it admits of a continuous extension $\psi : (\hat{R}, \hat{\tau}) \to (\hat{R}', \hat{\tau}')$.

26. The Wallman Compactification

In this section (R, τ) will always denote a T_1-space.

By 5.4 we know that the set of all nonnull closed filters of (R, τ) possesses minimal elements. We shall call such minimal filters ***ultraclosed filters***. In other words, a filter $\mathfrak{a}$ is ultraclosed if and only if, given any closed filter $\mathfrak{b}$ such that $\mathfrak{o} < \mathfrak{b} \leqslant \mathfrak{a}$, then $\mathfrak{b} = \mathfrak{a}$. Using Zorn's lemma we can immediately deduce that, given any closed filter $\mathfrak{b}$, there exists at least one ultraclosed filter $\mathfrak{a}$ such that $\mathfrak{a} \leqslant \mathfrak{b}$. Examples of ultraclosed filters are the principal filters $\ulcorner p \urcorner$ generated by the single points $p \in R$: For they are certainly ultrafilters and, by $(\mathbf{T}_1)$, they must also be closed. Note, however, that in general an ultraclosed filter need be no ultrafilter.

26.1 *If* $\mathfrak{a}$ *is an ultraclosed filter and* $\mathfrak{b}$ *a closed or open filter, then* $\mathfrak{a} \wedge \mathfrak{b} \neq \mathfrak{o}$ *implies* $\mathfrak{a} \leqslant \mathfrak{b}$.

Proof: First, if $\mathfrak{b}$ is a closed filter, so is $\mathfrak{a} \wedge \mathfrak{b}$, Hence $\mathfrak{o} < \mathfrak{a} \wedge \mathfrak{b} \leqslant \mathfrak{a}$ implies $\mathfrak{a} \wedge \mathfrak{b} = \mathfrak{a}$ and $\mathfrak{a} \leqslant \mathfrak{b}$. Secondly, let $\mathfrak{b}$ be open. For every open set $B \in \mathfrak{b}$ we see that $\mathfrak{d} = \mathfrak{a} \wedge \overline{\mathbf{C}B}$ is a closed filter and $\mathfrak{d} < \mathfrak{a}$. Hence $\mathfrak{d} = \mathfrak{o}$ and $\mathfrak{a} \leqslant \overline{B}$, which again implies $\mathfrak{a} \leqslant \mathfrak{b}$. ∎

If $\mathfrak{a}$ is a given filter, then, by 5.4, the intersection of all open filters $\mathfrak{b}$ such that $\mathfrak{a} \leqslant \mathfrak{b}$, is an open filter which we shall designate by $\mathfrak{a}^0$. For any point p of the space (R, τ) we then have $(\overline{p})^0 = \tau p$.

26.2 *If* $\mathfrak{a}$ *is an ultraclosed and* $\mathfrak{b}$ *a closed filter, then* $\mathfrak{a}^0 \wedge \mathfrak{b} \neq \mathfrak{o}$ *implies* $\mathfrak{a} \leqslant \mathfrak{b}$. *If both* $\mathfrak{a}_1$ *and* $\mathfrak{a}_2$ *are ultraclosed filters, then* $\mathfrak{a}_1^0 \leqslant \mathfrak{a}_2^0$ *implies* $\mathfrak{a}_1 = \mathfrak{a}_2$.

Proof: Suppose $\mathfrak{a} \leqslant \mathfrak{b}$ is false. Then, by 26.1, $\mathfrak{a} \wedge \mathfrak{b} = \mathfrak{o}$. Hence there exists a closed set $B \in \mathfrak{b}$ such that $\mathfrak{a} \leqslant \overline{\mathbf{C}B}$. Since $\mathbf{C}B$ is open, this implies $\mathfrak{a}^0 \leqslant \overline{\mathbf{C}B}$ and thus $\mathfrak{a}^0 \wedge \mathfrak{b} = \mathfrak{o}$.

From $\mathfrak{a}_1^0 \leqslant \mathfrak{a}_2^0$ we deduce $\mathfrak{a}_2^0 \wedge \mathfrak{a}_1^0 \neq \mathfrak{o}$ and from what has preceded, $\mathfrak{a}_2 \leqslant \mathfrak{a}_1$. ∎

Let A^* denote the set of all ultraclosed filters in the space (R, τ). By $(\mathbf{T}_1)$, $\overline{p} \in A^*$ for all $p \in R$. Hence the set $T^* = \{\mathfrak{a}^0 : \mathfrak{a} \in A^*\}$ contains all the neighborhoods filters of all the points of R and is, in fact, composed of all nonnull open filters. T^* is thus a trace system and, by Section 25, we know that there must exist a T^*-extension of (R, τ). It is known as the *Wallman extension* of the T_1-space (R, τ) and is generally denoted by $\omega(R, \tau)$. If $\mathfrak{a}_1, \mathfrak{a}_2 \in A^*$ and $\mathfrak{a}_1^0 = \mathfrak{a}_2^0$, then by 26.2 we have $\mathfrak{a}_1 = \mathfrak{a}_2$. Thus the points of the Wallman extension can be represented by the ultraclosed filters.

26.3 $\omega(R, \tau)$ *is a completely compact* T_1*-space.*

Proof: Let $\mathfrak{A}$ be a T^*-selection system. If $\mathfrak{v}$ is an ultrafilter of R, then there exists an ultraclosed filter $\mathfrak{a}$ such that $\mathfrak{a} \leqslant \bar{\mathfrak{v}}$. From $\mathfrak{a}^0 \wedge \mathfrak{v} = \mathfrak{o}$ and the fact that $\mathfrak{o}$ is open we could deduce $\mathfrak{a}^0 \wedge \bar{\mathfrak{v}} = \mathfrak{o}$, but this contradicts $\mathfrak{a} \leqslant \bar{\mathfrak{v}}$. Hence $\mathfrak{v} \leqslant \mathfrak{a}^0$ and $\mathfrak{v} \leqslant \overline{A}$ for a certain $A \in \mathfrak{A}$. By 3.13 this implies the existence of a finite covering $\mathfrak{A}_1$ of R, $\mathfrak{A}_1$ being contained in $\mathfrak{A}$. But $\mathfrak{A}_1$ is even a T^*-selection system: For if not, there would exist an ultraclosed filter $\mathfrak{a}$ such that $\mathfrak{a}^0 \wedge \overline{\mathbf{C}A} \neq \mathfrak{o}$ for all $A \in \mathfrak{A}_1$. But $\mathbf{C}A$ is closed and by 26.2 we would have $\mathfrak{a} \leqslant \overline{\mathbf{C}A}$—the finiteness of A would them imply $\mathfrak{a} = \mathfrak{o}$, a contradiction. Hence, by 25.4, $w(R, \tau)$ is completely compact.

If $\mathfrak{a}_1^0, \mathfrak{a}_2^0 \in T^*$ and $\mathfrak{a}_1^0 \leqslant \mathfrak{a}_2^0$ then, by 26.2, we must have $\mathfrak{a}_1 = \mathfrak{a}_2$ and, all the more so, $\mathfrak{a}_1^0 = \mathfrak{a}_2^0$. Hence any two distinct filters in T^* are noncomparable, and by 25.3, $\omega(R, \tau)$ must be a T_1-space. ∎

26.4 *Every continuous mapping φ of (R, τ) into a completely compact T_2-space (R', τ') admits of a continuous extension to the Wallman extension $\omega(R, \tau)$.*

Proof: Let $\mathfrak{a}$ be an ultraclosed filter of (R, τ). Then $\varphi\mathfrak{a}$ possesses an adherent point p' in (R', τ'). Since $\varphi^{-1}\tau'p'$ is an open filter and $\mathfrak{a} \wedge \varphi^{-1}\tau'p' \neq \mathfrak{o}$, by 26.1 we also have $\mathfrak{a} \leqslant \varphi^{-1}\tau'p'$ and even $\mathfrak{a}^0 \leqslant \varphi^{-1}\tau'p'$. This implies $\varphi\mathfrak{a}^0 \leqslant \tau'p'$, and hence we see that the images of all filters of the trace system of the Wallman extension are convergent. Since (R', τ') is a completely compact T_2-space it is also a T_3-space and our contention must follow from 25.2. ∎

The separation properties $(\mathbf{T}_2)$, $(\mathbf{T}_3)$, and $(\mathbf{T}_{3a})$ do not, in general, carry over from (R, τ) to $\omega(R, \tau)$—for the Wallman extension is always a completely compact T_1-space and hence these separation properties are equivalence to the normality of $\omega(R, \tau)$.

26.5 *$\omega(R, \tau)$ is normal if and only if (R, τ) is normal.*

Proof: If $\mathfrak{a}_1$ and $\mathfrak{a}_2$ are two distinct ultraclosed filters of R, then $\mathfrak{a}_1 \wedge \mathfrak{a}_2 = \mathfrak{o}$. If (R, τ) is normal, then we have $\mathfrak{a}_1^0 \wedge \mathfrak{a}_2^0 = \mathfrak{o}$. By 25.3, $\omega(R, \tau)$ must then be a T_2-space and thereby also normal.

Conversely, let $\omega(R, \tau)$ be normal and let A, B be two disjoint sets, closed in (R, τ). Let A and B denote the closures of A and B in $\omega(R, \tau)$. Let the point $\hat{p}$ of $\omega(R, \tau)$ be represented by the ultraclosed filter $\mathfrak{a}$. Then $\hat{p} \in \bar{A}$ implies $\mathfrak{a}^0 \wedge \ulcorner A \urcorner \neq \mathfrak{o}$. Since $\ulcorner A \urcorner$ is a closed filter in (R, τ), we have, by 26.2, $\mathfrak{a} \leqslant \ulcorner A \urcorner$. Hence $\bar{A} \cap \bar{B} = \emptyset$ and there must exist disjoint neighborhoods of $\bar{A}$ and $\bar{B}$ in $\omega(R, \tau)$. Their intersections with R are then disjoint neighborhoods of A and B, and (R, τ) must also be normal. ∎

From now on let $(\hat{R}, \hat{\tau}) = \omega(R, \tau)$ and let us denote by $\mathfrak{a}_{\hat{p}}$ the ultraclosed filter associated uniquely with the point $\hat{p} \in \hat{R}$. Then we have $\operatorname{Tr}(\hat{\tau}\hat{p}) = \mathfrak{a}_{\hat{p}}^0$. The above proof implies also: If A is a closed subset of (R, τ), then $\mathfrak{a}_{\hat{p}} \leqslant \ulcorner A \urcorner$ is a consequence of $\hat{p} \in h_{\hat{\tau}}(A)$. Conversely, if $\mathfrak{a}_{\hat{p}} \leqslant \ulcorner A \urcorner$ then it must follow that

$$\hat{\tau}\hat{p} \wedge \ulcorner A \urcorner \geqslant \mathfrak{a}_{\hat{p}}^0 \wedge \ulcorner A \urcorner \geqslant \mathfrak{a}_{\hat{p}} > 0$$

and hence $\hat{p} \in h_{\hat{\tau}}(A)$. For any two closed subsets A, B of (R, τ) we therefore have $h_{\hat{\tau}}(A \cap B) = h_{\hat{\tau}}(A) \cap h_{\hat{\tau}}(B)$.

For open subsets U of (R, τ) we have (cf. p. 166)

$$\complement_{\hat{R}} s(U) = \complement_{\hat{R}}\{\hat{p} : \mathfrak{a}_{\hat{p}}^0 \leqq \ulcorner U \urcorner\} = \{\hat{p} : \mathfrak{a}_{p}^0 \wedge \ulcorner \complement_R U \urcorner \neq \mathfrak{o}\} = h_{\hat{\tau}}(\complement_R U).$$

Since the sets of the form $s(U)$ constitute a basis of $(\hat{R}, \hat{\tau})$, every closed subset of $(\hat{R}, \hat{\tau})$ is the intersection of sets of the form $h_{\hat{\tau}}(\complement_R U)$, hence also the intersection of sets of form $h_{\tau}(A)$ with sets A closed in (R, τ).

The properties we have just enumerated will be shown to be characteristic of the Wallman compactification.

26.6 *A completely compact T_1-space (R', τ'), containing (R, τ) as subspace, is homeomorphic to the Wallman compactification $\omega(R, \tau) = (\hat{R}, \hat{\tau})$ if and only if it satisfies the following conditions:*

(a) *For any two sets A and B closed in (R, τ), we have*

$$h_{\tau'}(A \cap B) = h_{\tau'}(A) \cap h_{\tau'}(B).$$

(b) *Every closed subset of (R', τ') is the intersection of sets of the form $h_{\tau'}(A)$ where A is a closed set in (R, τ).*

Proof: By what we have previously proved, $(\hat{R}, \hat{\tau})$ itself satisfies properties (a) and (b). Let $\mathfrak{A}$ be the system of all closed subsets of (R, τ). For each $p' \in R'$ we see, by (b) and $(\mathbf{T}_1)$, that

$$\{p'\} = \cap \{h_{\tau'}(A) : A \in \mathfrak{A}, p' \in h_{\tau'}(A)\},$$

and, by (a), the $\{A : A \in \mathfrak{A}, p' \in h_{\tau'}(A)\}$ is a filter basis generating in R a closed filter $\mathfrak{a}_p \neq \mathfrak{o}$. The filter $\mathfrak{a}_{p'}$ is even ultraclosed: For let $\mathfrak{b}$ be a closed filter such that $\mathfrak{o} < \mathfrak{b} \leqslant \mathfrak{a}_{p'}$. Since (R', τ') is completely compact we have

$$\emptyset \neq \cap \{h_{\tau'}(B) : B \in \mathfrak{b}\} \subseteq \cap \{h_{\tau'}(A) : A \in \mathfrak{a}_{p'}\} = \{p\} \qquad (11.19)$$

For every closed set $B \in \mathfrak{b}$ we consequently have $p' \in h_{\tau'}(B)$, then $B \in \mathfrak{a}_{p'}$, and finally $\mathfrak{b} = \mathfrak{a}_{p'}$. From (b) we deduce that if $p' \neq q'$ there exists an $A \in \mathfrak{A}$ such that $p' \in h_{\tau'}(A)$ and $q' \notin h_{\tau'}(A)$, whence $\mathfrak{a}_{p'} \neq \mathfrak{a}_{q'}$. If, conversely, $\mathfrak{a}$ is an ultraclosed filter of R, then there exists a $p' \in R'$ such that $p' \in \cap\{h_{\tau'}(A) : A \in \mathfrak{a}\}$. This implies $\mathfrak{a}_{p'} \leqslant \mathfrak{a}$ and hence $\mathfrak{a}_{p'} = \mathfrak{a}$ (26.1). The mapping $p' \to \mathfrak{a}_{p'}$ is therefore 1-1, from R' onto the set of all ultraclosed filters of R. But this last mentioned set is in 1-1 correspondence with the points $\hat{p} \in \hat{R}$, and hence there exists a 1-1 mapping φ from R' onto the Wallman compactification, φ reducing to the identity on R. For $A \in \mathfrak{A}$, the condition $p' \in h_{\tau'}(A)$ is equivalent to $\mathfrak{a}_{p'} \leqslant \overline{A}$, which condition is in turn equivalent to $\varphi p' \in h_{\hat{\tau}}(A)$. For all $A \in \mathfrak{A}$ we therefore have $\varphi h_{p'}(A) = h_{\hat{\tau}}(A) = h_{\hat{\tau}}(\varphi A)$. By (b), this implies that $\varphi h_{\tau'}(M') = h_{\tau}(\varphi M')$ for all subsets M' of R' and φ is therefore a homeomorphism of (R', τ') onto $(\hat{R}, \hat{\tau})$. ∎

Notes and Exercises

26A *Exercise:* In a completely compact space every ultraclosed filter is a principal filter. Every completely compact T_1-space is its own Wallman compactification.

26B Let (R, τ) be a T_1-space. For every natural number $n \geqslant 2$ let S_n^* be the set of all filters of R satisfying the following property: Given any n closed subsets of (R, τ) whose intersection is empty, then at least one of them does not belong to $\mathfrak{a}$.

Exercises

1. The system T_n^* of all maximal filters in S_n^* is a trace system.
2. The T_n^*-extension of (R, τ) is a completely compact T_1-space.
3. $T^* = \cap \{T_n^* : n = 2, 3, ...\}$ is the trace system of the Wallman extension $\omega(R, \tau)$. Hence $\omega(R, \tau)$ is a subspace of all T_n^*-extensions of (R, τ) (cf. 25B).

27. The Stone-Čech Compactification

In this section we shall study a type of extension of a space which is intimately linked to the extension of continuous maps. As a special case we shall obtain an especially practical and important type of compactification.

Let (R, τ) denote a space which can be probed by the T_3-space (R', τ'). By 18.2, τ must be a T_3-topology also. Let C be the set of all continuous mappings from (R, τ) into (R', τ') and let Λ be the set of all mappings of C into R'. For each $\lambda \in \Lambda$ and $\varphi \in C$ we see that $\varphi^{-1}\tau'\lambda\varphi$ and also, by 5.4,

$$\mathfrak{a}_\lambda = \bigwedge \{\varphi^{-1}\tau'\lambda\varphi : \varphi \in C\}$$

are both open filters of (R, τ). For the mapping $\epsilon_p \in \Lambda$ defined by $\epsilon_p\varphi = \varphi p$ ($p \in R$, cf. p. 156) we then have, by 18.1,

$$\mathfrak{a}_{\epsilon_p} = \bigwedge \{\varphi^{-1}\tau'\varphi p : \varphi \in C\} = \tau p.$$

If Λ_0 denotes the set of all $\lambda \in \Lambda$ such that $\mathfrak{a}_\lambda \neq \mathfrak{o}$, then

$$T^* = \{\mathfrak{a}_\lambda : \lambda \in \Lambda_0\}$$

is a trace system. Hence there exists a T^*-extension of (R, τ) which we shall henceforth always denote by $(\hat{R}, \hat{\tau})$.

To each point $\hat{p} \in \hat{R}$ we can uniquely associate the map $\lambda_{\hat{p}} \in \Lambda_0$ stipulated by the condition $\mathrm{Tr}\,(\hat{\tau}\hat{p}) = \mathfrak{a}_{\lambda\hat{p}}$. If $\hat{\varphi}$ is a continuous mapping of $(\hat{R}, \hat{\tau})$ into (R', τ'), then φ, the restriction of $\hat{\varphi}$, is a mapping belonging to C. Since $\varphi\,\mathrm{Tr}\,(\hat{\tau}\hat{p}) \leqslant \tau'\lambda_{\hat{p}}\varphi$ and $\varphi\,\mathrm{Tr}\,(\hat{\tau}\hat{p}) \leqslant \tau'\hat{\varphi}\hat{p}$, we have $\tau'\lambda_{\hat{p}}\varphi \wedge \tau'\hat{\varphi}\hat{p} \neq \mathfrak{o}$. By 8.6, this implies $\tau'\lambda_{\hat{p}}\varphi = \tau'\hat{\varphi}\hat{p}$.

27.1 *Every $\varphi \in C$ possesses a continuous extension to the space $(\hat{R}, \hat{\tau})$. The space $(\hat{R}, \hat{\tau})$ is a T_3-space and it can be probed by (R', τ'). If we associate to each $\varphi \in C$ its unique extension $\hat{\varphi}$, then $\{\hat{\varphi} : \varphi \in C\}$ is a probing system.*

Proof: Since $\varphi\,\mathrm{Tr}\,(\hat{\tau}\hat{p}) \leqslant \tau'\lambda_{\hat{p}}\varphi$, the images of all filters in T^* must themselves be convergent filters. By 25.2 every $\varphi \in C$ must therefore possess a continuous extension $\hat{\varphi}$ to $(\hat{R}, \hat{\tau})$. This extension is, in general, not unique, but we suppose that it has been fixed once for all. The sets $s(U) = \{\hat{q} : \hat{q} \in \hat{R}, \operatorname{Tr}(\hat{\tau}\hat{q}) \leqslant \overline{U}\}$ where $U \in \operatorname{Tr}(\hat{\tau}\hat{p})$ constitute a basis of $\hat{\tau}\hat{p}$ (cf. p. 166). The condition $U \in \operatorname{Tr}(\hat{\tau}\hat{p})$ itself implies the existence of a finite number of mappings $\varphi_1, \ldots, \varphi_n \in C$ and a finite number of open neighborhoods $U'_\nu \in \tau'\lambda_{\hat{p}}\varphi_\nu (\nu = 1, \ldots, n)$ such that $\varphi_1^{-1}U'_1 \cap \ldots \cap \varphi_n^{-1}U'_n \subseteq U$. From $\hat{q} \in \hat{\varphi}_1^{-1}U'_1 \cap \ldots \cap \hat{\varphi}_n^{-1}U'_n$ we then deduce $\operatorname{Tr}(\hat{\tau}\hat{q}) \leqslant \overline{\varphi_1^{-1}U'_1} \wedge \ldots \wedge \overline{\varphi_n^{-1}U'_n} \subseteq \overline{U}$, hence $\hat{q} \in s(U)$. We then have

$$\hat{\varphi}_1^{-1}U'_1 \cap \ldots \cap \hat{\varphi}_n^{-1}U'_n \subseteq s(U)$$

and thus

$$\hat{\tau}\hat{p} = \bigwedge \{\hat{\varphi}^{-1}\tau'\lambda_{\hat{p}}\varphi : \varphi \in C\} = \bigwedge \{\hat{\varphi}^{-1}\tau'\hat{\varphi}\hat{p} : \varphi \in C\}.$$

By 18.1, $(\hat{R}, \hat{\tau})$ can therefore be probed by (R', τ'), with $\{\hat{\varphi} : \varphi \in C\}$ being a probing system. By 18.2 $(\hat{R}, \hat{\tau})$ is also a T_3-space. ∎

27.2 *If (R', τ') is completely compact, so is $(\hat{R}, \hat{\tau})$. If axiom* $(\mathbf{T}_2)$ *holds in (R, τ) and (R', τ'), then $(\hat{R}, \hat{\tau})$ is also a T_2-space.*

Proof: Let (R', τ') be completely compact. For each $\varphi \in C$ let $\hat{\varphi}$ be a continuous extension of φ to $(\hat{R}, \hat{\tau})$ (27.1). If $\hat{\mathfrak{v}}$ is an ultrafilter of $\hat{R}$, then for every $\varphi \in C$, $\hat{\varphi}\hat{\mathfrak{v}}$ converges to a point p'_φ of R'. We can define a mapping $\lambda \in \Lambda$ by $\lambda\varphi = p'_\varphi$ and then $\hat{\mathfrak{v}} \leqslant \bigwedge \{\hat{\varphi}^{-1}\tau'\lambda\varphi : \varphi \in C\}$. This last filter is open in $(\hat{R}, \hat{\tau})$ and different from $\mathfrak{o}$. Since R is dense in $(\hat{R}, \hat{\tau})$, the same thing is true for the trace $\mathfrak{a}_\lambda$ of this filter. Consequently $\lambda \in \Lambda_0$ and $\mathfrak{a}_\lambda$ represents a point $\hat{p} \in \hat{R}$. Since $\bigwedge \{\hat{\varphi}^{-1}\tau'\lambda\varphi : \varphi \in C\} = \bigwedge \{\hat{\varphi}^{-1}\tau'\hat{\varphi}\hat{p} : \varphi \in C\} = \hat{\tau}\hat{p}$ we have $\hat{\mathfrak{v}} \leqslant \hat{\tau}\hat{p}$ and hence $\hat{\mathfrak{v}}$ converges. Hence $(\hat{R}, \hat{\tau})$ is likewise completely compact.

Let us now suppose that axiom $(\mathbf{T}_2)$ holds in (R, τ) and (R', τ'). Given any two distinct mappings $\lambda_1, \lambda_2 \in \Lambda_0$ we can find a $\varphi \in C$ such that $\lambda_1\varphi \neq \lambda_2\varphi$, and thus $\mathfrak{a}_{\lambda_1} \wedge \mathfrak{a}_{\lambda_2} \leqslant \varphi^{-1}(\tau'\lambda_1\varphi \wedge \tau'\lambda_2\varphi) = \mathfrak{o}$. By 25.3, $(\hat{R}, \hat{\tau})$ is also a T_2-space. ∎

The space $(\hat{R}, \hat{\tau})$ possesses other extension properties.

27.3 *Suppose that the completely compact space (R'', τ'') can also be probed by (R', τ'). Then every continuous mapping $\psi : (R, \tau) \to (R'', \tau'')$ possesses a continuous extension to $(\hat{R}, \hat{\tau})$.*

Proof: By 18.2, (R'', τ'') is a T_3-space. For every continuous mapping φ'' of (R'', τ'') into (R', τ'), $\varphi = \varphi''\psi$ is a mapping onto C. For the fixed point $\hat{p} \in \hat{R}$ we have $\varphi''\psi \operatorname{Tr}(\hat{\tau}\hat{p}) = \varphi \operatorname{Tr}(\hat{\tau}\hat{p}) \leqslant \tau'\lambda_{\hat{p}}\varphi$. The filter $\psi \operatorname{Tr}(\hat{\tau}\hat{p})$ possesses an adherent point p'' in (R'', τ''). Since $\tau''p'' \wedge \psi \operatorname{Tr}(\hat{\tau}\hat{p}) \neq \mathfrak{o}$, we must also have $\tau'\varphi''p'' \wedge \tau'\lambda_{\hat{p}}\varphi \geqslant \varphi''\tau''p'' \wedge \varphi''\psi \operatorname{Tr}(\hat{\tau}\hat{p}) \neq \mathfrak{o}$, and hence by 8.6, $\tau'\varphi''p'' = \tau'\lambda_{\hat{p}}\varphi$. Since (R'', τ'') can be probed by (R', τ') we deduce from $\varphi''\psi \operatorname{Tr}(\hat{\tau}\hat{p}) \leqslant \tau'\varphi''p''$ that $\psi \operatorname{Tr}(\hat{\tau}\hat{p})$ converges to p''. Our contention then follows from 25.2. ∎

An important example of extension $(\hat{R}, \hat{\tau})$ occurs when (R', τ') is the unit interval [0, 1] on the real line. If we further assume that (R, τ) is a T_1-space then it will also be completely regular—with these hypotheses the extension $(\hat{R}, \hat{\tau})$ is known as the *Stone-Čech extension or compactification* and designated by $\mathfrak{B}(R, \tau)$. The use of the word compactification is justified by the next theorem, which is a direct consequence of previous results.

27.4 *The Stone-Čech extension or compactification* $\mathfrak{B}(R, \tau)$ *of a completely regular space* (R, τ) *is a completely compact and normal space* (27.1, 27.2). *Every continuous mapping from* (R, τ) *into* [0, 1] *possesses a unique continuous extension to* $\beta(R, \tau)$ (27.3).

27.5 *Let the completely compact* T_2*-space* $(\tilde{R}, \tilde{\tau})$ *be an extension of* (R, τ). *Then there exists a continuous mapping from* $\beta(R, \tau)$ *onto* $(\tilde{R}, \tilde{\tau})$ *which reduces to the identity map on R. The space* $(\tilde{R}, \tilde{\tau})$ *is homeomorphic to* $\beta(R, \tau)$ *if and only if every continuous mapping from* (R, τ) *into* [0, 1] *possesses a continuous extension to* $(\tilde{R}, \tilde{\tau})$; *hence there must even exist a homeomorphism between these two spaces which reduces to the identity mapping on R.*

Proof: Following our previous usage, let τ' denote the topology on [0, 1] and let C denote the set of all continuous mappings of (R, τ) into [0, 1].

The identity on R is a continuous mapping from (R, τ) into $(\tilde{R}, \tilde{\tau})$. Since $(\tilde{R}, \tilde{\tau})$ is a completely compact T_2-space it is also a T_{3a}-space (18.12), and hence, by 27.4, there exists a continuous extension $\hat{\psi}$ to $\beta(R, \tau) = (\hat{R}, \hat{\tau})$. The set $\hat{\psi}\hat{R}$ is completely compact (19.1) and, by $(\mathbf{T}_2)$, also closed (11.7). Since R is dense in $(\tilde{R}, \tilde{\tau})$, we have $\hat{\psi}\hat{R} \supseteq h_{\tilde{\tau}}(R) = \tilde{R}$, whence $\hat{\psi}\hat{R} = \tilde{R}$.

Let us now assume that every $\varphi \in C$ possesses a continuous extension $\tilde{\varphi}$ onto $(\tilde{R}, \tilde{\tau})$. If $\tilde{x}$ is a point in $\tilde{R}$, then $\lambda\varphi = \tilde{\varphi}\tilde{x}$ defines a mapping $\lambda \in \Lambda$. We have

$$\operatorname{Tr}(\tilde{\tau}\tilde{x}) \leqslant \bigwedge \{\varphi^{-1}\tau'\tilde{\varphi}\tilde{x} : \varphi \in C\} = \bigwedge \{\varphi^{-1}\tau'\lambda\varphi : \varphi \in C\} = \mathfrak{a}_\lambda .$$

From $\hat{\psi}\hat{p} = \hat{\psi}\hat{q} = \tilde{x}$ it follows (for $\hat{\psi}$ is the identity mapping on R) that $\operatorname{Tr}(\hat{\tau}\hat{p}) \leqslant \operatorname{Tr}(\tilde{\tau}\tilde{x}) \leqslant \mathfrak{a}_\lambda$ and $\operatorname{Tr}(\hat{\tau}\hat{q}) \leqslant \operatorname{Tr}(\tilde{\tau}\tilde{x}) \leqslant \mathfrak{a}_\lambda$. By $(\mathbf{T}_2)$ and 25.3 we now have $\operatorname{Tr}(\hat{\tau}\hat{p}) = \mathfrak{a}_\lambda = \operatorname{Tr}(\hat{\tau}\hat{q})$ and hence $\hat{p} = \hat{q}$, which implies that $\hat{\psi}$ is a 1-1 mapping. By 19.2, $\hat{\psi}$ must even be a homeomorphism. ∎

To terminate, in the case of a completely regular space (R, τ) we shall examine the relationship between the Stone-Čech compactification $\beta(R, \tau)$ and the Wallman compactification $\omega(R, \tau)$.

27.6 *There exists a continuous mapping of* $\omega(R, \tau)$ *onto* $\beta(R, \tau)$, *which reduces to the identity mapping on R. The spaces* $\omega(R, \tau)$ *and* $\beta(R, \tau)$ *are homeomorphic if and only if* (R, τ) *is normal; there even exists a homeomorphism between these two spaces which reduces to the identity mapping on R.*

Proof: Since $\beta(R, \tau)$ is a completely compact T_2-space, the identity of R, being a continuous mapping from (R, τ) into $\beta(R, \tau)$, must possess a continuous extension φ to $\omega(R, \tau)$ (26.4). The φ-image of $\omega(R, \tau)$ in $\beta(R, \tau)$ is completely compact (19.1) and, by (T_2), also closed (11.7). Since this image contains R it must coincide with $\beta(R, \tau)$.

By 26.4, every continuous mapping from (R, τ) into $[0, 1]$ possesses a continuous extension to $\omega(R, \tau)$. If (R, τ) is normal, then $\omega(R, \tau)$ is a completely compact T_2-space (26.5) and hence, by 27.5, homeomorphic to $\beta(R, \tau)$. Conversely, if $\omega(R, \tau)$ and $\beta(R, \tau)$ are homeomorphic, then 27.4 implies that $\omega(R, \tau)$ is normal and 26.5 implies that (R, τ) is normal. ∎

For further characterizations of the Stone-Čech compactification, cf. 27A and 28.5.

Notes and Exercises

27A (cf. 18B) Let $B(R, \tau)$ be the ring of all bounded, continuous, real-valued mappings on a completely regular space (R, τ).[1] Let $\hat{B}(R, \tau)$ be the analogous ring on the Stone-Čech compactification $\beta(R, \tau)$.

Exercises

1. The rings $B(R, \tau)$ and $\hat{B}(R, \tau)$ are isomorphic.

2. There exists a 1-1 mapping of $\beta(R, \tau)$ onto the set of all maximal ideals of $B(R, \tau)$.

3. For each $\hat{p} \in \beta(R, \tau)$ let $I(\hat{p})$ be the maximal ideal in $B(R, \tau)$ associated with $\hat{p}$ in (2) above. If M is a subset of $\beta(R, \tau)$, then $\hat{p} \in \bar{M}$ is equivalent to $\cap \{I(\hat{q}) : \hat{q} \in M\} \subseteq I(\hat{p})$.

27B Let (R, τ) be the space of all rational numbers under the topology (order topology).

Exercises

1. (R, τ) can be probed by the discretely topologized set $\{0, 1\}$. Hence there exists an extension $(\hat{R}, \hat{\tau})$ with respect to $\{0, 1\}$ (cf. the beginning of this section).

2. $(\hat{R}, \hat{\tau})$ is homeomorphic to the Stone-Čech extension $\beta(R, \tau)$.

28. Embeddings

If the space (R, τ) is homeomorphic to a subspace of the space (R', τ'), we say that (R, τ) is *embeddable* in (R', τ'). Any homeomorphism of (R, τ) onto a subspace of (R', τ') is called an *embedding mapping*, or simply an *embedding*.

[1] We say that a mapping of R into the reals is *bounded* if φR is a bounded subset of the real line.

Embedding theorems serve to characterize new spaces by establishing under what conditions these new spaces can be embedded in certain standard, well-known spaces.

We remind the reader that $(R', \tau')^I$ was defined as the set of all mapping from I into R', this set of mappings being topologized by pointwise convergence; i.e., $(R', \tau')^I$ is the product space of $|I|$ copies of the space (R', τ'). We shall call $(R', \tau')^I$ the *I-cube* over the space (R', τ'). If I_1 and I_2 are two sets of indices having the same cardinality, then the I_1-cube and the I_2-cube over any space must be homeomorphic. Hence we can also speak of the *k-cube*, where k is the cardinality of I. The following results refer to embeddability into such cubes.

28.1 *Suppose the T_1-space (R, τ) can be probed by the space (R', τ'), and let Φ be a probing system. Then (R, τ) can be embedded in the Φ-cube over (R', τ'), and $\psi = \times \{\varphi : \varphi \in \Phi\}$ is a map effectuating this embedding.*

Proof: By 22.9, $\psi = \times \{\varphi : \varphi \in \Phi\}$ is a continuous mapping of (R, τ) into $(R', \tau')^\Phi$. By $(\mathbf{T}_1)$, for two distinct points $p, q \in R$ we must have $\overline{p} \wedge \tau q = \mathfrak{o}$. Hence there exists a $\varphi \in \Phi$ such that $\overline{\varphi p} \wedge \tau' \varphi q = \mathfrak{o}$ (18.1), whence $\varphi p \neq \varphi q$ and also $\psi p \neq \psi q$. Thus ψ is a 1-1 mapping. To each "index" $\varphi \in \Phi$ there corresponds a projection mapping π_φ. We have $\varphi = \pi_\varphi \psi$ and $\psi \varphi^{-1} p' = \psi R \cap \pi_\varphi^{-1} p'$. If we let τ'^Φ denote the topology on $(R', \tau')^\Phi$, we then have

$$\psi \tau p = \psi \bigwedge_{\varphi \in \Phi} \varphi^{-1} \tau' \varphi p = \bigwedge_{\varphi \in \Phi} \psi \varphi^{-1} \tau' \varphi p = \overline{\psi R} \wedge \bigwedge_{\varphi \in \Phi} \pi_\varphi^{-1} \tau' \pi_\varphi \psi p = \overline{\psi R} \wedge \tau'^\Phi \psi p.$$

Hence ψ is even a homeomorphism of (R, τ) onto the subspace of (R', τ') determined by ψR.

Conversely, the k-cube over (R', τ') can be probed by (R', τ'): The projection mappings constitute a probing system of cardinality k (22.6). Every subspace of the k-cube and every space homeomorphic to such a subspace, i.e., every space embeddable in the k-cube, can therefore also be probed by (R', τ') and admits of a probing system of cardinality $\leqslant k$. ∎

28.2 *Suppose the T_1-space (R, τ) can be probed by the space (R', τ') and suppose also that* Bd $(R, \tau) = k \geqslant \aleph_0$. *Then (R, τ) can be embedded in the k-cube over (R', τ'). If, in addition,* Bd $(R, \tau) >$ Bd (R', τ'), *then k is the smallest cardinal number with this property.*

Proof: By 18.3, (R, τ) possesses a probing system Φ with $|\Phi| \leqslant k$. By 28.1, (R, τ) is embeddable in the Φ-cube, and hence certainly in the k-cube, over (R', τ'). Conversely, if (R, τ) is embeddable in the k'-cube over (R', τ'), then by 28.1 there exists a probing system Φ' with $|\Phi'| \leqslant k'$. By 18.3 we have

$$\aleph_0 \leqslant k = \mathrm{Bd}\,(R, \tau) \leqslant |\Phi'| \cdot \mathrm{Bd}\,(R', \tau') = \max\{|\Phi'|, \mathrm{Bd}\,(R', \tau')\}.$$

If, in addition, Bd $(R, \tau) >$ Bd (R', τ'), then $k \leqslant |\Phi'| \leqslant k'$. ∎

If (R, τ) is the unit interval $[0, 1]$ in the real line, we have the following special result.

28.3 *A space* (R, τ) *is completely regular if and only if it can be embedded in a cube over* $[0, 1]$; *if so, it can then be embedded in the k-cube, where* $k = \mathrm{Bd}\,(R, \tau)$. *If* $\mathrm{Bd}\,(R, \tau) > \aleph_0$, *then k is the smallest cardinal possessing this property.*

Proof: The space $[0, 1]$ is completely regular, as is every cube over $[0, 1]$ and every subspace of such a cube (22.6). Conversely, every completely regular space is a T_1-space and can be probed by $[0, 1]$. Hence our first contention follows from 28.1. A T_1-space with a finite basis is itself finite and discretely topologized; it can therefore be embedded in the 1-cube over $[0, 1]$. If, on the other hand, $\mathrm{Bd}\,(R, \tau) \geqslant \aleph_0$, then (R, τ) is, by 28.2, embeddable in the k-cube over $[0, 1]$, where $k = \mathrm{Bd}\,(R, \tau)$. But the interval $[0, 1]$ possesses a countable basis and thus our last contention follows from 28.2. ∎

We know that in a space with countable basis the notions of countable regularity and of regularity become identical (10.5, 18.4). Hence the above theorem also implies

28.4 (Urysohn's Embedding Theorem) *A space with countable basis is regular if and only if it can be embedded in the* $\aleph_0$*-cube over* $[0, 1]$.

Since the $\aleph_0$-cube over $[0, 1]$ is homeomorphic to the unit cube of real Hilbert Space (22A), we can also state Urysohn's Embedding Theorem in the language of the real Hilbert cube.

Theorem 28.1 can be used to obtain a new characterization of the Stone-Čech compactification: Let the space (R, τ) be completely regular, and let C be the set of all continuous mappings of (R, τ) into $[0, 1]$. Then $\psi = \times\, (\varphi : \varphi \in C\}$ is an embedding of (R, τ) into the C-cube W over $[0, 1]$, and the closure $\overline{\psi R}$ taken in W is a subspace $(\tilde{R}, \tilde{\tau})$ of W.

28.5 *The Stone-Čech extension* $\beta(R, \tau)$ *is homeomorphic to* $(\tilde{R}, \tilde{\tau})$.

Proof: By 27.4, each $\varphi \in C$ possesses a unique continuous extension $\hat{\varphi}$ to $\beta(R, \tau)$. Hence $\hat{C} = \{\hat{\varphi} : \varphi \in C\}$ is identical with the set of all continuous mappings from $\beta(R, \tau)$ into $[0, 1]$ and is thereby a probing system. Now $|\hat{C}| = |C|$, and by 28.1, $\hat{\psi} = \times\, \{\hat{\varphi} : \varphi \in C\}$ is both an embedding of $\beta(R, \tau)$ into W and an extension of ψ. The image space $\hat{\psi}\beta(R, \tau)$ is a completely compact (19.1) and closed subspace of W, because W is a T_2-space (11.7). Since R is dense in $\beta(R, \tau)$, so is ψR dense in $\hat{\psi}\beta(R, \tau)$ and thus $\hat{\psi}\beta(R, \tau) = (\tilde{R}, \tilde{\tau})$. ∎

Because of its relatively well-known structure the k-cube over $[0, 1]$ is an especially important, standard space useful in embedding of completely regular spaces. Nevertheless, in such problems it is often more useful to use a "cube-basis" more general than $[0, 1]$, thereby reducing the dimension k of the cube in question. The basis spaces we have in mind are, intuitively speaking, copies of the unit interval [0, 1], "emanating from a fixed point, in the form of a star." We can describe this form of space more precisely as follows: Let I be an indexing set and for each $\iota \in I$ let $(E_\iota, \tau_\iota) = [0, 1]$. We can define an equivalence over the sum set $\sum\{E_\iota : \iota \in I\}$ if we identify, i.e., place in one equivalence class, all the zero points of all the intervals E_ι. The resulting quotient space $\sum(E_\iota, \tau_\iota)/\rho$ is known as the *star space* associated with the indexing set I. The star space contains as subspaces all the intervals E_ι, those subspaces being known as *rays* of the star space. Every star space is evidently a T_2-space. We also have

28.6 *Every star space possesses an $\aleph_0$-discrete basis* (13e).

Proof: The interval [0, 1] possesses a countable basis $\mathfrak{B}$. Every set $B \in \mathfrak{B}$ can also be thought of as a ray E_ι—when we think of B as a ray we shall henceforth denote it by B_ι. If $0 \notin B$, then $\mathfrak{C}_B = \{B_\iota : \iota \in I\}$ is a discrete system of open sets of the star space. If, however, $0 \in B$, then let us suppose that $\mathfrak{C}_B$ is constituted by the one set $\cup B_\iota$, $\cup B_\iota$ being, of course, an open set in the star space. Then $\mathfrak{C} = \cup(\mathfrak{C}_B : B \in \mathfrak{B}\}$ is in either case a basis of the star space, and since $\mathfrak{B}$ is countable, $\mathfrak{C}$ is an $\aleph_0$-discrete system of sets. ∎

28.7 *A normal space (R, τ) can be embedded in the k-cube ($k \geqslant \aleph_0$) over a star space if and only if (R, τ) possesses a k-discrete basis.*

Proof: Let I be an indexing set such that $|I| = k$, let W be the k-cube over the star space S, and let $\pi_\iota(\iota \in I)$ be the projection mappings from W into S. By 28.6, S possesses a basis which is the union of a countable number of discrete systems of set $\mathfrak{B}_n$. For the fixed indices $n_1, \ldots, n_r$ and $\iota_1, \ldots, \iota_r \in I$ the sets of the form $\pi_{\iota_1}^{-1}B_{n_1} \cap \ldots \cap \pi_{\iota_r}^{-1}B_{n_r}$, where $B_{n_\rho} \in \mathfrak{B}_{n_\rho}$, constitute a discrete system in W. The union of all these systems is a basis of W, which, since $k \geqslant \aleph_0$, is also a k-discrete system of sets. Hence every space embeddable in W is also a space with k-discrete basis.

Conversely, let $\mathfrak{B}$ be a basis of the normal space (R, τ) such that $\mathfrak{B}$ equals the union of the discrete systems $\mathfrak{B}_\iota = (B_\kappa^\iota : \kappa \in K_\iota\}$, $\iota \in I$ and $|I| = k$. For any two indices $\iota_1, \iota_2 \in I$ and for any index $\kappa \in K_{\iota_1}$, let

$$V_{\iota_1,\iota_2}^\kappa = \cup\{\bar{B} : B \in \mathfrak{B}_{\iota_2}, \bar{B} \subseteq B_\kappa^{\iota_1}\}.$$

Since $\mathfrak{B}_{\iota_2}$ is a discrete system of sets, the sets $V_{\iota_1,\iota_2}^\kappa$ are closed by 13.1. Hence there exists a continuous mapping: $\varphi_{\iota_1,\iota_2}^\kappa : (R, \tau) \to [0, 1]$, taking the value 1 on $V_{\iota_1,\iota_2}^\kappa$ and the value 0 on $\complement B_\kappa^{\iota_1}$ (18.6). Now let S denote the star space associated with the indexing set $\cup K_\iota$. For each fixed couple (ι_1, ι_2) and any $p \in R$, there exists at most one index $\kappa \in K_{\iota_1}$, such that $\varphi_{\iota_1,\iota_2}^\kappa p \neq 0$. We can define a mapping $\psi_{\iota_1,\iota_2} : R \to S$ by the following rule: If $\varphi_{\iota_1,\iota_2}^\kappa p \neq 0$ for the index $\kappa \in K_\iota$, let $\psi_{\iota_1,\iota_2}p$ be the point in the ray E_κ of S stipulated by $\varphi_{\iota_1,\iota_2}^\kappa p$. If such an index κ does not exist, let $\psi_{\iota_1,\iota_2}p$ be the common zero point of all the rays of S. Since each $p \in R$ possesses a neighborhood U where U has a nonempty intersection with at most one set belonging to $\mathfrak{B}_\iota$, the mapping ψ_{ι_1,ι_2} must agree on U with one of the continuous mappings $\varphi^\kappa_{\iota_1,\iota_2}$ (conceived of as a mapping into the ray E_κ), and hence ψ_{ι_1,ι_2} must also be continuous. If now $\mathfrak{v}$ is an ultrafilter of R such that $\overline{\mathfrak{v} \wedge \tau p} = \mathfrak{o}$, then there exists an ι_1, and a $B^{\iota_1}_\kappa \in \mathfrak{B}_{\iota_1}$ with $p \in B^{\iota_1}_\kappa$ and $\mathfrak{v} \leqslant \overline{\complement B^{\iota_1}_\kappa}$. Then there also exists an ι_2 and a $B \in \mathfrak{B}_{\iota_2}$ such that $p \in B$ and $\bar{B} \subseteq B^{\iota_1}_\kappa$. Consequently $p \in V^\kappa_{\iota_1,\iota_2}$, and $\psi_{\iota_1,\iota_2}p$ is the point 1 in the ray E_κ. But since $\mathfrak{v} \leqslant \overline{\complement B^{\iota_1}_\kappa}$ we must have $\varphi^\kappa_{\iota_1,\iota_2}\mathfrak{v} = \overline{\mathfrak{o}}$ and $\psi_{\iota_1,\iota_2}\mathfrak{v}$ cannot converge to $\psi_{\iota_1,\iota_2}p$. Hence $\Psi = \{\psi_{\iota_1}, \psi_{\iota_2} : \iota_1, \iota_2 \in I\}$ is a probing system of continuous mappings from (R, τ) into the star space $S(18.1)$ with $|\Psi| = k$. By 28.1, we see that (R, τ) can be embedded in the k-cube over S. ∎

We cannot deduce that (R, τ) is normal even though we know it can be embedded in a cube over a star space. Nevertheless, for $\aleph_0$-cubes we can sharpen the previous theorem to obtain a result which will be decisive for the characterization of metric spaces (cf. Section 31).

28.8 *A space (R, τ) is embeddable in the $\aleph_0$-cube over a star space if and only if it is regular and possesses an $\aleph_0$-discrete basis.*

Proof: Every star space is evidently regular. The same holds then for the $\aleph_0$-cube over a star space (22.6). Every space embeddable in such a cube is therefore also regular (19.4) and possesses, by 28.7, an $\aleph_0$-discrete basis.[2] Conversely let (R, τ) be a regular space with $\aleph_0$-discrete basis $\mathfrak{B}$. Then every open covering of (R, τ) possesses an $\aleph_0$-discrete refinement contained in $\mathfrak{B}$. Hence by 13.11 (R, τ) must be paracompact, and by 13.3 it must also be normal. Our contention now follows from 28.7. ∎

Notes and Exercises

28A *Exercise:* The space of all rational numbers under the ordinary topology is embeddable in the $\aleph_0$-cube over the discretely topologized set $\{0, 1\}$ and hence also embeddable in the Cantor discontinuum (22.III).

28B *Exercise:* Every regular space with $\aleph_0$-discrete basis is even completely normal. But the k-cube over $[0, 1]$ where $k > \aleph_0$ is not normal.

(The first contention of this exercise is a direct consequence of the metrizability theorem 31.4, to be proved later; it can also be shown directly.)

[2] In 28.7 we did not need normality to prove this contention.

29. Representation Theorems

Alongside of embeddings as a way of characterizing spaces we have another powerful general method—that of expressing a given space as being homeomorphic to a quotient of a standard space.

Definition 29a: *A space* (R', τ') *is said to be a* representation *of the space* (R, τ) *whenever there exists a continuous mapping* φ *of* (R, τ) *onto* (R', τ') *such that* τ' *is identical to the quotient topology generated by* φ. *In this case* φ *is also known as the* representation mapping.

If (R', τ') is a representation of (R, τ) with representation map φ, then, by 21.8, we have the homeomorphism $(R', \tau') \simeq (R, \tau)/\tilde{\varphi}$. Conversely, if ρ is an equivalence relation over R, then $(R, \tau)/\rho$ is a representation of (R, τ), the cannonical map serving as representation map.

29.1 *Suppose the space* (R, τ) *is completely compact and that* φ *is a continuous mapping of* (R, τ) *onto the* T_2*-space* (R', τ'). *Then* (R', τ') *is a representation of* (R, τ) *with* φ *as representation mapping.*

Proof: By 19.2, φ is both a continuous and closed mapping. Hence, by 21.6, τ' must be identical with the quotient topology generated by φ. ∎

29.2 *Let* W *be the* $\aleph_0$*-cube over the discretely topologized set* $\{0, 1\}$. *Then the* $\aleph_0$*-cube over* $[0, 1]$, *like every closed subset of* W *itself, is a representation of* W.

Proof: Let $\pi_n (n \in \mathbf{N})$ be the projection mappings of W onto $\{0, 1\}$. For each $x \in W$ let

$$\xi_n = \sum_{\nu=1}^{\infty} 2^{-\nu} \pi_{2^{n-1}(2\nu-1)} x \qquad (n \in \mathbf{N})$$

then $\varphi x = \times \{\xi_n : n \in \mathbf{N}\}$ defines a mapping of W onto the $\aleph_0$-cube over $[0, 1]$. Let x be a fixed point of W. Then the set of all $y \in W$ whose first $2^{n-1}(2k - 1)$ coordinates agree with those of x forms a neighborhood U of x. Hence φ is also a continuous mapping. Since the $\aleph_0$-cube over $[0, 1]$ is a T_2-space, by 29.1 it must be a representation of W.

Let us now suppose that A is a closed subset of W. If p is one of the two numbers, 0 or 1, let $p_c = 1 - p$. For fixed $x \in W$, let us define $y_n = \pi_n x$ provided there exists an $a \in A$ such that $\pi_\nu a = y_\nu$ for $\nu < n$ and such that $\pi_n a = \pi_n x$. Otherwise, put $y_n = (\pi_n x)_c$. We can then define a mapping φ of W into W by $\varphi x = \times (y_n : n \in \mathbf{N}\}$. From the construction of φ it follows that for each $n \in \mathbf{N}$ the image point φx has the same first n coordinates as some point $a \in A$. Since A is closed, we must therefore have $\varphi x \in A$. And since $\varphi a = a$ for all $a \in A$, φ must be a mapping from W onto A. If two points have the same first n coordinates, then the same holds true for their images. Hence we finally have shown that φ is continuous, and our second contention now also follows from 29.1. ∎

Representation results can be used to completely characterize all spaces which can effectively be represented as quotients of a given standard space. The following result is such a characterization, the standard space being the Cantor discontinuum.

29.3 *A T_2-space (R, τ) is a representation of the Cantor discontinuum (and hence also a representation of the $\aleph_0$-cube over the discretely topologized set $\{0, 1\}$) if and only if it possesses a countable basis and is completely compact.*

Proof: By 22.III we know that the $\aleph_0$-cube W over $\{0, 1\}$ is homeomorphic to the Cantor discontinuum, and hence any representation of one of these homeomorphic spaces is also a representation of the other.

Since W is completely compact, if (R, τ) is a representation of W then (R, τ) must be completely compact also (19.1). By 22.3,

$$\mathfrak{B} = \{\pi_n^{-1}0 : n \in \mathbf{N}\} \cup \{\pi_n^{-1}1 : n \in \mathbf{N}\}$$

is a subbasis of W. Since $|\mathfrak{B}| \leqslant \aleph_0$, using 19.7 we see that

$$\mathrm{Bd}\,(R, \tau) \leqslant \mathrm{Bd}\,(W) \leqslant \aleph_0 .$$

Conversely, let (R, τ) be a completely compact T_2-space with countable basis. By 11.17 (R, τ) is regular and therefore, by Urysohn's Embedding Theorem 28.4, we can think of (R, τ) as a completely compact, hence closed subset of the $\aleph_0$-cube over $[0, 1]$. But by 29.2, the $\aleph_0$-cube over $[0, 1]$ is a representation of W. If φ is a mapping effectuating this representation, then $\varphi^{-1}R$ is a closed subset of W and, by 29.2 again, $\varphi^{-1}R$ is a representation of W. Hence there exists a continuous mapping ψ of W onto $\varphi^{-1}R$, and $\varphi\psi$ is a continuous mapping of W onto R. By 29.1, (R, τ) must be a representation of W. ∎

Definition 29b: *A T_2-space (R, τ) is called a* continuous curve, *or a* Peano curve, *if there exists a continuous mapping from* $[0, 1]$ *onto* (R, τ). *Any space homeomorphic to* $[0, 1]$ *will be called an* arc.

By 29.1, every continuous curve is a representation of the interval $[0, 1]$. If B is an arc and φ a homeomorphism of $[0, 1]$ onto B, then we call $\varphi 0$ and $\varphi 1$ the *endpoints* of the arc. The definition of end points is really independent of the choice of homeomorphism φ.

A further representation theorem treats of the characterization of continuous curves. We first need, however, some preparatory results on arcs.

29.4 *A space (R, τ) is homeomorphic to one of the real intervals* $[0, 1]$, $[0, 1[$ *or* $]0, 1[$ *if and only if it satisfies the following conditions:*

(1) *(R, τ) is a separable T_1-space containing at least two points.*

(2) *(R, τ) is both connected and locally connected.*

(3) *Given any three distinct, nonempty, connected proper subsets of (R, τ), there are always two of them which do not themselves cover R.*

Proof: The above listed intervals are all totally ordered, connected, and separable. By (15.5) they must possess properties (1)-(3). Conversely, if (R, τ) is a space with properties (1)-(3), then by 15.5 its topology is none other than the order topology associated with a total ordering of R. But R is a connected T_1-space containing at least two points and hence the cardinality of R must be infinite. Our contention now follows from 16.4. ∎

Let a and b be two fixed points of R. If in the previous theorem we replace condition (3) by the following one

(3') *R itself is the only connected subset of R, containing both a and b,*

then a proper, connected subset of R can never contain both a and b. Hence (3') implies (3). But in the total ordering of R, a and b must also be initial and terminal points, respectively: If not, then $[a, b]$ or $[b, a]$ would be a connected, proper subset of R, containing a and b. We therefore have the following characterization of arcs:

29.5 *A space (R, τ) is an arc with end points a and b ($a, b \in R$, $a \neq b$) if and only if it is a connected and locally connected separable T_1-space such that R itself is the only connected subset of R containing a and b.*

29.6 (R. L. Moore) *Let the completely compact T_2-space be locally connected and let it possess a countable basis. Then if a and b are two points of an open, connected subset Q of R, we can find an arc B contained in Q having a and b as its end points.*

Proof: Let $\mathfrak{B}$ be a countable-basis of (R, τ). Since R is completely compact, $\mathfrak{B}$ must contain finite coverings of R; since $|\mathfrak{B}| \leqslant \aleph_0$, there can be at most a countable number of such coverings of R—suppose they have been listed as $\mathfrak{C}_1, \mathfrak{C}_2, \ldots$.

$\mathfrak{B}_0 = \{Q\}$ is a minimal connection of a and b (cf. 15b). We can obtain a sequence $\mathfrak{B}_n$ of minimal connections of a and b by the following recursive method: Suppose $\mathfrak{B}_n = \{V_1, \ldots, V_r\}$. For $\rho = 1, \ldots, r - 1$, let x_ρ denote a point in $V_\rho \cap V_{\rho+1}$ and also let $x_0 = a$, $x_r = b$. Given any $y \in V_\rho$ we can find an open neighborhood $U_y \in \tau y$ such that $\bar{U}_y \subseteq V_\rho$ and such that U_y is also contained in a member set of $\mathfrak{C}_{n+1}$. And given any open covering

$$\mathfrak{H}_\rho = \{U_y : y \in V_\rho\}$$

of V_ρ, by 15.7 we can find a minimal connection $\mathfrak{W}_\rho$ of $x_{\rho-1}$ and x_ρ such that $\mathfrak{W}_\rho \ll \mathfrak{H}_\rho$. Let $\mathfrak{B}_{n+1}$ be a minimal connection of a and b contained in $\mathfrak{W}_1 \cup \ldots \cup \mathfrak{W}_r$.

As a consequence of the very construction of $\mathfrak{B}_n$ we have

(1) $\mathfrak{B}_n \ll \mathfrak{C}_n$ and (2) $\overline{\mathfrak{B}_{n+1}} \ll \mathfrak{B}_n$.

If $m < n$, $\mathfrak{B}_m = \{V_{m,1}, ..., V_{m,r}\}$ and $\mathfrak{B}_n = \{V_{n,1}, ..., V_{n,s}\}$. Hence if $V_{n,\sigma} \subseteq V_{m,\rho+2}$, using the minimality of our connections, we see that we also have

(3) $V_{n,\sigma'} \cap V_{m,\rho} = \emptyset$ for all $\sigma' \geqslant \sigma$. If we now put $S_n = \cup \{V : V \in \mathfrak{B}_n\}$ then, by (2), we first have $\overline{S_{n+1}} \subseteq S_n$, and then we see that

$$B = \cap S_n = \cap \bar{S}_n$$

must be a subset of Q containing a and b. We have to show that B is an arc with end points a and b and for this purpose we shall verify that the conditions of 29.5 hold. In any case, B is a T_1-space. By 10.7 and 10.4 we have

$$\mathrm{Sd}\,(\tau_B) \leqslant \mathrm{Bd}\,(\tau_B) \leqslant \mathrm{Bd}\,(\tau) \leqslant \aleph_0 .$$

And now, by 14.6, we see that B must be a connected set.

Let us suppose further that $p \in B$, and let U be an open neighborhood in τp. In order to prove local connectedness we must exhibit a connected set K and a neighborhood $V \in \tau p$ such that $V \cap B \subseteq K \subseteq U \cap B$. To begin, there exist open neighborhoods $U_1, U_2 \in \tau p$ with $\bar{U}_2 \subseteq U_1$ and $\bar{U}_1 \subseteq U$. Let K_n be the component of $U_1 \cap S_n$ containing the point p. Then $\overline{K_{n+1}} \subseteq \bar{K}_n \subseteq \bar{U}_1 \subseteq U$, and, by 14.6, $K = \cap \bar{K}_n$ must be a connected subset of $U \cap B$. The system $\mathfrak{L} = \{U_2, \complement p \cap U_1, \complement \bar{U}_2\}$ is an open covering of R, for which we can find an m such that $\mathfrak{C}_m \ll \mathfrak{L}$. Let V be a set belonging to $\mathfrak{B}_m$ and containing p. Then $V \in \tau p$. If q is an arbitrary point of $V \cap B$, then, for each $n > m$, $\mathfrak{B}_m$ contains a minimal connection $\mathfrak{W}_n$ of p and q. By (1) and (2) we have $\mathfrak{W}_n \ll \mathfrak{B}_m \ll \mathfrak{C}_m \ll \mathfrak{L}$ and by (3) we have $Z_n = \cup \{W : W \in \mathfrak{W}_n\} \subseteq U_1$. Since Z_n is connected and $p \in Z_n$ we also see that $Z_n \subseteq K_n$ and thus $q \in K_n$. Hence $V \cap B \subseteq K_n$ and then $V \cap B \subseteq K$.

Finally, let p denote a point in B distinct from a and b. If $\mathfrak{B}_n = \{V_1, ..., V_r\}$, let

$$X_n = \cup \{V_\rho : V_\rho \in \mathfrak{B}_n , p \notin V_\sigma \quad \text{for} \quad \sigma \leqq \rho\}$$

and

$$Y_n = \cup \{V_\rho : V_\rho \in \mathfrak{B}_n , p \notin V_\sigma \quad \text{for} \quad \sigma \geqq \rho\}.$$

Then $X = B \cap \cup X_n$ and $Y = B \cap \cup Y_n$ are two disjoint sets, open in B. From $q \in B$ and $q \neq p$ we deduce the existence of open neighborhoods $U_1, U_2 \in \tau p$ such that $\bar{U}_1 \subseteq U_2$ and $q \notin U_2$. As above, with the open covering $\mathfrak{L} = \{U_1, \complement p \cap U_2, \complement \bar{U}_1\}$ of R we can associate an m such that $\mathfrak{C}_m \ll \mathfrak{L}$. By (3) we deduce that q lies in X or in Y. In particular, $a \in X$ and $b \in Y$. We have $\complement_B p = X \cup Y$, and thus $\complement_B p$ is not connected. Hence B itself is the only connected subset of B containing the points a and b. ∎

29.7 (H. Hahn-S. Mazurkiewicz) *A T_2-space (R, τ) is a continuous curve if and only if it is completely compact, connected, locally connected, and possesses a countable basis.*

Proof: Let (R, τ) satisfy the conditions of this theorem. By the Urysohn Embedding Theorem 28.4 we can consider (R, τ) as a subspace of the $\aleph_0$-cube W over $[0, 1]$. By 29.3, there exists a continuous mapping φ from the Cantor discontinuum D onto (R, τ). D is a subset of $[0, 1]$. By 29.1, it only remains to show that φ can be extended continuously to $[0, 1]$.

The complement $\complement D$ of D in $[0, 1]$ is the union of pairwise disjoint open intervals. Let $]a, b[$ be one such interval, and let n be the largest natural number such that $|b - a| \leqslant 1/n$. The points a and b lie in D. Let us first suppose $\varphi a \neq \varphi b$. For $\epsilon > 0$ and $\nu = 1, \ldots, n$ let $V_{\epsilon,\nu}$ denote the open interval

$$]\pi_\nu \varphi a - \epsilon, \pi_\nu \varphi b + \epsilon[\quad \text{or} \quad]\pi_\nu \varphi b - \epsilon, \pi_\nu \varphi a + \epsilon[.$$

If we further set $V_{\epsilon,\nu} = [0, 1]$ for $\nu > n$, then $V = \times\, V_{\epsilon,\nu}$ is an open subset of W containing φa and φb. By 29.6, φa and φb can be connected by an arc in R. Let η denote the inf of all $\epsilon > 0$ for which V_ϵ contains such an arc. For $V_{a,b} = V_{\eta+1/n}$ we then have: There exists a homeomorphism $\varphi_{a,b}$ from $[a, b]$ into $V_{a,b} \cap R$ such that $\varphi_{a,b} a = \varphi a$ and $\varphi_{a,b} b = \varphi b$. Examining the other possibility where $\varphi a = \varphi b$—let $\varphi_{a,b}$ denote that mapping of $[a, b]$ into R, which sends all points in $[a, b]$ onto the unique point φa. The extension ψ of φ being sought must now be defined as follows: For $p \in D$, let us suppose $\psi p = \varphi p$. If $q \in \complement D$ then there must exist a uniquely determined interval $]a, b[$ contained in $\complement D$ with $q \in]a, b[$. Let us put $\psi q = \varphi_{a,b} q$. Since $q \in \complement D$, it follows that ψ coincides with the corresponding $\varphi_{a,b}$ on a suitable neighborhood of q, and hence the continuity of ψ at q follows from that of $\varphi_{a,b}$. It therefore only remains to prove the continuity of ψ at points of D.

Suppose $p \in D$ and let U be a neighborhood belonging to $\tau\psi p = \tau\varphi p$. Let us put $W_{\nu,n} =]\pi_\nu \varphi p - 1/n, \pi_\nu \varphi p + 1/n[$ for $\nu = 1, \ldots, n$, $W_{\nu,n} = [0, 1]$ for $\nu > n$, and $U_n = R \cap \times\, W_{\nu,n}$. Then there exists an n such that $U_n \subseteq U$. The component V of U_{3n} which contains ψp is, by 15.1, a connected and open neighborhood belonging to $\tau\psi p$. By the continuity of φ there exists an interval $J =]a', b'[$ where $a' < p < b'$ and $\varphi(J \cap D) \subseteq V$. We can even assume that $|b' - a'| < \frac{1}{3} n$. The interval J is a neighborhood of p in $[0, 1]$. If $]a, b[$ is an interval contained in $J \cap \complement D$, then we surely have $|b - a| < \frac{1}{3} n$ and $\varphi a, \varphi b \in V$. By 29.6 there must exist in V an arc with end points φa and φb. For the number η belonging to $]a, b[$ we therefore have $\eta < \frac{1}{3} n$ and thus

$V_{a_v b} \subseteq V_{2n/3} \subseteq U_n \subseteq U$. This implies that $\psi[a, b] = \varphi_{a,b}[a, b] \subseteq U$, whence $\psi J \subseteq U$. This shows that ψ is continuous at p.

Conversely, let us now assume that the T_2-space (R, τ) is a continuous curve; i.e., there exists a continuous map φ from $[0, 1]$ onto R. Since $[0, 1]$ is completely compact, connected, and locally connected, by 19.1, 19.3, and 19.5 we see that the same must hold for (R, τ). And since $[0, 1]$ possesses a countable basis, so does (R, τ), by 19.7. ∎

In the last part of this section we shall discuss another type of connectedness, a type which plays a major role especially in algebraic topology.

Definition 29c: *A subset W of a topological space is called a* path *whenever there exists a continuous mapping* φ *of the real interval onto W.*[3] *The points* $\varphi 0$ *and* $\varphi 1$ *are known respectively as the initial and terminal points of the path.*

Every continuous curve is a path. Conversely, in a T_2-space every path is a continuous curve.

Definition 29d: *Two points p, q of a topological space R are said to be* path-wise connected *if there exists a path in R having p as initial point and q as terminal point.*

We shall say that a subset M of R is pathwise connected *whenever any two of its points are pairwise connected by a path lying completely in M.*

The property of being pathwise connected is manifestly an equivalence relation among points, the associated classes being called *path components* of the space in question. The path components decompose the space into maximal path-connected subsets. Every path-connected subset is contained in a maximal path subset. The path components of a space need be neither open nor closed. In fact, the closure of a path-connected set need not itself be path-connected (cf. 29C, 29E).

29.8 *Every path-connected set is connected.*

Proof: This follows from 14.8 (4), for, by 19.3, every path is a connected set. ∎

Hence in any topological space the path components are contained in the components of the space. In particular, every path connected space is connected, but the converse is false (cf. 29C).

29.9 *The following two conditions are equivalent:*

(a) *All path components in the space are open* (*and hence also closed*).

(b) *Every point of the space possesses a path connected neighborhood.*

If these conditions are satisfied, then the path components and the components of the space become identical.

[3] Some authors consider the mapping φ itself to be the path.

Proof: Condition (a) implies that every path component is a path connected neighborhood of all its points. Conversely, suppose (b) holds and let K be a path component. Given any $p \in K$ we can therefore find a path-connected neighborhood $V \in \tau p$. Since K is the largest path-connected subset containing p, we have $U \subseteq K$, which shows that K is open.

Let us now suppose that (a) holds; let Z be a component and K a path component contained in Z. If Z were to contain other path components then K and the union of all other path components contained in Z would constitute a decomposition of Z into two open sets, contradicting the openness of Z. Hence $K = Z$. ∎

29.10 *A space is path connected if and only if it is connected and conditions* (a) *and* (b) *of the above theorem are satisfied.*

Proof: If a space is path-connected, then by 29.8 it is also connected and thus a path-connected neighborhood of each of its points; i.e., condition (b) is satisfied. Conversely if (a) or (b) holds for the connected space R, then by 29.9 R itself must be a path component and hence path-connected. ∎

Definition 29e: *A topological space is said to be* locally pathwise-connected *whenever the neighborhoods filter of every point of the space possesses a basis consisting of locally pathwise-connected sets* (cf. 29D).

By 29.8 we see that every locally pathwise-connected space is also locally connected. Since in a locally pathwise-connected space condition (b) of 29.9 is manifestly verified, its path components and its components must be identical. In particular, a locally pathwise-connected space is pathwise-connected if and only if it is connected.

Definition 29f: *A topological space* (R, τ) *is said to be* locally euclidian *if, given any point* $p \in R$ *we can find an* $n \in \mathbf{N}$ *and a neighborhood* $U \in \tau p$ *such that* U *is homeomorphic to a subset of* $\mathbf{R}^n$.

Since the spaces $\mathbf{R}^n$ (cf. 6C) are manifestly locally pathwise-connected, every locally Euclidian space is also locally pathwise-connected.

Notes and Exercises

29A Theorem 29.6 can be proved from 16.4. The notations we use in what follows are those used in the proof already given of 29.6. Let p and q be two distinct points of B. Suppose that $\mathfrak{B}_n = \{V_1, ..., V_n\}$. Whenever there exists indices ρ_1, ρ_2 such that $1 \leqslant \rho_1 \leqslant \rho_2 \leqslant r$ and $p \in V_{\rho_1}, q \in V_{\rho_2}$, then we shall write $p \leqslant {}_n q$ to abbreviate this state of affairs.

Exercise

1. Let us define $p \leqslant q$ if and only if $p \leqslant {}_n q$ for all n from a certain index onward. Then $\leqslant$ is a total order of B, having a as initial and b as terminal element.

2. The topology τ_B, induced in B by τ, is the order topology for $\leqslant$.

3. Hence 16.4 implies 29.6 [34, p. 123].

29B By 29.7, for every natural number n the n-cube W_n over $[0, 1]$ is a Peano curve. A representation mapping can be obtained as follows: The numbers $x \in [0, 1]$ can be put into 1-1 correspondence with all decimals of the form $0, \xi_1\xi_2\xi_3 \ldots$ if we agree and stick to a convention for representing terminating decimals as periodic ones. For $\nu = 1, \ldots, n$, let $\varphi_\nu x$ be the decimal $0, \eta_1^\nu\eta_2^\nu\eta_3^\nu \ldots$ where $\eta_\kappa^\nu = \xi_{\nu+(\kappa-1)n}$.

Exercise: $\varphi x = \varphi_1 x \times \ldots \times \varphi_n x$ defines a continuous mapping φ from $[0, 1]$ onto W_n.

A more intuitive construction of a continuous mapping of $[0, 1]$ onto W_2 has been given by G. Peano: Decompose W_2 into q congruent subsquares and construct the broken line S_1 out of the appropriate diagonals of these subsquares, as shown in the accompanying figure. Similarly, subdivide the segment $[0, 1]$ into q congruent subintervals. Then we can map these subintervals in a 1-1 linear fashion onto the 9 pices of S_1, defining in this way a topological mapping φ_1 of $[0, 1]$ onto S_1. If we now repeat our procedure, subdividing each of the above 9 subsquares into 9 congruent sub-subsquares and forming within each square, as before, a broken line made up of 9 parts, we ultimately construct a broken line S_2 made up of 9^2 parts, every ninth cornerpoint of which coincides with one of the corner points of S_1. Again we can find a topological, piecewise linear map φ_2 of $[0, 1]$ onto S_2, etc. We can thus form a sequence $(\varphi_n)_{n \in \mathbf{N}}$ of such mappings.

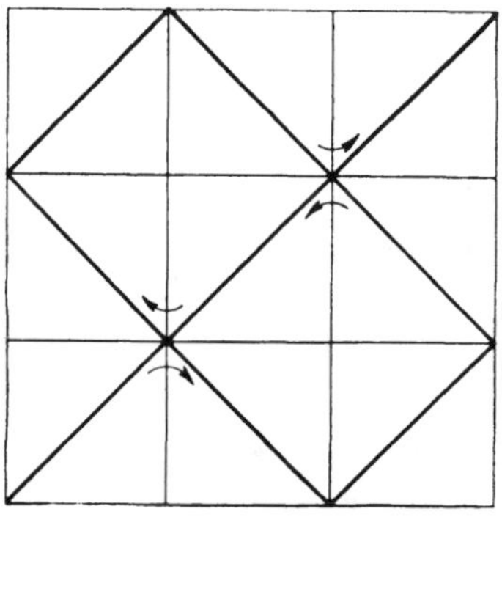

Exercise: The sequence $(\varphi_n)_{n \in \mathbf{N}}$ converges uniformly to a continuous mapping φ from $[0, 1]$ onto W_2.

29C *Exercise:* The set M defined in 15C forms a connected, but not path-connected, subspace of the plane $\mathbf{R}^2$. This subspace possesses two path components, one of which is open, the other closed. The closure of the open path component is not path-connected. The union of M with the line $y = 1$ determines a path-connected but not locally connected subspace of $\mathbf{R}^2$.

29D *Exercise:* The following statements are pairwise equivalent:

1. The space (R, τ) is locally path-connected.

2. The path components of open subspaces of (R, τ) are always open.

3. For every $p \in R$ and every $U \in \tau p$ there exists a $V \in \tau p$ where $V \subseteq U$ and such that: Every point of V can be path-connected to p in U.

In the literature, property (3) is often taken as the definition of local path-connectedness. Property (2) implies: In a locally path-connected space the neighborhoods filter of any point possesses a basis composed of open and path-connected sets.

29E *Exercise:* Let the space R satisfy the condition (b) in 29.9. If M is a path-connected subset of R, then M is also path-connected.

29F If in 29d we replace the word "path" by the word "arc" we say that the points p and q are *arc-connected.* Arc-connected points are manifestly path-connected.

Exercise: In a T_2-space path-connected points are also arc-connected. (In a T_2-space every path W is a continuous curve. If in 29.6 we set $R = Q = W$, then our contention follows from 29.7.)

29G A locally Euclidean and connected T_2-space is said to be a *manifold.* The definition of a locally Euclidean space (29f) associates to each point p a Euclidean space of dimension $n = n(p)$, where n in general does not depend on p. The following theorem then holds (we do not prove this result):

If a nonempty, open subset of $\mathbf{R}^n$ *is homeomorphic to an open subset of* $\mathbf{R}^m$, *then necessarily* $n = m$.

Exercise: Using the above theorem, one can show:

1. In a locally Euclidean space the number $n(p)$ is uniquely determined by the point p.

2. The number $n(p)$ is constant on each component of a locally Euclidean space.

Thus we see that all points of a manifold possesses the same number $n(p)$, which we can consequently define to be the *dimension* of the manifold.

VI

Metric and Uniform Spaces

Metric spaces are one of the most important types of topological spaces. The use of a metric is doubly valuable: on the one hand, a metric makes for quicker, more formal manipulations and deductions, and on the other hand, it enables us to think in terms of distance and thereby give intuitive, geometric interpretation to all concepts and theorems. Hence the great importance of metrization theorems, which allow us to characterize those topologies which can arise from a metric. The theory of uniform spaces represents an important, very useful generalization of the theory of metric spaces, concerning itself with questions stemming from the classical concepts of uniform continuity and uniform convergence.

30. Metric Spaces

We already have defined metric spaces, in Section 6, but for the sake of completeness we now repeat all definitions and results previously given.

A *metric* on the fundamental set R is a mapping $\delta: R \times R \to \mathbf{R}$ verifying the following postulates:

($\mathbf{M}_1$) $\delta(p, q) \geqslant 0$ and $\delta(p, p) = 0$.

($\mathbf{M}_2$) $\delta(p, q) = \delta(q, p)$ (symmetry).

($\mathbf{M}_3$) $\delta(p, q) \leqslant \delta(p, r) + \delta(r, q)$ (triangle inequality).

($\mathbf{M}_4$) If $\delta(p, q) = 0$, then $p = q$.

If only the first three postulates above are satisfied, δ is said to be a *pseudometric*. It is the couple (R, δ) which truly constitutes the *metric* or *pseudometric space*. Every pseudometric δ induces a topology τ_δ: For every real number $\epsilon > 0$ the set

$$U_\epsilon(p) = \{q : \delta(p, q) < \epsilon\}$$

is called the *ϵ-neighborhood* of p. (If we wish to emphasize the pseudometric δ in question, we can write $U_\epsilon^\delta(p)$ instead of $U_\epsilon(p)$.) Then $\mathfrak{U}_p = \{U_\epsilon(p) : \epsilon > 0\}$ is a filter basis, generating the filter $\overline{\mathfrak{U}_p} = \tau_\delta p$, which we define to be the *topology* τ_δ *induced by* δ. By abuse of language we often speak of (R, δ) rather than (R, τ_δ), as a topological space. Conversely, if (R, τ) is a topological space, whose topology can be induced by the pseudo metric δ, then we say that "δ belongs to τ."

We recall that $\bar{\mathbf{R}}$ denotes the real line "completed by $-\infty$ and $+\infty$," while $\bar{\mathbf{R}}^+ = \{x : x \in \bar{\mathbf{R}}, x \geqslant 0\}$. If M is a subset of the pseudometric space (R, δ), then the following sup, formed in $\bar{\mathbf{R}}^+$, is called the *diameter* of M:

$$\delta(M) = \sup\{\delta(p, q) : p, q \in M\}$$

We see that $\delta(\varnothing) = 0$. If $\delta(M) < +\infty$, M is said to be a bounded set (more precisely, δ-bounded). The inf, formed in $\bar{\mathbf{R}}$, defined by

$$\delta(M, N) = \inf\{\delta(p, q) : p \in M, q \in N\}$$

is called the *distance* of the sets M and N. We see that $\delta(\varnothing, N) = \delta(M, \varnothing) = +\infty$. Rather than write $\delta(\{p\}, N)$, we simply write $\delta(p, N)$. Since $\delta(\{p\}, \{q\}) = \delta(p, q)$ the two uses of δ, as distance between sets and as distance between points, can lead to no confusion. In (R, δ) we see that

$$\bar{M} = \{p : \delta(p, M) = 0\} \qquad \text{and} \qquad \mathring{M} = \{p : \delta(p, \complement M) > 0\},$$

while $U \in \tau_\delta p$ is equivalent to $\delta(p, \complement U) > 0$. $U_\epsilon(M) = \{p : \delta(p, M) < \epsilon\}$ is an open set, called the ϵ-neighborhood of the set M.

Every pseudometric space is paracompact (13.13) and satisfies the first countability axiom (10.II). Every $\aleph_0$-compact pseudometric space is completely compact (11.16). For the topology induced by the pseudometric δ we have Sd (τ_δ) = Bd (τ_δ) (10.11). The pseudometric δ is a metric if and only if (R, δ) is a T_1-space. Every pseudometric space is a T_3-, T_4-, and T_5-space. Every metric space is completely normal (8.12).

Since a metric space is normal it is also completely regular. For pseudometric spaces we can not quite make this conclusion. We do, however, have the analogous result:

30.1 *Every pseudometric space* (R, δ) *is a* T_{3a}*-space.*

Proof: Let $\mathfrak{v}$ be an ultrafilter not converging in (R, δ) to p. Hence there exists an $\epsilon > 0$ such that $\mathfrak{v} \leqslant \overline{\complement U_\epsilon(p)}$. Then $\varphi x = \delta(p, x)$ defines a mapping $\varphi : R \to \mathbf{R}$. We have $\varphi p = 0$ and $\varphi\mathfrak{v} \wedge \overline{]\to\epsilon, +\epsilon[} = \mathfrak{o}$; i.e., $\varphi\mathfrak{v}$ does not converge in $\mathbf{R}$ to φp. But φ is a continuous map: For if $y \in U_\eta(x)$, then, by the usual application of the triangle inequality, we have

$$|\varphi x - \varphi y| = |\delta(p, x) - \delta(p, y)| \leqslant \delta(x, y) < \eta$$

(cf. 18.1). ∎

The concepts of "boundedness," "diameter," and "distance" can be directly carried over from sets to filters.

Definition 30a: *Let* $\mathfrak{a}$ *and* $\mathfrak{b}$ *be filters in the pseudometric space* (R, δ). *Then the numbers, formed in* $\bar{\mathbf{R}}$, *and defined by*

$$\delta(\mathfrak{a}) = \inf\{\delta(A) : A \in \mathfrak{a}\}$$

and

$$\delta(\mathfrak{a}, \mathfrak{b}) = \sup\{\delta(A, B) : A \in \mathfrak{a},\ B \in \mathfrak{b}\}$$

are called, respectively, the diameter *of* $\mathfrak{a}$ *and the* distance *between* $\mathfrak{a}$ *and* $\mathfrak{b}$. *We say that the filter* $\mathfrak{a}$ *is* bounded *if* $\delta(\mathfrak{a}) < +\infty$.

We see immediately that $\delta(\mathfrak{o}) = 0$, $\delta(\mathfrak{o}, \mathfrak{a}) = \delta(\mathfrak{a}, \mathfrak{o}) = +\infty$, $\delta(\ulcorner M \urcorner) = \delta(M)$, and $\delta(\ulcorner M \urcorner, \ulcorner N \urcorner) = \delta(M, N)$.

30.2 *For any two filters* $\mathfrak{a}$ *and* $\mathfrak{b}$, *distinct from* $\mathfrak{o}$, *we have*

$$\max\{\delta(\mathfrak{a}), \delta(\mathfrak{b}), \delta(\mathfrak{a}, \mathfrak{b})\} \leqslant \delta(\mathfrak{a} \vee \mathfrak{b}) \leqslant \delta(\mathfrak{a}) + \delta(\mathfrak{b}) + \delta(\mathfrak{a}, \mathfrak{b}).$$

Proof: Since $\delta(A) \leqslant \delta(A \cup B)$, we have

$$\delta(\mathfrak{a}) = \inf\{\delta(A) : A \in \mathfrak{a}\} \leqslant \inf\{\delta(A \cup B) : A \in \mathfrak{a},\ B \in \mathfrak{b}\} = \delta(\mathfrak{a} \vee \mathfrak{b})$$

and likewise $\delta(\mathfrak{b}) \leqslant \delta(\mathfrak{a} \vee \mathfrak{b})$. Suppose $A, A' \in \mathfrak{a}$ and $B, B' \in \mathfrak{b}$. Since $\mathfrak{a} \neq \mathfrak{o}$ and $\mathfrak{b} \neq \mathfrak{o}$ there must exist points a, b such that $a \in A \cap A'$ and $b \in B \cap B'$. Hence $\delta(A, B) \leqslant \delta(a, b) \leqslant \delta(A' \cup B')$ and therefore $\delta(\mathfrak{a}, \mathfrak{b}) \leqslant \delta(\mathfrak{a} \vee \mathfrak{b})$. In similar fashion we see that $\delta(\mathfrak{a} \vee \mathfrak{b}) \leqslant \delta(A) + \delta(B) + \delta(A, B)$ for all $A \in \mathfrak{a}$ and $B \in \mathfrak{b}$, whence $\delta(\mathfrak{a} \vee \mathfrak{b}) \leqslant \delta(\mathfrak{a}) + \delta(\mathfrak{b}) + \delta(\mathfrak{a}, \mathfrak{b})$. ∎

30.3 *In the pseudometric space* (R, δ) *the following statements about the filter* $\mathfrak{a} \neq \mathfrak{o}$ *are all equivalent:*

(1) $\mathfrak{a}$ *converges to* p.

(2) $\delta(\mathfrak{a} \vee \ulcorner p \urcorner) = 0$.

(3) $\delta(\mathfrak{a}) = 0$ *and* $\delta(\mathfrak{a}, \ulcorner p \urcorner) = 0$.

Proof

(1) → (2): We have $\delta(\tau_\delta p) = \inf\{\delta(U_\epsilon(p)) : \epsilon > 0\} = 0$. From $\mathfrak{a} \leqslant \tau_\delta p$ we deduce $\mathfrak{a} \vee \ulcorner p \urcorner \leqslant \tau_\delta p$ and hence we also have $\delta(\mathfrak{a} \vee \ulcorner p \urcorner) = 0$.

(2) → (3): From 30.2 we see that $\delta(\mathfrak{a}) \leqslant \delta(\mathfrak{a} \vee \ulcorner p \urcorner)$ and $\delta(\mathfrak{a}, \ulcorner p \urcorner) \leqslant \delta(\mathfrak{a} \vee \ulcorner p \urcorner)$.

(3) → (1): By (3), given any $\epsilon > 0$ we can find an $A \in \mathfrak{a}$ such that $\delta(A) < \epsilon/3$ and $\delta(p, A) < \epsilon/3$. Hence there exists an $a' \in A$ with $\delta(p, a') < 2\epsilon/3$, and thus for any $a \in A$ we have $\delta(p, a) \leqslant \delta(p, a') + \delta(a', a) < \epsilon$. This implies $A \subseteq U_\epsilon(p)$, whence $\mathfrak{a}$ must converge to p. ∎

Definition 30b: *A filter in a pseudometric space* (R, δ) *is called a* Cauchy filter (*more exactly:* δ*-Cauchy filter*) *if* $\delta(\mathfrak{a}) = 0$.

30.4 *In a pseudometric space every convergent filter is a Cauchy filter. Conversely, if* $\mathfrak{a}$ *is a Cauchy filter and* M *a completely compact* (*k-compact*) *set, then from* $\mathfrak{a} \wedge \overset{\sqcap}{M} \neq \mathfrak{o}$ (*and from* $\mathrm{Rk}(\mathfrak{a}) \leqslant k$) *we can deduce the convergence of the filter* $\mathfrak{a}$.

Proof: From 30.3 we see that the convergence of the filter $\mathfrak{a}$ implies $\delta(\mathfrak{a}) = 0$. Conversely, from $\mathfrak{a} \wedge \overset{\sqcap}{M} \neq \mathfrak{o}$ (and $\mathrm{Rk}(\mathfrak{a}) \leqslant k$) we deduce the existance of an adherent point $p \in M$ to $\mathfrak{a}$. We have $p \in \bar{A}$, hence $\delta(p, A) = 0$ for all $A \in \mathfrak{a}$ and thus $\delta(\mathfrak{a}, \overset{\sqcap}{p}) = 0$. If, in addition, $\mathfrak{a}$ is a Cauchy filter then $\delta(\mathfrak{a}) = 0$, whence, by 30.3, $\mathfrak{a}$ must converge to p. ∎

The above theorem is a generalization of the well-known *Cauchy convergence criterion* from real analysis: Let $(a_n)_{n \in \mathbf{N}}$ be a sequence of real numbers with associated Fréchet-filter $\mathfrak{a}$. Then the condition $\delta(\mathfrak{a}) = 0$ is equivalent to the existence, for every $\epsilon > 0$, of a terminating section A_n of the sequence such that $\delta(A_n) < \epsilon$. This can itself happen if and only if for all $\nu, \mu \geqslant n$ we have $| a_\nu - a_\mu | < \epsilon$. Since every Cauchy filter is bounded and every bounded interval on the real line is completely compact, the validity of Cauchy convergence criterion on the real line is an immediate consequence of 30.4.

In any pseudometric space (R, δ) we can define an equivalence relation $\overset{\frown}{\delta}$ by

$$\overset{\frown}{\delta} = \{(p, q) : \delta(p, q) = 0\}.$$

If κ denotes the canonical mapping from R onto $R/\overset{\frown}{\delta}$, then $\kappa p = \kappa p'$ and $\kappa q = \kappa q'$ imply

$$\delta(p', q') \leqslant \delta(p', p) + \delta(p, q) + \delta(q, q') = \delta(p, q)$$

and similarly, $\delta(p, q) \leqslant \delta(p', q')$, whence $\delta(p, q) = \delta(p', q')$. If we therefore define $\delta'(\kappa p, \kappa q) = \delta(p, q)$, we obtain a pseudometric δ' on $R/\overset{\frown}{\delta}$.

30.5 κ *is a continuous, open, and closed mapping from* (R, τ_δ) *onto* $(R, \tau_\delta)/\overset{\frown}{\delta}$.

δ' *is a metric on* $R/\overset{\frown}{\delta}$.

We have $(R/\overset{\frown}{\delta}, \tau_{\delta'}) = (R, \tau_\delta)/\overset{\frown}{\delta}$.

Proof: From the very definition of δ' we see immediately that $\tau_{\delta'}$ is really the quotient topology $(\tau_\delta)_\kappa$. Furthermore, $\delta(p, q) = 0$ if and only if $\tau_\delta p = \tau_\delta q$. Our contention then follows from 21.11. ∎

The above theorem enables us to pass from a pseudometric space to a metric space by taking quotients. The quotient set $R/\overset{\frown}{\delta}$, bearing the metric δ', shall henceforth be denoted by $(R, \delta)/\overset{\frown}{\delta}$.

Every pseudometric induces a unique topology, but several distinct pseudometrics can induce the same topology. This situation leads us to

Definition 30c: *Two pseudometrics on R are said to be* topologically equivalent *if they induce the same topology on R.*

If δ is a metric on R while δ' is a pseudometric equivalent to δ, then δ' must also be a metric, for $\tau_{\delta'} = \tau_\delta$ is a T_1-topology.

30.6 *Given any pseudometric δ on R such that $\delta(R) > 0$, then we can find a pseudometric δ^* with respect to which the diameter of R equals any prescribed number d verifying $0 < d < +\infty$.*

Proof: Suppose $\delta(R) < +\infty$, then

$$\delta^*(p, q) = \frac{d}{\delta(R)} \delta(p, q)$$

defines a pseudometric δ^* equivalent to δ such that $\delta^*(R) = d$. Suppose that, on the other hand, $\delta(R) = +\infty$. Then $\delta'(p, q) = \min\{1, \delta(p, q)\}$ defines a pseudometric on R such that $\delta'(R) = 1$. For $0 < \epsilon \leqslant 1$ the ϵ-neighborhoods under δ and δ' coincide. Hence δ and δ' are topologically equivalent. ∎

If $\delta(R) < +\infty$ there may be no pseudometric, topologically equivalent to δ, with respect to which the diameter of R is $+\infty$ (cf. 30A).

Theorem 30.6 tells us that "diameter," "distance," "boundedness," etc., are not topological concepts: They depend essentially on the pseudometric at hand and can be different for two topologically equivalent pseudometrics.[1] But note that if a set has a diameter equal to zero, then this property is a topological invariant: $\delta(M) = 0$ is equivalent to $\overline{M} \leqslant \tau_\delta p$ for all $p \in M$. On the other hand, being a Cauchy filter is not a topological property:

30.I Let R be the set of all real numbers p where $|p| > 1$. Let δ be the absolute value on R, (i.e., $\delta(p, q) = |p - q|$. Then we can define a second metric δ' by

$$\delta'(p, q) = \begin{cases} |p - q| & \text{for} \quad pq > 0 \\ |p - q| - 2 & \quad pq < 0. \end{cases}$$

If $0 < \epsilon \leqslant |p| - 1$, then $U_\epsilon^\delta(p) = U_\epsilon^{\delta'}(p)$. Hence δ and δ' are topologically equivalent. Let $A_n = \{p : 1 < |p| < 1 + 1/n\}$, $\mathfrak{A} = \{A_n : n \in \mathbf{N}\}$ and $\mathfrak{a} = \overline{\mathfrak{A}}$; then $\delta(\mathfrak{a}) = 2$ and $\delta'(\mathfrak{a}) = 0$. We see that $\mathfrak{a}$ is a Cauchy filter with respect to δ' but not with respect to δ.

Definition 30d: *A mapping $\varphi : R \to R'$ is said to be* metric-preserving *or* distance-preserving *from the pseudometric space (R, δ) into the pseudometric space (R', δ') if $\delta(p, q) = \delta'(\varphi p, \varphi q)$ for all $p, q \in R$.*

[1] There exist other notions of boundedness which are topologically invariant.

A metric preserving 1-1 *mapping from* (R, δ) *onto* (R', δ') *is called an* isometry. *Whenever an* isometry *exists between* (R, δ) *and* (R', δ') *we say that these two spaces are* isometric.

The composition of two metric preserving maps, or two isometries, gives as product a map of the same nature. The relation of isometry between pseudometric spaces is an equivalence in the class of all such spaces.

30.7 *Every distance preserving map is continuous. If such a map is onto a pseudometric space, then it must also be an open and closed map. Every isometry is a homeomorphism.*

Proof: Let φ be a metric preserving mapping from (R, δ) into (R', δ'). For each point $p \in R$ we then have $\varphi U_\epsilon^\delta(p) \subseteq U_\epsilon^{\delta'}(\varphi p)$ and thus $\varphi \tau_\delta p \leqslant \tau_{\delta'} \varphi p$. If φ is a mapping from R onto R' we even have $\varphi U_\epsilon^\delta(p) = U_\epsilon^{\delta'}(\varphi p)$ and $U_\epsilon^\delta(p) = \varphi^{-1} U_\epsilon^{\delta'}(\varphi p)$: For $\varphi q \in U_\epsilon^{\delta'}(\varphi p)$ implies $\delta(p, q) = \delta'(\varphi p, \varphi q) < \epsilon$, whence $q \in U_\epsilon^\delta(p)$. We see that $\tau_\delta = {}_\varphi(\tau_{\delta'})$, and hence, by 20.1, φ is an open and closed mapping. Thus an isometry is both a continuous and 1-1 mapping and consequently a homeomorphism. ∎

30.8 *Let* φ *be a metric preserving mapping from* (R, δ) *onto* (R', δ'). *If* δ *is a metric then so is* δ' *and* φ *is an isometry.*

Proof: From $p \neq q$ it follows that $\delta'(\varphi p, \varphi q) = \delta(p, q) > 0$, hence $\varphi p \neq \varphi q$. Thus φ is a 1-1 mapping and thereby an isometry. Conversely, if $p' \neq q'$, then $\delta'(p', q') = \delta(\varphi^{-1}p', \varphi^{-1}q') > 0$, which implies that δ' is also a metric. ∎

30.9 *If* φ *is a metric-preserving mapping of* (R, δ) *onto* (R', δ'), *then the metric spaces* $(R, \delta)/\bar{\delta}$ *and* $(R', \delta')/\bar{\delta}'$ (cf. 30.5) *are isometric.*

Proof: $\delta(p, q) = 0$ is equivalent to $\delta'(\varphi p, \varphi q) = 0$. Hence φ maps the equivalence classes under $\bar{\delta}$ onto the equivalence classes of $\bar{\delta}'$. If κ and κ' denote the canonical maps from R onto $R/\bar{\delta}$ and R' onto $R'/\bar{\delta}'$, respectively, then $\psi = \kappa' \varphi \kappa^{-1}$ is a metric-preserving map of $(R, \delta)/\bar{\delta}$ onto $(R', \delta')/\bar{\delta}'$ and hence, by 30.8, an isometry. ∎

By metric theorems, concepts, properties, etc., we mean theorems, concepts, properties, etc., which remain invariant under metric-preserving mappings. "Boundedness" and "Cauchy filter" are metric concepts; "distance" and "diameter" are likewise metric invariants.

Notes and Exercises

30A *Exercise:* Every completely compact subset of a pseudometric space is bounded.

Hence if the pseudometric space (R, δ) is completely compact and δ' is a pseudometric equivalent to δ, then $\delta'(R) < +\infty$.

30B Suppose that the pseudometric space (R, δ) obeys $\delta(R) = 1$, although $\delta(p, q) < 1$ for all $p, q \in R$.

Exercise: By

$$\delta'(p, q) = \frac{\delta(p, q)}{1 - \delta(p, q)}$$

we can define a pseudometric δ' on R, topologically equivalent to δ and such that $\delta'(R) = +\infty$.

30C Let C^* be the set of all Cauchy filters of the pseudometric space (R, δ).

Exercise: Given any filter $\mathfrak{a}$ of C^* there exists in C^* exactly one maximal filter $\mathfrak{b}$ containing $\mathfrak{a}$.

30D *Exercise:* Every pseudometric δ is a continuous map from $(R, \delta) \times (R, \delta)$ into $\mathbf{R}$.

31. Metrization

In this section we shall seek to determine those topological spaces which can be considered as metric spaces.

Definition 31a: *A space (R, τ) and its topology are said to be* (pseudo-) metrizable *if there exists a* (pseudo-) metric *on R belonging to τ (i.e., δ induces the topology τ).*

The property of (pseudo-) metrizability is manifestly topological: For if φ is a topological mapping of (R, τ) into (R', τ') and δ' is a (pseudo-) metric of R' belonging to τ', then we can define a (pseudo-) metric δ on R belonging to τ by $\delta(p, q) = \delta'(\varphi p, \varphi q)$. We also see that if a topological space is (pseudo-) metrizable so are all its subspaces.

31.1 *A produce space $(R, \tau) = \times \{(R_n, \tau_n) : n \in \mathbf{N}\}$ is (pseudo-) metrizable if and only if all its coordinate spaces are (pseudo-) metrizable. If δ_n is a (pseudo-) metric of R_n belonging to τ_n and $\delta_n(R_n) \leqslant 1$, then we can define a (pseudo-) metric on the product space belonging to τ by*

$$\delta(\xi, \eta) = \sum_{n-1}^{\infty} \frac{1}{2^n} \delta_n(\pi_n \xi, \pi_n \eta)$$

Proof: If (R, τ) is a pseudometrizable space, then so is the subspace

$$T_{\xi, n} = \{\eta : \pi_\nu \eta = \pi_\nu \xi \quad \text{for} \quad \nu \neq n\}.$$

For a fixed $\xi \in R$, the space (R_n, τ_n) is homeomorphic to $T_{\xi,\eta}$ (22.5) and hence also (pseudo-) metrizable.

Conversely, let δ_n be a pseudometric on R_n belonging to τ_n. By 30.6 we can assume $\delta_n(R_n) \leqslant 1$. Then the above series (*) is convergent and thereby defines a metric δ on R. Since $\delta_n(\tau_n\xi, \tau_n\eta) \leqslant 2^n\delta(\xi, \eta)$ we have $\pi_n\tau_\delta\xi \leqslant \tau_n\pi_n\xi$, which implies $\tau_\delta \leqslant \tau$. Conversely, given any $\epsilon > 0$ there exists a natural number n such that $2^{-n} < \epsilon/2$. If, for $\nu = 1, ..., n$, we have $\pi_\nu\eta \in U_{\epsilon/2}(\pi_\nu\xi)$, we must then also have

$$\delta(\xi, \eta) = \sum_{\nu=1}^{\infty} \frac{1}{2^\nu} \delta_\nu(\pi_\nu\xi, \pi_\nu\eta) < \frac{\epsilon}{2}\sum_{\nu=1}^{n} 2^{-\nu} + \frac{\epsilon}{2}\sum_{\nu=n+1}^{\infty} 2^{-(\nu-n)} < \epsilon.$$

The neighborhood $\bigcap \{\pi_\nu^{-1}U_{\epsilon/2}(\pi_\nu\xi) : \nu = 1, ..., n\}$ belonging to $\tau\xi$ is therefore contained in $U_\epsilon^\delta(\xi)$, whence $\tau \leqslant \tau_\delta$ and finally $\tau = \tau_\delta$. This tells us that δ is a pseudometric belonging to τ. If δ_n is a metric for all n, then τ_n must be a T_1-topology for all n—hence, by 22.6, τ is also a T_1-topology and therefore δ must be a metric. ∎

31.2 *The intersection of a countable number of (pseudo-) metrizable topologies is itself (pseudo-) metrizable.*

Proof: Let the topologies τ_n on R ($n \in \mathbf{N}$) be (pseudo-) metrizable. By 31.1, the product space $\times\ (R, \tau_n)$ must also possess this property. By 22.10, the space $(R, \bigwedge \tau_n)$ is embeddable in $\times\ (R, \tau_n)$ and hence (pseudo-) metrizable. ∎

31.3 (Urysohn's Metrization Theorem) *A space with a countable basis is metrizable if and only if it is regular.*

Proof: The interval [0, 1] is metrizable and hence, by 31.1, so is the $\aleph_0$-cube over [0, 1]. By Urysohn's embedding theorem 28.4 we see that every regular space with a countable basis is embeddable in the above cube and hence also metrizable. Conversely, by 8.12, we see that every metrizable space is completely normal and hence surely regular. ∎

By 19.7, we have an immediate corollary to the above theorem:

Let (R, τ) and (R', τ') be two completely compact T_2-spaces such that there exists a continuous mapping of (R, τ) onto (R', τ'). Then if (R', τ') is metrizable, so is (R, τ).

The Urysohn Metrization proof only applies to spaces with a countable basis. We can prove a more general result in a similar way using the theorem on embeddability in the $\aleph_0$-cube over a star space.

31.4 (Metrization Theorem of R. H. Bing) *A space is metrizable if and only if it is regular and possesses an $\aleph_0$-discrete basis.*

Proof: By 28.8, every regular space R with an $\aleph_0$-discrete basis can be embedded in the $\aleph_0$-cube over the star space S. We can define a metric belonging to the topology on S by

$$\delta(p, q) = \begin{cases} |p - q| & \\ & , \quad \text{if } p, q \text{ lie in the} \\ |p| + |q| & \end{cases} \begin{matrix} \text{same ray} \\ \\ \text{different rays.} \end{matrix}$$

If S is metrizable, then, by 31.1, so is the $\aleph_0$-cube over S, and hence R itself is metrizable.

Conversely, let the space (R, τ) be metrizable (pseudometrizable). Then it must also be regular (a T_3-space) by 8.12 and paracompact by 13.13. Let δ be a metric (pseudometric) belonging to τ. For fixed $\epsilon > 0$, $\mathfrak{H}_\epsilon = \{U_\epsilon^\delta(p) : p \in R\}$ is an open neighborhood of (R, τ). By 13.11, there exists an $\aleph_0$-discrete open covering $\mathfrak{K}_\epsilon$ of R such that $\mathfrak{K}_\epsilon \ll \mathfrak{H}_\epsilon$. Then $\mathfrak{B} = \cup \{\mathfrak{K}_{1/n} : n \in \mathbf{N}\}$ is an $\aleph_0$-discrete system of open sets. If U is a neighborhood belonging to τp, then there exists an n such that $\overline{U_{1/n}^\delta(p)} \subseteq U$. Let us suppose that $V \in \mathfrak{K}_{n/2}$ and $p \in V$. Then $\delta(V) \leqslant 1/n$ and $V \subseteq \overline{U_{1/n}^\delta(p)}$. Hence $\mathfrak{B}$ is a basis of (R, τ). ∎

The last part of the above proof holds if (R, τ) is only assumed pseudometrizable, i.e., when δ is only a pseudometric. Hence every pseudometrizable space is a T_3-space with $\aleph_0$-discrete basis. The following theorem shows that these properties are also characteristic ones.

31.5 *The space* (R, τ) *is pseudometrizable if and only if it is a* T_3*-space with* $\aleph_0$*-discrete basis.*

Proof: Let (R, τ) be a T_3-space with $\aleph_0$-discrete basis. By 21.11 the quotient space $(R, \tau)/\rho$, where the equivalence is given by $\rho = \{(p, q) : \tau p = \tau q\}$, must be regular while the associated canonical map κ is continuous and open. Hence $\kappa\mathfrak{B}$ is a basis of the quotient space. If M is an open subset of (R, τ) then $M = \kappa^{-1}\kappa M$. Hence for any two open sets B and U, if

$$\varnothing = B \cap U = \kappa^{-1}(\kappa B \cap \kappa U)$$

then we must also have $\kappa B \cap \kappa U = \varnothing$. Consequently we see that κ maps any discrete system of open sets onto a system of the same nature—$\kappa\mathfrak{B}$ is thus an $\aleph_0$-discrete basis of $(R, \tau)/\rho$. By 31.4 there exists a metric δ' belonging to the quotient topology τ_κ. By $\delta(p, q) = \delta'(\kappa p, \kappa q)$ we can define a pseudometric on R belonging to τ'. For by 21.11 and 30.5 we have $T = {}_\kappa(\tau_\kappa) = {}_\kappa(\tau_{\delta'}) = \tau_\delta$. ∎

The formulation of the last two theorems is still valid if we replace "discrete" by "locally finite":

31.6 (Metrization Theorem of J. Nagata and Y. U. M. Smirnov) *A space is metrizable* (*pseudometrizable*) *if and only if it is regular* (T_3*-space*) *with an* $\aleph_0$*-locally finite basis.*

Proof: Every $\aleph_0$-discrete system is also $\aleph_0$-locally finite. By 31.4 and 31.5 it only remains to show that the existence of an $\aleph_0$-locally finite basis implies that of an $\aleph_0$-discrete basis.

Let $\mathfrak{B} = \bigcup \{\mathfrak{B}_n : n \in \mathbf{N}\}$ be a basis of the T_3-space (R, τ), where the $\mathfrak{B}_n$ are assumed to be locally finite systems of sets. Given any open covering of R we can then find a refinement contained in $\mathfrak{B}$ such that this refinement is also $\aleph_0$-locally finite. By 13.11, (R, τ) must therefore be paracompact. For any two natural numbers n and m and for $B \in \mathfrak{B}_n$ we must have

$$V^B_{n,m} = \bigcup \{\overline{B'} : B' \in \mathfrak{B}_m, \overline{B'} \subseteq B\}.$$

The sets $V^B_{n,m}$ are closed by 13.1. Since $V^B_{n,m} \subseteq B$, $\mathfrak{B}_{n,m} = \{V^B_{n,m} : B \in \mathfrak{B}_n\}$ is also a locally finite system of sets. Hence every union of sets belonging to $\mathfrak{B}_{n,m}$, in particular the union $V_{n,m}$ of all sets in $\mathfrak{B}_{n,m}$, must be closed. This means that for every point $p \in R$,

$$H_{n,m}(p) = \bigcap \{B : p \in V^B_{n,m}\} \cap \complement \bigcup \{V^B_{n,m} : p \notin V^B_{n,m}\}$$

must be an open set. Thus $\mathfrak{H}_{n,m} = \{H_{n,m}(p) : p \in R\} \cup \{\complement V_{n,m}\}$ is an open covering of R, for which, by 13.11, there exists an open and $\aleph_0$-discrete covering $\mathfrak{K}_{n,m}$ such that $\mathfrak{K}_{n,m} \ll \mathfrak{H}_{n,m}$. $\mathfrak{K} = \bigcup \{\mathfrak{K}_{n,m} : n, m \in \mathbf{N}\}$ is likewise an $\aleph_0$-discrete system of sets. It only remains to show that $\mathfrak{K}$ is even a basis of (R, τ).

Let us therefore denote by U an open neighborhood in τp. Then there exists an n and a $B \in \mathfrak{B}_n$ such that $p \in B$ and $B \subseteq U$. By $(\mathbf{T_3})$, there also exists an m and a $B' \in \mathfrak{B}_m$ such that $p \in B'$ and $\overline{B'} \subseteq B$. Hence $p \in V^B_{n,m}$. Finally, there exists a $K \in \mathfrak{K}_{n,m}$ such that $p \in K$. Since $\mathfrak{K}_{n,m} \ll \mathfrak{H}_{n,m}$, $K \subseteq H_{n,m}(q)$ for a $q \in R$. If $q \notin V^B_{n,m}$, then $p \notin H_{n,m}(q)$ and therefore $p \notin K$, a contradiction! Hence $H_{n,m}(q) \subseteq B$ and $K \subseteq B \subseteq U$, which means that $\mathfrak{K}$ is a basis of (R, τ). ∎

We conclude with another characterization of paracompact T_3-spaces. Let (R, τ) be a space and k a cardinal number such that $k \geqslant |\mathfrak{C}|$, where $\mathfrak{C}$ denotes any discrete system of open subsets of (R, τ); for instance, we can take $k = \operatorname{Bd}(R, \tau)$ or $k = |R|$. Let I be an indexing set of cardinality k and W the $\aleph_0$-cube over star space S belonging to I.

31.7 *For a T_3-space (R, τ) the following properties are pairwise equivalent:*

(1) *(R, τ) is paracompact.*

(2) *Given any open covering $\mathfrak{H}$ of (R, τ) we can find a continuous map $\psi : (R, \tau) \to W$ and an open covering $\mathfrak{L}$ of ψR such that $\psi^{-1}\mathfrak{L} = \{\psi^{-1}L : L \in \mathfrak{L}\} \leqslant \mathfrak{H}$.*

(3) *Given any open covering $\mathfrak{H}$ of (R, τ) there exists a pseudometric δ on R such that $\tau \leqslant \tau_\delta$ and an open covering of (R, δ) such that $\mathfrak{K} \ll \mathfrak{H}$.*

(4) *A filter $\mathfrak{a}$ possesses an adherent point in (R, τ) if and only if, given any metric space (R', δ') and any continuous mapping φ of (R, τ) onto (R', δ'), then $\varphi\mathfrak{a}$ possesses an adherent point in (R', δ').*

Proof

(1) → (2): (R, τ) is a paracompact T_3-space and, hence, by 13.3, it must also be a T_4-space. By 13.11 we can associate with $\mathfrak{H}$ an $\aleph_0$-discrete open covering $\mathfrak{K}$ such that $\mathfrak{K} \ll \mathfrak{H}$. Suppose $\mathfrak{K} = \cup \mathfrak{K}_n$ where the $\mathfrak{K}_n$ are discrete systems for $n \in \mathbf{N}$. If I is an indexing set of the type described above, we can assume that $\mathfrak{K}_n = \{K_{n,\iota} : \iota \in I\}$. (If $|\mathfrak{K}| < |I|$, we agree that for the "extra" indices, $K_{n,\iota} = \emptyset$.) By 13.15 $\mathfrak{K}$ admits of a contraction $\mathfrak{K}'$. For each $K_{n,\iota} \in \mathfrak{K}$ we can find a $K'_{n,\iota} \in \mathfrak{K}'$ such that $\overline{K'_{n,\iota}} \subseteq K_{n,\iota}$. Because of $(\mathbf{T}_4)$ there exists a continuous mapping $\varphi_{n,\iota} : (R, \tau) \to [0, 1]$, taking the value 1 on $\overline{K'_{n,\iota}}$ and the value 0 on $\complement K_{n,\iota}$. As in the proof of 28.7, let the map $\psi_n : R \to S$ be defined by the following: Let $\psi_n p$ be the point on the ιth ray of S determined by $\varphi_{n,\iota} p$, provided $\varphi_{n,\iota} p \neq 0$; if $\varphi_{n,\iota} p = 0$, let $\psi_n p$ be the common zero point of all the rays of S. In any case, ψ_n is a continuous map of (R, τ) into S and $\psi = \times\, \psi_n$ is a continuous map of (R, τ) into W.

Let $L_{n,\iota}$ be the set of all $\xi \in W$, for which $\pi_n \xi$ lies in the ιth ray of S without being the null point. For each $p \in R$ we can find a $K_{n,\iota} \in \mathfrak{K}$ such that $p \in K'_{n,\iota}$. Then $\pi_n \psi p = \psi_n p$ is the unit point 1 one the ιth ray, and hence $\psi p \in L_{n,\iota}$. Thus $\mathfrak{L} = \{L_{n,\iota} : n \in \mathbf{N}, \iota \in I\}$ is an open covering of ψR. Moreover $\psi^{-1} L_{n,\iota} = \{p : p \in R, \varphi_{n,\iota} p \neq 0\} \subseteq K_{n,\iota}$. This implies $\psi^{-1}\mathfrak{L} \ll \mathfrak{K} \ll \mathfrak{H}$.

(2) → (3): By (2), we can associate with $\mathfrak{H}$ a continuous map $\psi : (R, \tau) \to W$ and an open covering $\mathfrak{L}$ of ψR such that $\psi^{-1}\mathfrak{L} \ll \mathfrak{H}$. Being a subspace of W, ψR must be metrizable. Hence let δ' be a metric on ψR. We can define a metric δ on R by $\delta(p, q) = \delta'(\psi p, \psi q)$. Since ψ is continuous, we see that

$$\tau p \leqslant \psi^{-1} \tau_{\delta'} \psi p = \tau_\delta p.$$

And since $\mathfrak{L}$ is an open covering of $(\psi R, \delta')$, $\mathfrak{K} = \psi^{-1}\mathfrak{L}$ is an open covering of (R, δ).

(3) → (4): Like $\mathfrak{a}$, for every continuous φ, $\varphi\mathfrak{a}$ possesses an adherent point in the image space. Conversely, suppose that $\mathfrak{a}$ had no adherent point in (R, τ). Then $\mathfrak{H} = \{\complement \bar{A} : A \in \mathfrak{a}\}$ is an open covering of (R, τ). By (3) there exists a pseudometric δ on R with $\tau \leqslant \tau_\delta$ and an open covering $\mathfrak{K}$ of (R, δ) with $\mathfrak{K} \ll \mathfrak{H}$. Hence $\mathfrak{a}$ cannot have any adherent point in (R, δ). Since $\tau \leqslant \tau_\delta$, the canonical map κ of (R, δ) onto the metric space $(R, \delta)/\bar{\delta}$ (cf. 30.5) is also a continuous map from (R, τ) onto $(R, \delta)/\bar{\delta}$. From the very definition of $\bar{\delta}$ we see immediately that $\kappa\mathfrak{a}$ cannot possess an adherent point in $(R, \delta)/\bar{\delta}$.

(4) → (1): Let $\mathfrak{H}$ be an open covering of (R, τ) and let $\mathfrak{a}$ be the complementary filter of $\mathfrak{H}$ (cf. p. 28). Since $\mathfrak{a}$ does not possess any adherent point, there exists a continuous mapping φ of (R, τ) onto a metric space (R', δ') such that $\varphi\mathfrak{a}$ does not possess any adherent point. Then $\mathfrak{H}' = \{\complement \overline{\varphi A} : A \in \mathfrak{a}\}$ is an open covering of (R', δ'). Since every metric space is paracompact, there exists a locally finite open covering $\mathfrak{K}'$ of (R', δ') such that $\mathfrak{K}' \ll \mathfrak{H}'$. To each $K' \in \mathfrak{K}'$ we can associate an

$A_{K'} \in \mathfrak{a}$ such that $\varphi^{-1}K' \subseteq \varphi^{-1}\complement\overline{\varphi A_{K'}} \subseteq \complement A_{K'}$. In addition there exists a finite number of sets $H'_{K'}, \ldots, H^{(rK')}_{K'} \in \mathfrak{H}$ such that $\varphi^{-1}K' \subseteq \complement A'_K \subseteq H'_{K'} \cup \ldots \cup H^{r(K')}_{K'}$. Then

$$\mathfrak{K} = \varphi^{-1}K' \cap H^{\rho}_{K'} : K' \in \mathfrak{K}', \rho = 1, \ldots, r(K')\}$$

is an open covering of (R, τ) with $\mathfrak{K} \ll \mathfrak{H}$. For each $p \in R$ there exists a neighborhood $U' \in \tau_{\delta'}\varphi p$ which possesses a nonempty intersection with at most a finite number of $K' \in \mathfrak{K}'$. Then $\varphi^{-1}U'$ is a neighborhood of p possessing this same property with respect to $\mathfrak{K}$. Hence $\mathfrak{K}$ is locally finite and (R, τ) is paracompact. ∎

Notes and Exercises

31A *Exercise:* The k-cube over $[0, 1]$ is not metrizable for $k > \aleph_0$. (First Countability Axiom:).

31B For the metric space (R, τ) let us assume that $\delta(R) < +\infty$ and let $\mathfrak{A}$ denote the system of all closed subsets of (R, δ). Finally, for $A_1, A_2 \in \mathfrak{A}$ let

$$\delta'(A_1, A_2) = \inf\{\epsilon : \epsilon > 0,\ A_2 \subseteq U^{\delta}_{\epsilon}(A_1),\ A_1 \subseteq U^{\delta}_{\epsilon}(A_2)\}.$$

Exercise

1. δ' is a metric on $\mathfrak{A}$.

2. The map $\varphi : R \to \mathfrak{A}$ defined by $\varphi x = \{x\}$ is a metric 1-1 mapping of (R, δ) into $(\mathfrak{A}, \delta')$.

Let R be the set of all positive reals. For $p, q \in R$ let

$$\delta_1(p, q) = \min\{1, |p - q|\} \qquad \text{and} \qquad \delta_2(p, q) = \left|\frac{p}{1 + p} - \frac{q}{1 + q}\right|.$$

3. δ_1 and δ_2 are topologically equivalent metrics on R.

4. The metrics δ_1' and δ_2' on $\mathfrak{A}$ nevertheless induce distinct topologies. (Examine the set of all points of $\mathfrak{A}$ represented by positive integers) [24, p. 131].

32. Extensions and Completions of Metric Spaces

An arbitrary extension $(\hat{R}, \hat{\tau})$ of a metric space (R, δ) is in general not metrizable. We are therefore led to examine the following question: Under what conditions can we extend δ to a metric $\hat{\delta}$ of $\hat{R}$ in such a way that $\hat{\delta}$ induces the topology $\hat{\tau}$? The following two theorems are needed before we can give an answer to this question.

32.1 *Let M be a dense subset of a pseudometric space (R, δ). For every open subset A, for every open filter $\mathfrak{a}$, and for every two points $p, q \in R$ we have:*

$$\delta(A) = \delta(A \cap M), \qquad \delta(\mathfrak{a}) = \delta(\mathfrak{a} \wedge \overset{\sqcap}{M}), \qquad \delta(p, q) = \delta((\tau_\delta p \vee \tau_\delta q) \wedge \overset{\sqcap}{M}).$$

Proof: Since M is dense in (R, δ), given any $\epsilon > 0$ and any $p \in A$ we can find a $p_\epsilon \in A \cap M$ such that $\delta(p, p_\epsilon) < \epsilon/2$. Consequently

$$\delta(A \cap M) \leqslant \delta(A) = \sup\{\delta(p, q) : p, q \in A\} \leqslant \epsilon + \sup\{\delta(p_\epsilon, q_\epsilon) : p, q \in A\}$$
$$\leqslant \epsilon + \sup\{\delta(p', q') : p', q' \in A \cap M\} = \epsilon + \delta(A \cap M)$$

for each $\epsilon > 0$, which establishes the desired equality. Moreover, if $\mathfrak{A}$ is an open basis of $\mathfrak{a}$, then

$$\delta(\mathfrak{a}) = \inf\{\delta(A) : A \in \mathfrak{A}\} = \inf\{\delta(A \cap M) : A \in \mathfrak{A}\} = \delta(\mathfrak{a} \wedge \overset{\sqcap}{M}).$$

Lastly, we have $\delta(\tau_\delta p) = 0$, $\delta(p, q) \leqslant \delta(\tau_\delta p \vee \tau_\delta q)$, and $\delta(\tau_\delta p, \tau_\delta q) \leqslant \delta(p, q)$. by 30.2, we therefore see that

$$\delta(p, q) \leqslant \delta(\tau_\delta p \vee \tau_\delta q) \leqslant \delta(\tau_\delta p) + \delta(\tau_\delta q) + \delta(\tau_\delta p, \tau_\delta q) \leqslant \delta(p, q),$$

whence

$$\delta(p, q) = \delta(\tau_\delta p \vee \tau_\delta q) = \delta((\tau_\delta p \vee \tau_\delta q) \wedge \overset{\sqcap}{M}). \blacksquare$$

Definition 32a: *Let $\mathfrak{a}$ be a filter in the pseudometric space (R, δ) and let $\mathfrak{r}(\mathfrak{a})$ denote the filter generated by the basis $\{U_\epsilon(A) : A \in \mathfrak{a}, \epsilon > 0\}$.*

If $\mathfrak{r}(\mathfrak{a}) = \mathfrak{a}$, we shall say that the filter $\mathfrak{a}$ is round.

32.2

(1) *Every round filter is open.*

(2) *The filter $\mathfrak{r}(\mathfrak{a})$ is round.*

(3) *The neighborhoods filters $\tau_\delta p$ are round.*

(4) *$\mathfrak{r}(\mathfrak{a})$ is a Cauchy filter if and only if $\mathfrak{a}$ is a Cauchy filter.*

(5) *Suppose $\mathfrak{a} \leqslant \mathfrak{b}$, $\mathfrak{a}$ is round, and $\mathfrak{b}$ is a Cauchy filter. Then $\mathfrak{a} = \mathfrak{b}$.*

Proof

(1) If $\mathfrak{a}$ is round, the open sets $U_\epsilon(A)$, where $A \in \mathfrak{a}$ and $\epsilon > 0$, constitute a basis of $\mathfrak{a}$.

(2) This is a consequence of $U_\epsilon(U_\epsilon(A)) \subseteq U_{2\epsilon}(A)$.

(3) We have $\tau_\delta p = \mathfrak{r}(\overset{\sqcap}{p})$ and hence our contention follows from (2).

(4) Since $\delta(U_\epsilon(A)) \leqslant \delta(A) + 2\epsilon$, we have $\delta(\mathfrak{r}(\mathfrak{a})) = \delta(\mathfrak{a})$. Hence $\delta(\mathfrak{r}(\mathfrak{a})) = 0$ is equivalent to $\delta(\mathfrak{a}) = 0$.

(5) If $\delta(\mathfrak{b}) = 0$ and $\mathfrak{a} \leqslant \mathfrak{b}$ then $\mathfrak{b} \leqslant \overline{U_\epsilon(A)}$ for all $A \in \mathfrak{a}$ and all $\epsilon > 0$, whence $\mathfrak{b} \leqslant \mathfrak{r}(\mathfrak{a}) = \mathfrak{a}$. ∎

Definition 32b: *Let $(\hat{R}, \hat{\tau})$ be an extension of (R, τ) and δ a pseudometric belonging to τ. A pseudometric $\hat{\delta}$ of $\hat{R}$ is then said to be an* extension *of δ to $\hat{R}$ whenever $\hat{\delta}(p, q) = \delta(p, q)$ for all $p, q \in R$, while $\hat{\delta}$ is said to be an extension to $(\hat{R}, \hat{\tau})$ if $\hat{\delta}$ induces the topology $\hat{\tau}$.*

32.3 *Let $(\hat{R}, \hat{\tau})$ be an extension of (R, τ) with trace system T^* and let δ be a pseudometric on R such that $\tau_\delta \geqslant \tau$. Then δ admits of an extension $\hat{\delta}$ to $\hat{R}$ satisfying the condition $\tau_{\hat{\delta}} \geqslant \hat{\tau}$ if and only if T^* is composed entirely of δ-Cauchy filters. In this case $\hat{\delta}$ is uniquely determined by δ and we have*

$$(*) \qquad \hat{\delta}(\hat{p}, \hat{q}) = \delta(\mathrm{Tr}\,(\hat{\tau}\hat{p}) \vee \mathrm{Tr}\,(\hat{\tau}\hat{q})).$$

Proof: Let $\hat{\delta}$ be an extension verifying the above conditions. Then

$$\hat{\delta}(\hat{\tau}\hat{p}) \leqslant \hat{\delta}(\tau_{\hat{\delta}}\hat{p}) = 0 \qquad \text{and} \qquad \delta(\mathrm{Tr}\,(\hat{\tau}\hat{p})) = 0.$$

Hence T^* is made up of δ-Cauchy filters. In addition, by 32.1, the relation (*) holds and thus $\hat{\delta}$ is uniquely determined by δ.

Conversely, suppose T^* is made up entirely of δ-Cauchy filters. For the map $\hat{\delta} : \hat{R} \times \hat{R} \to \mathbf{R}$ defined by (*) above we then have $\hat{\delta}(\hat{p}, \hat{q}) \geqslant 0$, $\hat{\delta}(\hat{p}, \hat{q}) = \hat{\delta}(\hat{q}, \hat{p})$, and $\hat{\delta}(\hat{p}, \hat{q}) = \delta(\mathrm{Tr}\,(\hat{\tau}\hat{p})) = 0$. In order to prove the triangle inequality let us make the following abbreviations: $\mathrm{Tr}\,(\hat{\tau}\hat{p}) \vee \mathrm{Tr}\,(\hat{\tau}\hat{q}) = \mathfrak{a}$, $\mathrm{Tr}\,(\hat{\tau}\hat{p}) \vee \mathrm{Tr}(\hat{\tau}\hat{r}) = \mathfrak{b}$, and $\mathrm{Tr}\,(\hat{\tau}\hat{r}) \vee \mathrm{Tr}\,(\hat{\tau}\hat{q}) = \mathfrak{c}$. Then $\mathfrak{a} \leqslant \mathfrak{b} \vee \mathfrak{c}$. Using 30.2, we have

$$\hat{\delta}(\hat{p}, \hat{q}) = \delta(\mathfrak{a}) \leqslant \delta(\mathfrak{b} \vee \mathfrak{c}) \leqslant \delta(\mathfrak{b}) + \delta(\mathfrak{c}) + \delta(\mathfrak{b}, \mathfrak{c}) = \hat{\delta}(\hat{p}, \hat{r}) + \hat{\delta}(\hat{r}, \hat{q}).$$

Hence $\hat{\delta}$ is a pseudometric on $\hat{R}$. For the points $p, q \in R$ we have

$$\hat{\delta}(p, q) = \delta(\tau p \vee \tau q) = \delta(p, q),$$

from which we see that $\hat{\delta}$ is also an extension of δ. By 32.1, we have

$$\hat{\delta}(\hat{\tau}\hat{p}) = \delta(\mathrm{Tr}\,(\hat{\tau}\hat{p})) = 0$$

and therefore $\hat{\tau}\hat{p} \leqslant \tau_{\hat{\delta}}\hat{p}$. ∎

32.4 *Let (R, δ) be a pseudometric space and $(\hat{R}, \hat{\tau})$ a T^*-extension of (R, τ_δ).*

(1) *There exists a unique extension $\hat{\delta}$ of δ to $(\hat{R}, \hat{\tau})$ if and only if T^* is composed entirely of round Cauchy filters.*

(2) *For every pseudometric extension (R', δ') of (R, δ) with trace system T^* there exists a metric-preserving mapping of (R', δ') onto $(\hat{R}, \hat{\delta})$ which agrees with the identity on R. This property uniquely determines $(\hat{R}, \hat{\delta})$ up to isometry.*

(3) *$\hat{\delta}$ is a metric if and only if δ is a metric.*

Proof: If δ admits of the extension $\hat{\delta}$ to $(\hat{R}, \hat{\delta})$, then by 32.2, $\hat{\tau}\hat{p} = \tau_{\hat{\delta}}\hat{p}$ is a round Cauchy filter in $(\hat{R}, \hat{\delta})$ and Tr $(\hat{\tau}\hat{p})$ is likewise round in (R, δ). Conversely, suppose T^* is made up of only round Cauchy filters. Then, by 32.3, δ admits of an extension $\hat{\delta}$ to $\hat{R}$ with $\tau_{\hat{\delta}} \geqslant \hat{\tau}$. Consequently Tr $(\hat{\tau}\hat{p}) \leqslant$ Tr $(\tau_{\hat{\delta}}\hat{p})$. But Tr $(\hat{\tau}\hat{p})$ is a round filter and Tr $(\tau_{\hat{\delta}}\hat{p})$ is a Cauchy filter and hence, by 32.2(5), we must have Tr $(\hat{\tau}\hat{p}) =$ Tr $(\tau_{\hat{\delta}}\hat{p})$. Since $\hat{\tau}$ is the coarsest topology on $\hat{R}$ having the trace system T^* (25.1), we finally see that $\tau_{\hat{\delta}} = \hat{\tau}$.

(2) By 25.1 there exists a continuous mapping φ from $(R', \tau_{\delta'})$ onto $(\hat{R}, \hat{\tau})$, which reduces to the identity on R and satisfies the trace condition

$$\mathrm{Tr}\,(\tau_{\delta'}p) = \mathrm{Tr}\,(\hat{\tau}\varphi p').$$

For every such map we have

$$\delta'(p', q') = \delta(\mathrm{Tr}\,(\tau_{\delta'}p') \vee \mathrm{Tr}\,(\tau_{\delta'}q')) = \delta(\mathrm{Tr}\,(\hat{\tau}\varphi p') \vee \mathrm{Tr}\,(\hat{\tau}\varphi q')) = \hat{\delta}(\varphi p', \varphi q');$$

i.e., such maps are metric-preserving. Conversely, if φ is a metric-preserving mapping of (R', δ') onto $(\hat{R}, \hat{\delta})$ and equal to the identity on R, then, by 30.7, φ must be open and continuous. Consequently $\varphi\tau_{\delta'}p' = \hat{\tau}\varphi p'$ and

$$\mathrm{Tr}\,(\tau_{\delta'}p') = \mathrm{Tr}\,(\varphi\tau_{\delta'}p') = \mathrm{Tr}\,(\hat{\tau}\varphi p').$$

By exactly the same line of reasoning as in 25.1 we see that $(\hat{R}, \hat{\delta})$ is uniquely determined by the imposed conditions, up to isometry.

(3) If $\hat{\delta}$ is a metric, so is δ. Conversely, let us suppose that δ is a metric and that consequently (R, τ_δ) is a T_2-space. From Tr $(\hat{\tau}\hat{p}) \wedge$ Tr $(\hat{\tau}\hat{q}) \neq$ o it follows that

$$\delta(\mathrm{Tr}\,(\hat{\tau}\hat{p}) \vee \mathrm{Tr}\,(\hat{\tau}\hat{q})) \leqslant \delta(\mathrm{Tr}\,(\hat{\tau}\hat{p})) + \delta(\mathrm{Tr}\,(\hat{\tau}\hat{q})) + \delta(\mathrm{Tr}\,(\hat{\tau}\hat{p}), \mathrm{Tr}\,(\hat{\tau}\hat{q})) = 0.$$

Since Tr $(\hat{\tau}\hat{p})$ and Tr $(\hat{\tau}\hat{q})$ are round filters, by 32.2 we even have

$$\mathrm{Tr}\,(\hat{\tau}\hat{p}) = \mathrm{Tr}\,(\hat{\tau}\hat{p}) \vee \mathrm{Tr}\,(\hat{\tau}\hat{q}) = \mathrm{Tr}\,(\hat{\tau}\hat{q}).$$

By 25.3, $(\hat{R}, \hat{\tau})$ is therefore also a T_2-space and hence $\hat{\delta}$ is a metric. ∎

The extensions of a pseudometric space (R, δ) are in general not uniquely determined by their trace systems; not even if we only consider those extension spaces to which δ can also be extended. But if δ is a metric and we consider only metric extensions, we do have the following theorem.

32.5 *If the metric spaces (R', δ') and (R'', δ'') are two extensions of the metric space (R, δ) having the same trace system T^* then they must both be isometric.*

Proof: Since T^* is necessarily composed of round Cauchy filters, by 32.3 we know that δ can be extended to a metric $\hat{\delta}$ on the T^*-extension $(\hat{R}, \hat{\tau})$ of

(R, δ). Furthermore, there exists a metric-preserving map from (R', δ') onto $(\hat{R}, \hat{\delta})$ which must even be an isometry by 30.8. Hence (R', δ') and (R'', δ'') are both isometric to $(\hat{R}, \hat{\delta})$. ∎

Definition 32c: *A pseudometric space* (R, δ) *is said to be* complete *if each of its Cauchy filters converges. A pseudometric space* $(\hat{R}, \hat{\delta})$ *is said to be a* complete extension *of* (R, δ) *if* $(\hat{R}, \hat{\delta})$ *is an extension of* (R, δ) *and* $(\hat{R}, \hat{\delta})$ *is complete while* $\hat{\delta}$ *is also an extension of* δ *to* $\hat{R}$.

For example, by 30.4 we see that all completely compact pseudometric spaces are complete. The real line, although not completely compact, nevertheless is a completion of the metric space of rational numbers.

Let (R, δ) be a pseudometric space and T^* the set of all round Cauchy filters of (R, δ). By 32.2, T^* contains all neighborhoods filters $\tau_\delta p$ and is composed entirely of open filters. Hence T^* must be a trace system. By 32.3, δ can be extended to $\hat{\delta}$ on the T^*-extension $(\hat{R}, \hat{\tau})$. Let $\hat{\mathfrak{a}}$ be a Cauchy filter on $(\hat{R}, \hat{\delta})$. Then, by 32.2, $\mathfrak{r}(\mathfrak{a})$ must be a round, open Cauchy filter of $(\hat{R}, \hat{\delta})$. Hence $\mathfrak{a} = \mathfrak{r}(\mathfrak{a}) \wedge \ulcorner R \urcorner \neq \mathfrak{o}$ and $\mathfrak{a}$ is a round Cauchy filter of (R, δ) and thus a filter belonging to T^*. For the point $\hat{p} \in \hat{R}$ represented by $\mathfrak{a}$ we have $\hat{\tau}\hat{p} \wedge \hat{\mathfrak{a}} \neq \mathfrak{o}$, hence $\hat{\delta}(\hat{\tau}\hat{p} \vee \mathfrak{a}) \leqslant \hat{\delta}(\hat{\tau}\hat{p}) + \hat{\delta}(\hat{\mathfrak{a}}) = \mathfrak{o}$. This implies, using 32.2, that $\hat{\tau}\hat{p} = \hat{\tau}\hat{p} \vee \hat{\mathfrak{a}}$ and hence $\hat{\mathfrak{a}} \leqslant \hat{\tau}\hat{p}$. Every Cauchy filter of $(\hat{R}, \hat{\delta})$ is therefore convergent, and $(\hat{R}, \hat{\delta})$ is a complete extension of (R, δ), called the *completion* of (R, δ). The concept of completion is metric but not topological: topologically equivalent pseudometrics can lead to nonhomeomorphic completions!

32.6 *Every complete extension of a pseudometric space* (R, δ) *can be mapped in a metric-preserving fashion onto the completion* $(\hat{R}, \hat{\delta})$ *of* (R, δ). *This property characterizes* $(\hat{R}, \hat{\delta})$ *up to isometry.*

If δ *is a metric, then* $\hat{\delta}$ *is also a metric and* $(\hat{R}, \hat{\delta})$ *is the only metric, complete extension of* (R, δ) *possible, up to isometry.*[2]

Proof: Let (R', δ') be a complete extension of (R, δ). If $\mathfrak{a}$ is a round Cauchy filter of (R, δ), then $\mathfrak{a}$ converges in (R', δ') to a point p'. Hence $\mathfrak{a} \leqslant \mathrm{Tr}\,(\tau_{\delta'} p')$ and, by 32.2, $\mathfrak{a} = \mathrm{Tr}\,(\tau_{\delta'} p')$. Hence the trace system T^* of (R', δ') contains all round Cauchy filters. But since, conversely, T^* can only contain such filters, T^* must be identical with the set of all round Cauchy filters on (R, δ). Hence the T^*-extension of (R, δ) is the same as the completion of (R, δ) and our contention follows immediately from 32.3 and 32.4. ∎

The following definition serves to characterize those pseudometric spaces whose completion is completely compact.

Definition 32d: *A pseudometric space* (R, δ) *is said to be* precompact *if, given any* $\epsilon > 0$, *we can find a finite open covering* $\{H_1, \ldots, H_n\}$ *of* R *such that* $\delta(H_\nu) < \epsilon$ *for* $\nu = 1, \ldots, n$.

[2] Cf. 35.12 for the extensions of metric-preserving mappings.

We have used the term "precompact" in order to better concord with standard usage—if we had followed the logic of our prior nomenclature we would have had to use "pre-completely-compact." A frequently used synonym for "precompact" is "totally bounded."

32.7 *The completion $(\hat{R}, \hat{\delta})$ of a pseudometric space (R, δ) is completely compact if and only if (R, δ) is precompact.*

Proof: Let $(\hat{R}, \hat{\delta})$ be completely compact. Given $\epsilon > 0$, then $\{U_{\epsilon/3}(\hat{p}) : \hat{p} \in \hat{R}\}$ is an open covering of $(\hat{R}, \hat{\delta})$ which must contain a finite covering

$$\{U_{\epsilon/3}(\hat{p}_1), ..., U_{\epsilon/3}(\hat{p}_n)\}.$$

The sets $H_\nu = U_{\epsilon/3}(\hat{p}_\nu) \cap R$ then form a covering of R. In addition,

$$\delta(H_\nu) \leqslant \hat{\delta}(U_{\epsilon/3}(\hat{p}_\nu)) < \epsilon$$

and hence (R, δ) is precompact.

Conversely, suppose (R, δ) is precompact. We must show that every ultrafilter $\hat{\mathfrak{v}}$ of $\hat{R}$ converges in $(\hat{R}, \hat{\delta})$. By 32.2, $\mathfrak{r}(\hat{\mathfrak{v}})$ is an open filter. Hence $\mathfrak{d} = \mathfrak{r}(\hat{\mathfrak{v}}) \wedge \overline{R} \neq \mathfrak{o}$, and there exists an ultrafilter $\mathfrak{v}$ of R with $\mathfrak{v} \leqslant \mathfrak{d}$. Given $\epsilon > 0$, let $\{H_1, ..., H_n\}$ be a covering of R with $\delta(H_\nu) < \epsilon$. Then $\mathfrak{v} \leqslant \overline{H_\nu}$ for at least one ν and hence $\delta(\mathfrak{v}) < \epsilon$. Thus $\mathfrak{v}$ is a Cauchy filter of (R, δ), converging in $(\hat{R}, \hat{\delta})$ to a point $\hat{p}$. If $\hat{\mathfrak{v}} \wedge \hat{\tau}\hat{p} = \mathfrak{o}$, then we would have $\hat{\mathfrak{v}} \leqslant \overline{\complement U_\epsilon(\hat{p})}$ and $\mathfrak{r}(\hat{\mathfrak{v}}) \leqslant \overline{\complement U_{\epsilon/3}(\hat{p})}$ for an $\epsilon > 0$, which would contradict $\mathfrak{r}(\hat{\mathfrak{v}}) \wedge \hat{\tau}\hat{p} \geqslant \mathfrak{v}$. Hence $\hat{\mathfrak{v}} \leqslant \hat{\tau}\hat{p}$. ∎

To terminate, let us examine the question of when a topological space can be considered a complete pseudometric space. This question depends on the determination of all subspaces of a complete pseudometric space which themselves are complete under an equivalent pseudometric.

For each subset M of a pseudometric space (R, δ) let $\check{M}$ denote the set of all points $p \in R$ for which there exists a $q \in M$ with $\delta(p, q) = 0$. If δ is a metric, then evidently $\check{M} = M$.

32.8 *For subsets M of a complete pseudometric space (R, δ) the following statements are equivalent:*

(1) *There exists a pseudometric δ' on M, topologically equivalent to δ, such that (M, δ') is complete.*

(2) *$\check{M}$ is the intersection of a countable collection of open subsets of (R, δ).*[3]

Proof

(1) → (2): For each $n \in \mathbf{N}$ let T_n denote the set of all $p \in R$, for which we can find an open neighborhood $U_n(p) \in \tau_\delta p$ possessing the following properties:
(a) $\delta(U_n(p)) < 1/n$, (b) $U_n(p) \cap M \neq \varnothing$, (c) $\delta'(U_n(p) \cap M) < 1/n$.

[3] Such sets are known as G_δ-*sets*.

Then the sets T_n are open subsets of (R, δ) and $M \subseteq T_n$. Let p be a point in $D = \cap \{T_n : n \in \mathbf{N}\}$. By (a), $\{U_n(p) : n \in \mathbf{N}\}$ is a basis of $\tau_\delta p$. From (b) we see that $\tau_\delta p \wedge \overline{M} \neq \mathfrak{o}$ and from (c) we see that $\delta'(\tau_\delta p \wedge \overline{M}) = 0$. Hence $\tau_\delta p \wedge \overline{M}$ is a δ'-Cauchy filter and, since (M, δ') is complete, this filter conveges to a point $q \in M$. Since δ and δ' are topologically equivalent, we also have $\tau_\delta p \wedge \overline{M} \leqslant \tau_\delta q$. Using 30.2 we therefore have

$$\delta(p, q) \leqq \delta(\tau_\delta p \vee \tau_\delta q) \leqq \delta(\tau_\delta p) + \delta(\tau_\delta q) + \delta(\tau_\delta p, \tau_\delta q) = 0.$$

Since $q \in M$ we have $p \in \check{M}$ and thus $D \subseteq \check{M}$. On the other hand, from $M \subseteq T_n$ and the openness of the sets T_n, we deduce that $\check{M} \subseteq D$ and hence $\check{M} = \cap T_n$.

(2) → (1): Suppose $\check{M} = \cap \{T_n : n \in \mathbf{N}\}$ for open subsets T_n of (R, δ). We can then define a pseudometric δ_n of T_n by

$$\delta_n(p, q) = \min \left\{ 1, \delta(p, q) + \left| \frac{1}{\delta(p, \complement T_n)} - \frac{1}{\delta(q, \complement T_n)} \right| \right\}.$$

From $p \in T_n$ we first see that $\delta(p, \complement T_n) = \eta > 0$. If $\delta(p, q) < n/2$, then $\delta_n(p, q) \leqslant (1 + 2/\eta^2) \cdot \delta(p, q)$. Conversely, $\delta_n(p, q) < 1$ implies $\delta(p, q) \leqslant \delta_n(p, q)$. Hence δ and δ_n are topologically equivalent and induce the same topology on T_n and all the more so on $\check{M}$. By 31.2, the pseudometric δ' of $\check{M}$ induces the same topology as δ, where δ' is defined by

$$\delta'(p, q) \sum_{n=1}^{\infty} \frac{1}{2^n} \delta_n(p, q).$$

Let now $\mathfrak{a}$ be a Cauchy filter of (M, δ'). Since

$$\min \{1, \delta(\mathfrak{a})\} \leqslant \delta_1(\mathfrak{a}) \leqslant 2\delta'(\mathfrak{a}) = 0,$$

$\mathfrak{a}$ is then also a δ-Cauchy filter and hence converges to a point $p \in R$. If $p \notin \check{M}$ then p would not belong to T_n for an appropriate $n \in \mathbf{N}$. In every set $A \in \mathfrak{a}$ there would then be points x, y such that $\delta(y, \complement T_n) < \frac{1}{2}\delta(x, \complement T_n) < \frac{1}{2}$. We would then have $\delta_n(x, y) = 1$ and $\delta'(x, y) \geqslant 2^{-n}\delta_n(x, y) \geqslant 2^{-n}$, contradicting $\delta'(\mathfrak{a}) = 0$. Hence $p \in \check{M}$, and there must be a $q \in M$ such that $\delta(p, q) = 0$. Hence $\mathfrak{a}$ converges under δ to q also. But since δ and δ' are topologically equivalent, $\mathfrak{a}$ also converges under δ' to q. The pseudometric space (M, δ') is therefore complete. ∎

We are now in a position to answer our original question.

32.9 *For a space (R, τ) the following statements are pairwise equivalent:*

(1) *There exists on R a pseudometric δ_0, belonging to τ, for which (R, δ_0) is complete.*

(2) (R, τ) *is pseudometrizable, and for every embedding of* (R, τ) *in a pseudometric space* (R', δ'), $\check{R}$ *is the intersection of a countable number of open subsets of* (R', δ').

(3) (R, τ) *is pseudometrizable, and for any pseudometric of* R *belonging to* τ, *we have: In the completion* $(\hat{R}, \hat{\delta})$ *of* (R, δ), $\check{R}$ *is the intersection of a countable collection of open sets.*

Proof

(1) → (2): Condition (1) implies that (R, τ) must in any case be pseudometrizable. If (R, τ) is embedded in the pseudometric space (R', δ'), then we can consider R as a subset of the completion $(\hat{R}', \hat{\delta}')$ of (R', δ'). From 32.8 we see that $\check{R}$ is the intersection of a countable number of open subsets $\hat{T}_n$ of $(\hat{R}', \hat{\delta}')$. If we put $T_n = \hat{T}_n \cap R'$, then the sets T_n are open in (R', δ') and hence we also have, in (R', δ'), the relation $\check{R} = \bigcap T_n$.

(2) → (3): (3) is a special case of (2), obtained by putting $(R', \delta') = (\hat{R}, \hat{\delta})$.

(3) → (1): This follows immediately from 32.8. ∎

Notes and Exercises

32A *Exercise:* A pseudometric space (R, δ) is precompact if and only if, given any filter $\mathfrak{a}$ of R we can find a Cauchy-filter $\mathfrak{b}$ of (R, δ) such that $\mathfrak{b} \leqslant \mathfrak{a}$.

32B Let δ_1 and δ_2 be two pseudometrics on R and let $(\hat{R}_1, \hat{\delta}_1)$, $(\hat{R}_2, \hat{\delta}_2)$ be the completions of (R, δ_1) and (R, δ_2), respectively.

Exercise: The Cauchy filters of (R, δ_1) and (R, δ_2) are the same if, and only if, there exists a homeomorphism between $(\hat{R}_1, \hat{\delta}_1)$ and $(\hat{R}_2, \hat{\delta}_2)$ which reduces to the identity on R.

32C For $n \in \mathbf{N}$ let (R_n, δ_n) be a pseudometric space and let δ be the pseudometric on $R = \times R_n$ defined in 31.1. Let $(\hat{R}, \hat{\delta})$, $(\hat{R}_n, \hat{\delta}_n)$ be the completions of (R, δ) and (R_n, δ_n) respectively and let δ' be the pseudometric on $R' = \times \hat{R}_n$, also defined in 31.1.

Exercise: $\mathfrak{a}$ is a Cauchy filter of (R, δ) if and only if $\pi_n \mathfrak{a}$ is a Cauchy filter of (R_n, δ_n) for all n. The spaces $(\hat{R}, \hat{\delta})$ and (R', δ') are isometric.

32D *Exercise:* If the spaces (R_n, τ_n) $(n \in \mathbf{N})$ are complete pseudometric spaces, then so is their product space.

33. Uniform Structures

The concepts of real variable theory which involve the idea of uniformity (uniform continuity, uniform convergence) are based upon the possibility of comparing distances in a pseudometric space: We can say that the couple of points (p, q) of a pseudometric space (R, δ) are near to degree ϵ, or neighbors of degree ϵ, if $\delta(p, q) < \epsilon$. In arbitrary topological spaces, however we can not avail ourselves of this "distance language." Nevertheless we can still manage to speak of nearness, whether or not we have a pseudometric at hand.

Subsets of $R \times R$ have already been defined as synonymous to relations and been designated by small Greek letters of the alphabet. In what follows, however, it will be more appropriate for us to re-emphasize the "subset nature" of relations by using a notation different from that of small Greek letters. We shall designate subsets of $R \times R$ by capital Latin letters, while points of $R \times R$ will be noted as ordered couples of points of R.

The identity relation on R corresponds to the subset

$$D = D(R) = \{(p, p) : p \in R\};$$

it is called the *diagonal*. We define the inverse relation T^- to the relation T by $T^- = \{(p, q) : (q, p) \in T\}$. If $T = T^-$, T is said to be a *symmetric* set. If $\mathfrak{a}$ is a filter in $R \times R$, then $\mathfrak{a}^- = \{A^- : A \in \mathfrak{a}\}$ is also a filter of $R \times R$. A filter $\mathfrak{a}$ is called symmetric if $\mathfrak{a} = \mathfrak{a}^-$. If $A \in \mathfrak{a}$ then $S_A = A \cap A^- \in \mathfrak{a}$ and hence S_A must be a symmetric set. This tells us that a filter in $R \times R$ is symmetric only when it possesses a basis of symmetric sets. For any two subsets S and T of $R \times R$ let $S \circ T$ denote the set of all couples (p, q) for which there exists an $x \in R$ such that $(p, x) \in S$ and $(x, q) \in T$. This new definition corresponds to the old product of relations: note, however, the difference of order that occurs in this new notation for product of relations! If $\mathfrak{a}$ and $\mathfrak{b}$ are filters of $R \times R$, then $\{A \circ B : A \in \mathfrak{a}, B \in \mathfrak{b}\}$ is a filter basis, generating a filter denoted by $\mathfrak{a} \circ \mathfrak{b}$. For any subset A of $R \times R$ and for any point p of R, $A[p] = \{q : (p, q) \in A\}$ is a subset of R. We can likewise associate with the filter $\mathfrak{a}$ of $R \times R$ the filter $\mathfrak{a}[p] = \{A[p] : A \in \mathfrak{a}\}$ of R. We have $(A \circ B)[p] = \cup \{B[q] : q \in A[p]\}$.

Let (R, δ) be a pseudometric space. The subset

$$W_\epsilon = W_\epsilon^\delta = \{(p, q) : \delta(p, q) < \epsilon\}$$

of $R \times R$ is known as the *ϵ-uniform neighborhood* belonging to δ. The filter generated by the filter basis $\{W_\epsilon : \epsilon > 0\}$ is known as the uniform structure or uniformity of δ and denoted by $\mathfrak{w}_\delta$. This filter possesses the following properties:

(1) Since $\delta(p, p) = 0$ every *ε-uniform neighborhood* contains the diagonal and hence $\overline{D(R)} \leqslant \mathfrak{w}_\delta$.

(2) Since $W_\epsilon^- = W_\epsilon$, the filter $\mathfrak{w}_\delta$ is symmetric.

(3) $W_\epsilon \subseteq W_\epsilon \circ W_\epsilon \subseteq W_{2\epsilon}$ implies $\mathfrak{w}_\delta \circ \mathfrak{w}_\delta = \mathfrak{w}_\delta$.

Properties (1)-(3) have been expressed independently of any pseudometric and are therefore meaningful in any space.

Definition 33a: *A filter* $\mathfrak{w}$ *of* $R \times R$ *is said to be a* uniform structure *or* uniformity *of* R *whenever it possesses the following properties:*

(1) $\overline{D(R)} \leqslant \mathfrak{w}$.

(2) $\mathfrak{w}^- = \mathfrak{w}$.

(3) $\mathfrak{w} \circ \mathfrak{w} = \mathfrak{w}$.

The sets belonging to $\mathfrak{w}$ *are called* uniform neighborhoods *or* vicinities *while the couple* $(R, \mathfrak{w})$ *is called a* uniform space.

We see by the above that every pseudometric induces in a unique way a uniform structure. Hence every pseudometric space can also be viewed as a uniform space. But as we shall show below, the concept of uniform space is more general: Not every uniform space can be obtained from a pseudometric.

33.1 *Given any set* W *belonging to the filter* $\mathfrak{w}$, *we can find a symmetric set* W' *such that* $W' \circ W' \subseteq W$.

Proof: From $\mathfrak{w} \circ \mathfrak{w} = \mathfrak{w}$ we deduce that, given any $W \in \mathfrak{w}$ we can find sets $W_1, W_2 \in \mathfrak{w}$ with $W_1 \circ W_2 \subseteq W$. Since $\mathfrak{w}$ is symmetric, there also exists a symmetric set $W' \in \mathfrak{w}$ such that $W' \subseteq W_1 \cap W_2$. Hence

$$W' \circ W' \subseteq W_1 \circ W_2 \subseteq W. \blacksquare$$

Every uniform space can also be considered a topological space in the following sense: From $\overline{D(R)} \leqslant \mathfrak{w}$ we deduce that $p \in W[p]$ for all $W \in \mathfrak{w}$ and hence $\overline{p} \leqslant \mathfrak{w}[p]$. Given $W \in \mathfrak{w}$, let W' be a set belonging to $\mathfrak{w}$ such that $W' \circ W' \subseteq W$ (33.1). From $q \in W'[p]$ and $x \in W'[q]$ (and thereby $(p, q) \in W'$ and $(q, x) \in W'$) we deduce that $(p, x) \in W' \circ W' \subseteq W$; i.e., $W'[q] \subseteq W[p]$. Hence, by

$$\tau_{\mathfrak{w}} p = \mathfrak{w}[p],$$

we can define a topology $\tau_{\mathfrak{w}}$ on R [cf. p. 34, condition (2)], known as the *topology induced by* $\mathfrak{w}$. It is in this sense of induced topology that we can call $(R, \mathfrak{w})$ a topological space.

If δ is a pseudometric on R, then δ induces, on the one hand, a direct topology τ_δ on R. On the other hand, δ also induces a uniform structure $\mathfrak{w}_\delta$ and thereby a topology $\tau_{\mathfrak{w}_\delta}$. But since

$$W_\epsilon^\delta[p] = \{q : \delta(p, q) < \epsilon\} = U_\epsilon^\delta(p)$$

we have $\tau_\delta = \tau_{\mathfrak{w}_\delta}$.

33.2 *A uniform space* $(R, \mathfrak{w})$ *is a* T_1*-space if and only if*

$$D(R) = \cap \{W : W \in \mathfrak{w}\}.$$

Proof: This condition is equivalent to

$$\{p\} = \cap \{W[p] : W \in \mathfrak{w}\} = \cap \{U : U \in \tau_{\mathfrak{w}}p\}$$

for all $p \in R$. ∎

33.3 *Let* $\{\mathfrak{w}_\iota : \iota \in I\}$ *be a system of uniform structures on R. Then* $\mathfrak{w} = \bigwedge\{\mathfrak{w}_\iota : \iota \in I\}$ *is also a uniform structure on R, and* $\tau_{\mathfrak{w}_\iota} = \bigwedge \{\tau_{\mathfrak{w}} : \iota \in I\}$.

Proof: Since $\overline{D(R)} \leqslant \mathfrak{w}_\iota$, we also have $\overline{D(R)} \leqslant \mathfrak{w}$. The filters $\mathfrak{w}_\iota$ are symmetric and hence possess bases made up of symmetric sets. Every intersection of symmetric sets is again symmetric and hence $\mathfrak{w}$ possesses a basis of symmetric sets and must itself be a symmetric filter. From $W \in \mathfrak{w}$ we deduce that $W = W_1 \cap \ldots \cap W_n$ with $W_\nu \in \mathfrak{w}_{\iota_\nu}$. By 33.1 there must exist sets $W'_\nu \in \mathfrak{w}_{\iota_\nu}$ such that $W'_\nu \circ W'_\nu \subseteq W_\nu$. Then $W' = W'_1 \cap \ldots \cap W'_n$ is a set in $\mathfrak{w}$ with $W' \circ W' \subseteq W$, whence $\mathfrak{w} \circ \mathfrak{w} = \mathfrak{w}$. Thus $\mathfrak{w}$ is a uniform structure. Lastly, we have

$$\tau_{\mathfrak{w}}p = \mathfrak{w}[p] = \bigwedge \mathfrak{w}_\iota[p] = \bigwedge (\tau_{\mathfrak{w}_\iota}p) = (\bigwedge \tau_{\mathfrak{w}_\iota})\, p. \blacksquare$$

Uniform spaces stand between (and even help to delimit) topological and pseudometrizable spaces.

Definition 33b: *A uniform space* $(R, \mathfrak{w})$ *is said to be* uniformly (pseudo-) metrizable *if there exists a (pseudo-) metric* δ *on R with* $\mathfrak{w}_\delta = \mathfrak{w}$.

Uniform metrizability is a stronger property than metrizability: A uniform space can be metrizable (i.e., the topological space $(R, \tau_{\mathfrak{w}})$ is metrizable) yet there may not exist a unique metric belonging to $\tau_{\mathfrak{w}}$ which, in turn, induces the uniform structure $\mathfrak{w}$.

33.4 *A uniform space* $(R, \mathfrak{w})$ *is uniformly pseudometrizable if and only if* $\mathrm{Rk}\,(\mathfrak{w}) \leqslant \aleph_0$. *It is uniformly metrizable if, in addition,* $D(R) = \cap\{W : W \in \mathfrak{w}\}$.

Proof: If there exists a pseudometric δ on R with $\mathfrak{w}_\delta = \mathfrak{w}$, then the ϵ-vicinities $W^\delta_{1/n}$ constitute a countable basis of w. Conversely, suppose $\mathrm{Rk}\,(\mathfrak{w}) \leqslant \aleph_0$. Since $\mathfrak{w}$ is symmetrical there must exist a countable basis $\{W_n : n \in \mathbf{N}\}$ of $\mathfrak{w}$, made up entirely of symmetric sets. By 33.1, we may even assume

$$W_{n+1} \circ W_{n+1} \circ W_{n+1} \subseteq W_n .$$

For each couple (x, y) of $R \times R$ let

$$\eta(x, y) = \inf \{2^{-n} : (x, y) \in W_n\}.$$

Let us further define

$$\lambda(x_0, \ldots, x_r) = \sum_{\rho=1}^{r} \eta(x_{\rho-1}, x_\rho)$$

and

$$\delta(p, q) = \inf\{\lambda(x_0, \ldots, x_r)\},$$

where these infs are taken over all finite sets of points $\{x_0, \ldots, x_r\}$ such that $x_0 = p$ and $x_r = q$. Since $(p, p) \in D(R) \subseteq W_n$ for all n, we have $\eta(p, p) = 0$ and hence $\delta(p, p) = 0$. From the symmetry of W_n we have $\eta(p, q) = \eta(q, p)$ and then $\delta(p, q) = \delta(q, p)$. Lastly, the very definition of δ immediately implies that δ obeys the triangle inequality and therefore δ is a pseudometric on R. From $(p, q) \in W_n$ we deduce $\delta(p, q) \leqslant \eta(p, q)$, hence $W_n \subseteq W^{\delta}_{2^{-n}}$ and therefore $\mathfrak{w} \leqslant \mathfrak{w}_\delta$. If, conversely, we can establish the inequality

$$(*) \qquad \eta(x_0, x_r) \leqslant 2\lambda(x_0, \ldots, x_r)$$

then it would follow therefrom that $W^{\delta}_{2^{-(n+1)}} \subseteq W_n$ and hence $\mathfrak{w}_\delta \leqslant \mathfrak{w}$. This would imply $\mathfrak{w}_\delta = \mathfrak{w}$ and $(R, \mathfrak{w})$ would therefore be uniformly pseudometrizable.

We shall establish the inequality (*) by induction on r. For $r = 1$ this inequality is obvious. Let $\lambda(x_0, \ldots, x_r) = \alpha$ and let t be the largest index such that $\lambda(x_0, \ldots, x_t) \leqslant \alpha/2$. Then $\lambda(x_{t+1}, \ldots, x_r) \leqslant \alpha$ and $\eta(x_t, x_{t+1}) \leqslant \alpha$. From the inductive hypothesis it follows that $\eta(x_0, x_t) \leqslant \alpha$ and $\eta(x_{t+1}, x_t) \leqslant \alpha$. If n is the smallest natural number such that $2^{-(n+1)} \leqslant \alpha$, then (x_0, x_t), (x_t, x_{t+1}), and (x_{t+1}, x_r) lie in W_{n+1}. Since $W_{n+1} \circ W_{n+1} \circ W_{n+1} \subseteq W_n$, we deduce that $(x_0, x_r) \in W_n$ and therefore $\eta(x_0, x_n) \leqslant 2^{-n} \leqslant 2\alpha$.

Our contention on uniform metrizability now follows directly from 33.2. ∎

If $\{\delta_\iota : \iota \in I\}$ is a system of pseudometrics on the fundamental set R, then, by 33.3, $\mathfrak{w} = \bigwedge_\iota \mathfrak{w}_{\delta_\iota}$ is a uniform structure on R. Conversely, we shall now show that every uniform structure of R can be obtained in this way. Let, therefore, $\Delta(\mathfrak{w})$ denote the set of all pseudometrics δ on R such that $\mathfrak{w}_\delta \geqslant \mathfrak{w}$.

33.5 $$\mathfrak{w} = \bigwedge \{\mathfrak{w}_\delta : \delta \in \Delta(\mathfrak{w})\}.$$

Proof: Let W be a symmetric set in $\mathfrak{w}$. By 33.1 there must then exist a sequence $(W_n)_{n \in \mathbf{N}}$ of symmetric sets $W_n \in \mathfrak{w}$ such that $W_{n+1} \circ W_{n+1} \subseteq W_n$ and $W_0 = W$. Consequently $W_{n+1} \subseteq W_n$. The sets $\overline{W_n}$ therefore constitute a filter basis and generate a filter $\mathfrak{w}^*$ with $\mathfrak{w} \leqslant \mathfrak{w}^* \leqslant \overline{W}$. Since $\mathfrak{w} \leqslant \mathfrak{w}^*$ we have $\overline{D(R)} \leqslant \mathfrak{w}^*$. From the symmetry of the W_n we see that $\mathfrak{w}^{*-1} = \mathfrak{w}^*$. Since $W_{n+1} \circ W_{n+1} \subseteq W_n$, we finally see that $\mathfrak{w}^* \circ \mathfrak{w}^* = \mathfrak{w}^*$ and therefore $\mathfrak{w}^*$ is a uniform structure on R. Moreover, since $\mathrm{Rk}(\mathfrak{w}^*) \leqslant \aleph_0$, there must exist, by 33.4, a pseudometric δ on R with $\mathfrak{w}_\delta = \mathfrak{w}^* \geqslant \mathfrak{w}$, hence with $\delta \in \Delta(\mathfrak{w})$. Since $\mathfrak{w}_\delta = \mathfrak{w}^* \leqslant \overline{W}$ and since $\mathfrak{w}$ possesses a basis of symmetric sets, we see that $\mathfrak{w} \leqslant \bigwedge \{\mathfrak{w}_\delta : \delta \in \Delta(\mathfrak{w})\} \leqslant \bigwedge \{\overline{W} : W \in \mathfrak{w}\} = \mathfrak{w}$. ∎

Definition 33c: *A topological space* (R, τ) *and its topology are both said to be* uniformizable *if there exists on* R *a uniform structure* $\mathfrak{w}$, *belonging to* τ; i.e., if $\tau = \tau_{\mathfrak{w}}$.

33.6 *The following statements are pairwise equivalent:*

(1) (R, τ) *is uniformizable.*

(2) τ *is the intersection of pseudometrizable topologies of* R.

(3) (R, τ) *is a* T_{3a}-*space.*

Proof

(1) → (2): Let $\mathfrak{w}$ be a uniform structure of R belonging to τ. By 33.5 we then have $\mathfrak{w} = \bigwedge \{\mathfrak{w}_\delta : \delta \in \Delta(\mathfrak{w})\}$ and consequently $\tau = \tau_{\mathfrak{w}} = \bigwedge\{\tau_\delta : \delta \in \Delta(\mathfrak{w})\}$ (33.3).

(2) → (3): Since every pseudometrizable topology is, by 30.1, a T_{3a}-topology, the intersection of pseudometrizable topologies is also a T_{3a}-topology (18.14).

(3) → (1): Let Φ be the set of all continuous maps $\varphi : (R, \tau) \to [0, 1]$. For each $\varphi \in \Delta$ and each $\epsilon > 0$ let

$$W_{\varphi,\epsilon} = \{(p, q) : | \varphi p - \varphi q | < \epsilon\}.$$

The intersection of any finite collection of sets of the form $W_{\varphi,\epsilon}$ is a filter basis, generating a filter $\mathfrak{w}$ of $R \times R$. Since $D(R) \subseteq W_{\varphi,\epsilon}$ we have $\overline{D(R)} \leqslant \mathfrak{w}$. Since the sets $W_{\varphi,\epsilon}$ are symmetric, we have $\mathfrak{w}^- = \mathfrak{w}$. Finally, $W_{\varphi,\epsilon} \circ W_{\varphi,\epsilon} \subseteq W_{\varphi,2\epsilon}$, whence $\mathfrak{w} \circ \mathfrak{w} = \mathfrak{w}$. Thus we see that $\mathfrak{w}$ is a uniform structure on R. Given any $p \in R$, $\varphi \in \Phi$, and $\epsilon > 0$, we can find, because of the continuity of φ, a $U \in \tau p$ such that $\varphi U \subseteq U_\epsilon(\varphi p)$. This implies $U \subseteq W_{\varphi,\epsilon}[p]$, hence

$$\tau p \leqslant \mathfrak{w}[p] = \tau_{\mathfrak{w}} p$$

and thus $\tau \leqslant \tau_{\mathfrak{w}}$. Conversely, by $(\mathbf{T}_{3a})$, given any open neighborhood $U \in \tau p$, we can find a $\varphi \in \Phi$ such that $\varphi p = 0$ and $\varphi \complement U = \{1\}$. From $q \in W_{\varphi,1}[p]$ we therefore deduce that $\varphi q = | \varphi p - \varphi q | < 1$; hence $q \in U$ and $W_{\varphi,1}[p] \subseteq U$. Hence we have the converse relation, $\tau_{\mathfrak{w}} \leqslant \tau$ and therefore every T_{3a}-space is uniformizable. ∎

We have the following important corollary.

33.7 *Every uniform space is a* T_{3a}-*space*; *every uniform* T_1-*space is completely regular.*

A uniform structure $\mathfrak{w}$ on R also determines a unique topology on $R \times R$, namely, the product topology $\tau_{\mathfrak{w}} \times \tau_{\mathfrak{w}}$. It is therefore natural for us to study the topological properties of the vicinities filter $\mathfrak{w}$ in the product space $(R, \tau_{\mathfrak{w}}) \times (R, \tau_{\mathfrak{w}})$.

33.8 *The filter* $\mathfrak{w}$ *is open and closed in* $(R, \tau_{\mathfrak{w}}) \times (R, \tau_{\mathfrak{w}})$.

Proof: Given $W \in \mathfrak{w}$ we can find, by 33.1, a symmetric set $W' \in \mathfrak{w}$ with $W' \circ W' \circ W' \subseteq W$. We shall show that the interior $\overset{\circ}{W}$ of W, taken with respect to the product-topology $\tau_{\mathfrak{w}} \times \tau_{\mathfrak{w}}$, contains the vicinity W'. Let therefore (p, q) be any point of W'. Then $W'[p] \times W'[q]$ is a neighborhood of (p, q) in $(R, \tau_{\mathfrak{w}}) \times (R, \tau_{\mathfrak{w}})$. From $(x, y) \in W'[p] \times W'[q]$ it follows that

$$(p, x) \in W', \qquad (q, y) \in W',$$

and, by the symmetry of W', that $(x, p) \in W'$. Since $(p, q) \in W'$ we have $(x, y) \in W' \circ W' \circ W' \subseteq W$; hence $W'[p] \times W'[q] \subseteq W$. This implies $(p, q) \in \overset{\circ}{W}$ and thus $W' \subseteq \overset{\circ}{W}$. Therefore $\overset{\circ}{W} \in \mathfrak{w}$; i.e., $\mathfrak{w}$ possesses a basis of open sets. Likewise, from $(p, q) \in \overline{W'}$ we deduce $(W'[p] \times W'[q]) \cap W' \neq \emptyset$, and by similar reasoning to that above, (p, q) must belong to W. Hence, $\overline{W'} \subseteq W$ and $\mathfrak{w}$ also possesses a basis of closed sets. ∎

The uniform neighborhoods or vicinities in $\mathfrak{w}$ are, by the above theorem, neighborhoods of the diagonal in $(R, \tau_{\mathfrak{w}}) \times (R, \tau_{\mathfrak{w}})$. The converse result is, in general, not true: An arbitrary neighborhood of the diagonal need not be a uniform neighborhood of $\mathfrak{w}$. For example, let δ be the absolute-value metric on $\mathbf{R}$. Then the set $\{(x, y) : |x^2 - y^2| < 1\}$ is a neighborhood of the diagonal in $(\mathbf{R}, \tau_\delta) \times (\mathbf{R}, \tau_\delta)$ but this set contains no ϵ-uniform neighborhood belonging to δ. If, however, we restrict ourselves to completely compact uniform spaces, then the converse also holds.

33.9 *In a completely compact uniform space $(R, \mathfrak{w})$ the filter $\mathfrak{w}$ is composed of all the neighborhoods of the diagonal of $(R, \tau_{\mathfrak{w}}) \times (R, \tau_{\mathfrak{w}})$.*

Proof: If $(R, \tau_{\mathfrak{w}})$ is completely compact, so is the product space $(R, \tau_{\mathfrak{w}}) \times (R, \tau_{\mathfrak{w}})$ (22.6). Let V be an open neighborhood of the diagonal of this product space. Suppose $\mathfrak{a} = \mathfrak{w} \wedge \ulcorner \complement U \urcorner \neq \mathfrak{o}$, then we deduce the existence of an adherent point $(p, q) \in \complement U$ of $\mathfrak{a}$. Since U is open, there exists a $W \in \mathfrak{w}$ such that

$$W[p] \times W[p] \subseteq \bar{U}.$$

Since $\mathfrak{a} \leqslant \mathfrak{w}$, and since $\mathfrak{w}$ is closed by 33.8, we see that $\ulcorner (p, q) \urcorner \leqslant \bar{\mathfrak{a}} \leqslant \bar{\mathfrak{w}} = \mathfrak{w} \leqslant \ulcorner W \urcorner$; hence $(p, q) \in W$ and $q \in W[p]$. This implies that $(p, q) \in W[p] \times W[p] \subseteq U$ contradicting $(p, q) \in \complement U$. Hence $\mathfrak{w} \wedge \ulcorner \complement U \urcorner = \mathfrak{o}$ and $\mathfrak{w} \leqslant \ulcorner U \urcorner$. ∎

This theorem tells us that we can associate one and only one uniform structure with each completely compact T_{3a}-space (R, τ) (uniformizable by 33.6)—this uniform structure being the filter of all neighborhoods of the diagonal in $(R, \tau) \times (R, \tau)$. For arbitrary uniform spaces we do not have such a uniqueness theorem: There can exist different uniform structures in R which induce the same topology. Likewise, distinct pseudometrics of R can induce the same topology on R (cf. 33C).

Definition 33d: *Two uniform structures on R are said to be* topologically equivalent *if they induce the same topology on R. Two pseudometrics of R are said to be* uniformly equivalent *if they induce the same uniform structure on R.*

Uniformly equivalent pseudometrics are also topologically equivalent. Within every class of topologically equivalent uniform structures there exists a finest such structure; namely, the intersection of all the uniform structures of this class (33.3).

If M is a subset of the uniform space $(R, \mathfrak{w})$, then the trace filter $\mathfrak{w}_M$ of $\mathfrak{w}$ in $M \times M$ is manifestly a uniform structure of M. We call $\mathfrak{w}_M$ the *relative structure* of $\mathfrak{w}$ with respect to M and $(M, \mathfrak{w}_M)$ a (uniform) subspace of $(R, \mathfrak{w})$. The topology induced by $\mathfrak{w}_M$ on M is identical to the relative topology induced by $\tau_{\mathfrak{w}}$ on M.

Notes and Exercises

33A For every subset T of $R \times R$ and every subset M of R let us suppose $T[M] = \cup \{T[x] : x \in M\}$.

Exercise

1. $(S \circ T)[M] = S[T[M]]$.
2. If S is symmetric, then $S \circ T \circ S = \cup \{S[x] \times S[y] : (x, y) \in T\}$.
3. Let M be a subset of the uniform space $(R, \mathfrak{w})$. Then

$$\bar{M} = \cap \{W[M] : W \in \mathfrak{w}\}.$$

4. Let T be a subset of $R \times R$ and $\mathfrak{w}$ a uniform structure on R In $(R, \tau_{\mathfrak{w}}) \times (R, \tau_{\mathfrak{w}})$ we then have $\bar{T} = \cap \{W \circ T \circ W : W \in \mathfrak{w}\}$.

33B Let I be an indexing set. To each point x of the fundamental set R and to each index $\iota \in I$ let a subset $V_\iota(x)$ of R be associated, such that the following conditions are satisfied:

(a) $x \in V_\iota(x)$ *for all* $\iota \in I$.

(b) *Given* $\alpha, \beta \in I$ *we can find a* $\gamma \in I$ *such that* $V_\gamma(x) \subseteq V_\alpha(x) \cap V_\beta(x)$ *for all* $x \in R$.

(c) *Given* $\alpha \in I$ *there exists a* $\beta \in I$ *with the following property:* $y \in V_\beta(x)$ *implies* $x \in V_\alpha(y)$.

(d) *Given* $\alpha \in I$ *there exists a* $\beta \in I$ *such that: If* $x \in V_\beta(y)$ *and* $y \in V_\beta(z)$ *then* $x \in V_\alpha(z)$.

If, now, we put $W_\iota = \{(x, y) : y \in V_\iota(x)\}$, then W_ι is a subset of $R \times R$.

Exercise: $\mathfrak{W} = \{W_\iota : \iota \in I\}$ is a filter basis. The filter generated by $\mathfrak{W}$ is a uniform structure on R.

This description of uniform structure is due to A. Weil. Every uniform structure can be obtained in this way: For let $\mathfrak{w}$ be a uniform structure on R. If we think of $\mathfrak{w}$ as an indexing set, then to each "index" $W \in \mathfrak{w}$ we can associate the set $V_W(x) = W[x]$.

Exercise: The sets $V_W(x)$ satisfy conditions (a)-(d). The resulting uniform structure is identical to the initial structure $\mathfrak{w}$.

33C In $\mathbf{R} \times \mathbf{R}$, for every $\epsilon > 0$, let

$$W_\epsilon = \{(x, y) : | x - y | < \epsilon\} \qquad \text{and} \qquad W'_\epsilon = \{(x, y) : | e^x - e^y | < \epsilon\}.$$

Exercise: The filter bases $\{W_\epsilon : \epsilon > 0\}$ and $\{W'_\epsilon : \epsilon > 0\}$ each generate a uniform structure on $\mathbf{R}$. Call them $\mathfrak{w}$ and $\mathfrak{w}'$, respectively. These filters $\mathfrak{w}$ and $\mathfrak{w}'$ are noncomparable, yet they are nevertheless topologically equivalent uniform structures.

33D Let (R, τ) be a uniformizable space. And let $\mathfrak{w}$ be the collection of all neighborhoods W of the diagonal in $(R, \tau) \times (R, \tau)$ which satisfy the following condition: There exists a sequence $(U_n)_{n \in N}$ of neighborhoods of the diagonal such that $U_1 \subseteq W$ and $U_{n+1} \circ U_{n+1} \subseteq U_n$.

Exercise: $\mathfrak{w}$ is the finest uniform structure on R inducing the topology τ [24, p. 205].

34. Uniform Continuity and Uniform Convergence

Let $(R, \mathfrak{w})$ and $(R', \mathfrak{w}')$ be two uniform spaces, and let φ be a map from R into R'. We can speak of the continuity of φ in the sense of the induced topology. This continuity means that $\varphi\mathfrak{w}[p] \leqslant \mathfrak{w}'[\varphi p]$ for all $p \in R$. Given any $W' \in \mathfrak{w}'$ and any $p \in R$ there must exist a $W \in \mathfrak{w}$ such that $\varphi W[p] \subseteq W'[\varphi p]$. For fixed W', W still depends on p. If $\varphi W[p] \subseteq W'[\varphi p]$, where W remains the same for all p, then we can say that φ is uniformly continuous. Let $\varphi \times \varphi$ denote the mapping of $R \times R$ into $R' \times R'$ defined by $(\varphi \times \varphi)(p, q) = (\varphi p, \varphi q)$ (cf. Section 22). Then the above condition is equivalent to $(\varphi \times \varphi) W \subseteq W'$. We are thus led to:

Definition 34a: *A mapping* $\varphi : R \to R'$ *is said to be* uniformly continuous *with respect to the uniform structures* $\mathfrak{w}$ *of* R *and* $\mathfrak{w}'$ *of* R' *if* $(\varphi \times \varphi)\, \mathfrak{w} \leqslant \mathfrak{w}'$. *We shall denote this relationship by* $\varphi : (R, \mathfrak{w}) \to (R', \mathfrak{w}')$.

In the case where the uniform structures $\mathfrak{w}$ and $\mathfrak{w}'$ can be induced by the pseudometrics, δ and δ', respectively, then the condition for the uniform continuity of φ can be reformulated as follows: Given any $\epsilon > 0$, we can find an $\eta > 0$ such that $\delta(p, q) < \eta$ implies $\delta'(\varphi p, \varphi q) < \epsilon$. Every uniformly continuous map from $(R, \mathfrak{w})$ into $(R', \mathfrak{w}')$ is also a continuous map from $(R, \tau_{\mathfrak{w}})$ into $(R', \tau_{\mathfrak{w}'})$. But as for the converse, continuity does not in general imply uniform continuity, as the example of $\varphi x = 1/x$ shows, where $\varphi :]0, \rightarrow] \rightarrow \mathbf{R}$ (nevertheless, cf. 34.2).

34.1 *Let φ be a mapping of R into R' and let $\mathfrak{w}'$ be a uniform structure on R'. Then, ${}_{\varphi}\mathfrak{w}' = (\varphi \times \varphi)^{-1}\mathfrak{w}'$ defines a uniform structure on R.*

${}_{\varphi}\mathfrak{w}'$ is the coarsest uniform topology on R with respect to which φ is a uniformly continuous map into $(R', \mathfrak{w}')$ (i.e., $\varphi : (R, \mathfrak{w}) \rightarrow (R', \mathfrak{w}')$ always implies $\mathfrak{w} \leqslant {}_{\varphi}\mathfrak{w}'$.

The topology induced on R by ${}_{\varphi}\mathfrak{w}'$ is the topology ${}_{\varphi}(\tau_{\mathfrak{w}'})$ transported by φ^{-1} (cf. Section 20).

Proof: Since $(\varphi \times \varphi)\, D(R) \subseteq D(R')$, we have $\overline{D(R)} \leqslant {}_{\varphi}\mathfrak{w}'$. For arbitrary subsets S' and T' of $R' \times R'$ we also have

$$(\varphi \times \varphi)^{-1}S'^{-} = ((\varphi \times \varphi)^{-1}S')^{-}$$

and

$$(\varphi \times \varphi)^{-1}(S' \circ T') \supseteq ((\varphi \times \varphi)^{-1}S') \circ ((\varphi \times \varphi)^{-1}T').$$

This implies $({}_{\varphi}\mathfrak{w}')^{-} = {}_{\varphi}\mathfrak{w}'$ and ${}_{\varphi}\mathfrak{w}' \circ {}_{\varphi}\mathfrak{w}' = {}_{\varphi}\mathfrak{w}'$ and hence ${}_{\varphi}\mathfrak{w}'$ is a uniform structure on R. Since $(\varphi \times \varphi)_{\varphi}\mathfrak{w}' = (\varphi \times \varphi)\,(\varphi \times \varphi)^{-1}\mathfrak{w}' \leqslant \mathfrak{w}'$, we see that φ is a uniformly continuous map of $(R, {}_{\varphi}\mathfrak{w}')$ into $(R', \mathfrak{w}')$. Conversely, let $\mathfrak{w}$ be a uniform structure on R with $\varphi : (R, \mathfrak{w}) \rightarrow (R', \mathfrak{w}')$. Then $(\varphi \times \varphi)\, \mathfrak{w} \leqslant \mathfrak{w}'$ and consequently $\mathfrak{w} \leqslant (\varphi \times \varphi)^{-1}\mathfrak{w}' = {}_{\varphi}\mathfrak{w}'$. This means that ${}_{\varphi}\mathfrak{w}'$ is the coarsest uniform structure with this property. Lastly, we have

$$((\varphi \times \varphi)^{-1}W')\,[p] = \{q : (\varphi p, \varphi q) \in W'\} = \varphi^{-1}(W'[\varphi p]),$$

which implies

$${}_{\varphi}\mathfrak{w}'[p] = \varphi^{-1}\mathfrak{w}'[\varphi p] = \varphi^{-1}\tau_{\mathfrak{w}'}\varphi p = {}_{\varphi}(\tau_{\mathfrak{w}'})\, p.$$

This is the final contention of our theorem. ∎

The uniform structure ${}_{\varphi}\mathfrak{w}'$ is known as the (uniform) structure transported by φ^{-1}.

34.2 *Let $\mathfrak{w}$ be a uniform structure on R, the finest one which induces a given topology τ. And let $(R', \mathfrak{w}')$ be an arbitrary uniform space. Then every continuous map from (R, τ) into $(R', \tau_{\mathfrak{w}'})$ is also a uniformly continuous mapping from $(R, \mathfrak{w})$ into $(R', \mathfrak{w}')$.*

Proof: From $\varphi : (R, \tau) \rightarrow (R', \tau_{\mathfrak{w}'})$ we see that $\tau \leqslant {}_{\varphi}(\tau_{\mathfrak{w}'})$ (20.1). In addition, by 33.3 and 34.1 we see that $\mathfrak{w} \wedge {}_{\varphi}\mathfrak{w}'$ is a uniform structure on R which induces the topology $\tau \wedge {}_{\varphi}(\tau_{\mathfrak{w}'}) = \tau$, and hence $\mathfrak{w} \leqslant {}_{\varphi}\mathfrak{w}'$. We finally have $(\varphi \times \varphi)\, \mathfrak{w} \leqslant (\varphi \times \varphi)_{\varphi}\mathfrak{w}' \leqslant \mathfrak{w}'$. ∎

We have the following corollary:

34.3 *If a uniform space $(R, \mathfrak{w})$ is completely compact, then every continuous mapping of $(R, \tau_{\mathfrak{w}})$ into $(R', \tau_{\mathfrak{w}'})$ is also a uniformly continuous map of $(R, \mathfrak{w})$ into $(R', \mathfrak{w}')$.*

Proof: By 33.9, $\mathfrak{w}$ is the finest uniform structure on R inducing the topology $\tau_{\mathfrak{w}}$. ∎

The composition of two uniformly continuous maps gives as a product another uniformly continuous map. A uniformly continuous 1-1 map φ from $(R, \mathfrak{w})$ onto $(R', \mathfrak{w}')$ is called a *uniform isomorphism* whenever φ^{-1} is also uniformly continuous. We then have $(\varphi \times \varphi)\,\mathfrak{w} = \mathfrak{w}'$ and hence $\varphi\tau_{\mathfrak{w}} = \tau_{\mathfrak{w}'}\varphi$. Hence every uniform isomorphism is at the same time a homeomorphism. An isometry between two pseudometric spaces (R, δ) and (R', δ') is also a uniform isomorphism of the uniform spaces $(R, \mathfrak{w}_\delta)$ and $(R', \mathfrak{w}_{\delta'})$. Two uniform spaces are said to be *uniformly isomorphic* if there exists a uniform isomorphism between then. Properties which are preserved under all uniform isomorphisms are called *uniform invariants.*

We can generalize the situation in 34.1 as follows: Let $\{(R'_\iota, \mathfrak{w}'_\iota) : \iota \in I\}$ be a collection of uniform spaces, and for each $\iota \in I$ let φ_ι be a map from R into R'_ι. Then manifestly,

$$\mathfrak{w} = \bigwedge \{(\varphi_\iota \times \varphi_\iota)^{-1}\mathfrak{w}'_\iota : \iota \in I\} = \bigwedge \{{}_{\varphi_\iota}\mathfrak{w}'_\iota : \iota \in I\}$$

is the coarsest uniform structure of R with respect to which all mappings φ_ι are uniformly continuous. We can therefore use this process to define a uniform structure on the product of sets.

As in Section 22, for each $\kappa \in I$ let π_κ denote the projection map from $R = \times \{R_\iota : \iota \in I\}$ onto R_κ. Then $\pi_\kappa \times \pi_\kappa$ is a map from

$$R \times R = (\times R_\iota) \times (\times R_\iota)$$

onto $R_\kappa \times R_\kappa$. If now, for each $\iota \in I$, $\mathfrak{w}_\iota$ is a uniform structure on R_ι, then $\mathfrak{w} = \bigwedge \{(\pi_\iota \times \pi_\iota)^{-1}\mathfrak{w}_\iota : \iota \in I\}$ is the coarsest uniform structure on the product set R for which all projection mappings are uniformly continuous. This uniform structure $\mathfrak{w}$ is called the *product-structure* of the uniform structures $\mathfrak{w}_\iota$, while $(R, \mathfrak{w})$ is called the *uniform product space* of the uniform spaces $(R_\iota, \mathfrak{w}_\iota)$ and often also denoted by $\times (R_\iota, \mathfrak{w}_\iota)$.

The product structure $\mathfrak{w} = \bigwedge (\pi_\iota \times \pi_\iota)^{-1}\mathfrak{w}_\iota$ does not turn out to be the same as the product-filter $\times\, \mathfrak{w}_\iota$ defined in Section 22: For $\times\, \mathfrak{w}_\iota$ is a filter in the product set $\times (R_\iota \times R_\iota)$ while $\mathfrak{w}$ is a filter in $(\times R_\iota) \times (\times R_\iota)$. Nevertheless there exists a definite relation between these two filters. If we define the map $\theta : \times (R_\iota \times R_\iota) \to (\times R_\iota) \times (\times R_\iota)$ by $\theta \times (p_\iota, q_\iota) = (\times p_\iota, \times q_\iota)$, then $\mathfrak{w} = \theta \times \mathfrak{w}_\iota$.

34.4 *The product structure* $\mathfrak{w} = \theta \times \mathfrak{w}_\iota$ *induces on* $R = \times R_\iota$ *the product topology* $\times \tau_{\mathfrak{w}_\iota}$.

For mappings $\varphi_\iota : R'' \to R_\iota$ *and* $\psi_\iota : R_\iota \to R'_\iota$ *the conditions*

$$\times \varphi_\iota : (R'', \mathfrak{w}'') \to \times (R_\iota , \mathfrak{w}_\iota) \qquad \textit{and} \qquad \times \psi_\iota : \times R_\iota , \mathfrak{w}_\iota) \to \times (R''_\iota, \mathfrak{w}'_\iota)$$

are equivalent, respectively, to the conditions

$$\varphi_\iota : (R'', \mathfrak{w}'') \to (R_\iota , \mathfrak{w}_\iota) \qquad \textit{and} \qquad \psi_\iota : (R_\iota , \mathfrak{w}_\iota) \to (R'_\iota ; \mathfrak{w}'_\iota)$$

for all $\iota \in I$.

Proof: For an arbitrary point $p \in R$ and for each $W_\iota \in \mathfrak{w}_\iota$ we have

$$((\pi_\iota \times \pi_\iota)^{-1} W_\iota) [p] = \{q : (\pi_\iota p, \pi_\iota q) \in W_\iota\} = \pi_\iota^{-1} W_\iota[\pi_\iota p]$$

and for subsets M_ν of $R \times R$ we also have

$$(M_1 \cap \ldots \cap M_n) [p] = \{q : (p, q) \in M_1 \cap \ldots \cap M_n\} = M_1[p] \cap \ldots \cap M_n[p].$$

This implies that

$$\tau_{\mathfrak{w}} p = (\bigwedge (\pi_\iota \times \pi_\iota)^{-1} \mathfrak{w}_\iota)[p] = \bigwedge (\pi_\iota^{-1} \mathfrak{w}_\iota[\pi_\iota p]) = \bigwedge (\pi_\iota^{-1} \tau_{\mathfrak{w}_\iota} \pi_\iota p) = (\times \tau_{\mathfrak{w}_\iota})\, p.$$

From $\varphi_\iota : (R'', \mathfrak{w}'') \to (R_\iota , \mathfrak{w}_\iota)$, and hence $(\varphi_\iota \times \varphi_\iota)\, \mathfrak{w}'' \leqslant \mathfrak{w}_\iota$, we deduce that $(\times (\varphi_\iota \times \varphi_\iota))\, \mathfrak{w}'' \leqslant \times \mathfrak{w}_\iota$ and that

$$((\times \varphi_\iota) \times (\times \varphi_\iota))\, \mathfrak{w}'' = \vartheta \times (\varphi_\iota \times \varphi_\iota)\, \mathfrak{w}'' \leqq \vartheta \times \mathfrak{w}_\iota = \mathfrak{w};$$

i.e., $\times \varphi_\iota : (R'', \mathfrak{w}'') \to (R, \mathfrak{w})$. This implies, conversely, that

$$(\varphi_\kappa \times \varphi_\kappa)\, \mathfrak{w}'' = (\pi_\kappa \times \pi_\kappa)\, ((\times \varphi_\iota) \times (\times \varphi_\iota))\, \mathfrak{w}'' \leqslant (\pi_\kappa \times \pi_\kappa)\, \mathfrak{w} = \mathfrak{w}_\kappa ,$$

hence $\varphi_\kappa : (R'', \mathfrak{w}'') \to (R_\kappa , \mathfrak{w}_\kappa)$. The last of the contentions above follows in similar fashion. ∎

For every uniform structure $\mathfrak{w}$ on R we have, by 33.5, the equation

$$\mathfrak{w} = \bigwedge \{\mathfrak{w}_\delta : \delta \in \Delta(\mathfrak{w})\},$$

where $\Delta(\mathfrak{w})$ denotes the set of all pseudometrics δ on R with $\mathfrak{w} \leqslant \mathfrak{w}_\delta$. The system $\Delta(\mathfrak{w})$ can also be characterized using uniform continuity.

34.5 *Let* $\mathfrak{w}'$ *be the uniform structure induced on the real line* **R** *by the usual absolute value. Then* $\delta \in \Delta(\mathfrak{w})$ *if and only if* $\delta : (R, \mathfrak{w}) \times (R, \mathfrak{w}) \to (\mathbf{R}, \mathfrak{w}')$.

Proof: Suppose $\delta \in \Delta(\mathfrak{w})$, and hence $\mathfrak{w} \leqslant \mathfrak{w}_\delta$. Given $\epsilon > 0$ we can then find a $W \in \mathfrak{w}$ such that $W \subseteq W^\delta_{\epsilon/2}$. From $(p, q) \in W$ and $(x, y) \in W$ it follows that $\delta(p, q) < \epsilon/2$ and $\delta(x, y) < \epsilon/2$. We have

$$\delta(p, x) - \delta(q, y) \leqq \delta(p, q) + \delta(q, y) + \delta(y, x) - \delta(q, y) < \epsilon.$$

Similarly, we have $\delta(q, y) - \delta(p, x) < \epsilon$ and thus $|\delta(p, x) - \delta(q, y)| < \epsilon$. This means that δ is uniformly continuous. Conversely, suppose that the pseudometric δ is uniformly continuous; i.e., given $\epsilon > 0$, there exists a $W \in \mathfrak{w}$ with $|\delta(p, x) - \delta(q, y)| < \epsilon$ for all $(p, q) \in W$ and $(x, y) \in W$. If we now put $x = y = q$ we get $\delta(p, q) = |\delta(p, q) - \delta(q, q)| < \epsilon$. Hence $W \subseteq W_\epsilon^\delta$ and $\mathfrak{w} \leqslant \mathfrak{w}_\delta$. ∎

34.6 *Every uniform space is uniformly isomorphic to a subspace of some product of pseudometric spaces. Every uniform T_1-space is even uniformly isomorphic to a subspace of a product of metric spaces.*

Proof: Every uniform space $(R, \mathfrak{w})$ determines a product space

$$(R^*, \mathfrak{w}^*) = \times \{(R, \mathfrak{w}_\delta) : \delta \in \Delta(\mathfrak{w})\}$$

of the pseudometric spaces (R, δ), where $\delta \in \Delta(\mathfrak{w})$. For each $\delta \in \Delta(\mathfrak{w})$ let φ_δ be the identity on R. Then $\varphi = \times\, \varphi_\delta$ is a 1-1 mapping of R into R^*. If we denote the projection maps from R^* onto R by π_δ, then we have

$$(\varphi \times \varphi)\, \mathfrak{w} = (\varphi \times \varphi) \bigwedge \mathfrak{w}_\delta = \bigwedge\, (\varphi \times \varphi)\, \mathfrak{w}_\delta = \overline{\varphi R \times \varphi R} \wedge \bigwedge\, (\pi_\delta \times \pi_\delta)^{-1}\, \mathfrak{w}_\delta$$
$$= \overline{\varphi R \times \varphi R} \wedge \mathfrak{w}^*.$$

Hence φ is a uniform isomorphism of $(R, \mathfrak{w})$ onto the uniform subspace of $(R^*, \mathfrak{w}^*)$ determined by φR.

We can pass from the pseudometric space (R, δ) by the formation of quotients to the metric space $(R, \delta)/\tilde{\delta}$ (30.5), whose metric we shall denote by δ'. Let κ_δ denote the corresponding canonical map. Then $\delta'(\kappa_\delta p, \kappa_\delta q) = \delta(p, q)$. Letting $\mathfrak{w}_\delta$ and $\mathfrak{w}_{\delta'}$ denote the vicinities filters associated with δ and δ' respectively, we then have $(\kappa_\delta \times \kappa_\delta)\, \mathfrak{w}_\delta = \mathfrak{w}_{\delta'}$. If we put $\kappa = \times \{\kappa_\delta : \delta \in \Delta(\mathfrak{w})\}$, then $\psi = \kappa\varphi$ is a map from R into the product set $R'' = \times \{R/\tilde{\delta} : \delta \in \Delta(\mathfrak{w})\}$. From $\psi p = \psi q$ we deduce that $\delta(p, q) = \delta'(\kappa_\delta p, \kappa_\delta q) = 0$ and hence

$$(p, q) \in \bigcap \{W : W \in \mathfrak{w}_\delta\}$$

for all $\delta \in \Delta(\mathfrak{w})$. Since $\mathfrak{w} = \bigwedge \{\mathfrak{w}_\delta : \delta \in \Delta(\mathfrak{w})\}$ we have

$$(p, q) \in \bigcap \{W : W \in \mathfrak{w}\} \supseteq D(R).$$

Whenever $(R, \mathfrak{w})$ is a T_1-space this inclusion becomes an inequality and we have $p = q$, whence ψ is a 1-1 mapping. Lastly, let $\mathfrak{w}''$ be the product structure $\theta \times \{\mathfrak{w}_{\delta'} : \delta \in \Delta(\mathfrak{w})\}$ of the uniform structure $\mathfrak{w}_{\delta'}$ on $(R, \delta)/\delta$. As above, we have $(\psi \times \psi)\, \mathfrak{w} = \mathfrak{w}'' \wedge \overline{\psi R \times \psi R}$ and hence ψ is a uniform isomorphism of $(R, \mathfrak{w})$ onto the uniform subspace of $(R'', \mathfrak{w}'')$ determined by ψR. ∎

An even more important concept than uniform continuity is that of uniform convergence. We arrive at this concept with the aid of certain uniform structures on sets of maps.

Let $(R', \mathfrak{w}')$ be a uniform space and, as in Section 23, let $\hat{A} = R'^R$ be the set of all maps from R into R'. If then S is a subset of R and W' a uniform neighborhood in $\mathfrak{w}'$, then[4]

$$\langle S, W' \rangle = \{(\varphi, \psi) : (\varphi \times \psi)\, S \subseteq W'\}$$

defines a subset $\langle S, W' \rangle$ of $\hat{A} \times \hat{A}$. This set consists of all couples of mappings (φ, ψ) such that $(\varphi p, \psi p) \in W'$ for all $p \in S$. Since $(\varphi p, \varphi p) \in D(R') \subseteq W'$ we also have $D(\hat{A}) \subseteq \langle S, W' \rangle$. Furthermore, $\langle S, W' \rangle^- = \langle S, W'^- \rangle$. Also, to each W' we can, by 33.1, associate a $W_1' \in \mathfrak{w}'$ with $W_1' \circ W_1' \subseteq W'$. From $(\varphi, \chi) \in \langle S, W_1' \rangle$ and $(\chi, \psi) \in \langle S, W_1' \rangle$ one deduces that $(\varphi p, \chi p) \in W_1'$ and $(\chi p, \psi p) \in W_1'$; hence $(\varphi p, \psi p) \in W_1' \circ W_1' \subseteq W'$ for all $p \in S$ and therefore $(\varphi, \psi) \in \langle S, W' \rangle$. This implies $\langle S, W_1' \rangle \circ \langle S, W_1' \rangle \subseteq \langle S, W' \rangle$ and, evidently, also $\langle S, W_1' \rangle \subseteq \langle S, W_1' \rangle \circ \langle S, W_1' \rangle$. These partial results, taken together, tell us that the filter basis $\{\langle S, W' \rangle : W' \in \mathfrak{w}'\}$ generates a vicinities filter $\hat{\mathfrak{w}}_S$ of $\hat{A}$.

Definition 34b: *The uniform structure $\hat{\mathfrak{w}}_S$ of $\hat{A}$ is called the* structure of uniform convergence *on S (with respect to the uniform structure $\mathfrak{w}'$ of R'). The topology induced by $\hat{\mathfrak{w}}_S$ on $\hat{A}$ is called the* topology of uniform convergence *on S, while a filter convergent under this topology is said to* converge uniformly on S.

In particular, if $S = R$ then $\hat{\mathfrak{w}}_R$ is simply called the structure of uniform convergence. If $\mathfrak{S}$ is a collection of subsets of R, then, by 33.3,

$$\hat{\mathfrak{w}}_{\mathfrak{S}} = \bigwedge \{\hat{\mathfrak{w}}_S : S \in \mathfrak{S}\}$$

is also a uniform structure on $\hat{A}$. It is called the structure of uniform convergence over the sets in $\mathfrak{S}$.

34.I Let R and R' both be the real line **R**, bearing the usual uniform structure (induced by the absolute value δ). Also let S be the interval (a, b) of **R** and let $(\varphi_n)_{n \in \mathbf{N}}$ be a sequence of real-valued functions. To say that the Fréchet filter $\hat{\mathfrak{a}}$ of this sequence converges uniformly on S to the function ψ is equivalent to saying that, given $\epsilon > 0$, we can find a terminating section A_n of the sequence such that $A_n \subseteq \langle S, W_\epsilon^\delta \rangle$. This condition itself is equivalent to saying that for all $\nu \geqslant n$ and all $p \in [a, b]$ we always have the inequality $|\varphi_\nu p - \psi p| < \epsilon$. Hence we see that when $S = [a, b]$ our new "abstract" uniform convergence on S agrees with the classical concept of uniform convergence of the sequence of functions $(\varphi_n)_{n \in \mathbf{N}}$ over the interval $[a, b]$.

[4] Here $\varphi \times \psi$ denotes a map from R into $R' \times R'$.

34.II Let $\mathfrak{S}$ denote the system of all finite subsets of R. Then the topology of uniform convergence induced by $\hat{\mathfrak{w}}_{\mathfrak{S}}$ on the sets of $\mathfrak{S}$ is identical to the topology of pointwise convergence (with respect to the topology induced by the uniform structure on the image space R'; cf. p. 156).

34.7 *The topology induced on the sets of $\mathfrak{S}$ by the structure of uniform convergence in R'^R is a T_1-topology (and hence completely regular) if and only if the uniform structure $\mathfrak{w}'$ of the image space R' induces a T_1-topology and $\mathfrak{S}$ is a covering of R.*[5]

Proof: From $(\varphi, \psi) \in \langle S, W' \rangle$ for all $S \in \mathfrak{S}$ and $W' \in \mathfrak{w}'$, we deduce that $(\varphi p, \psi p) \in \bigcap \{W' : W' \in \mathfrak{w}\}$ for all $p \in \bigcup \{S : S \in \mathfrak{S}\}$. If $\mathfrak{w}$ induces a T_1-topology, then by 32.2 we have $\bigcap \{W' : W' \in \mathfrak{w}'\} = D(R')$, whence $\varphi p = \psi p$. Moreover, if $\mathfrak{S}$ is a covering of R, then $\varphi p = \psi p$ for all $p \in R$ and therefore $\varphi = \psi$. This means that the intersection of all sets in $\hat{\mathfrak{w}}_{\mathfrak{S}}$ is equal to the diagonal $D(\hat{A})$, and, by 33.2 again, $\hat{\mathfrak{w}}_{\mathfrak{S}}$ induces a T_1-topology.

Conversely, if $\mathfrak{w}'$ does not induce a T_1-topology, then there exist two distinct points $p', q' \in R'$ with $(p', q') \in \bigcap \{W' : W' \in \mathfrak{w}'\}$; the maps φ and ψ defined by $\varphi R = \{p'\}$ and $\psi R = \{q'\}$ then satisfy the relation $(\varphi, \psi) \in \langle S, W' \rangle$ for all $S \in \mathfrak{S}$ and $W' \in \mathfrak{w}'$. Hence the intersection of the filter sets of $\hat{\mathfrak{w}}_{\mathfrak{S}}$ contains the diagonal $D(\hat{A})$ properly, and $\hat{\mathfrak{w}}_{\mathfrak{S}}$ cannot induce a T_1-topology. If, on the other hand, $\mathfrak{S}$ is not a covering of R, then there exists a $p \in R$ such that $p \notin \bigcup \{S : S \in \mathfrak{S}\}$. Let x' and y' be distinct points of R'. Then $\varphi R = \{x'\}$ and $\psi \,\mathbf{C}p = \{x'\}$, $\psi p = y'$ define two mappings on $\hat{A}$ with $(\varphi, \psi) \in \langle S, W' \rangle$ for all $S \in \mathfrak{S}$ and $W' \in \mathfrak{w}'$. As above, we conclude that $(\varphi, \psi) \in \langle S, W' \rangle$ for all $S \in \mathfrak{S}$ and $W' \in \mathfrak{w}'$. ∎

If δ' is a pseudometric on R' with $\delta'(R') < +\infty$ and S is again a subset of R, then

$$\delta'_S(\varphi, \psi) = \sup \{\delta'(\varphi p, \psi p) : p \in S\}$$

defines a pseudometric δ'_S on $\hat{A}$. If δ' itself is a metric, then (for $S = R$) we see that δ'_R is also a metric. The pseudometric δ'_S induces in $\hat{A}$ a uniform structure $\hat{\mathfrak{w}}_{\delta', S}$. Let $\Delta_b(\mathfrak{w}')$ denote the set of all pseudometrics $\delta' \in \Delta(\mathfrak{w}')$ for which $\delta'(R') < +\infty$. Then we have the following result.

34.8 $$\hat{\mathfrak{w}}_S = \bigwedge \{\hat{\mathfrak{w}}_{\delta', S} : \delta' \in \Delta_b(\mathfrak{w}')\}.$$

Proof: Suppose we have been given $\epsilon > 0$. Let $\delta' \in \Delta_b(\mathfrak{w}')$. Since $\mathfrak{w}' \leqslant \mathfrak{w}_{\delta'}$ there exists a $W' \in \mathfrak{w}'$ such that $W' \subseteq W^{\delta'}_{\epsilon/2}$. From $(\varphi, \psi) \in \langle S, W' \rangle$ we deduce that $(\varphi p, \psi p) \in W'$; hence $\delta'(\varphi p, \psi p) < \epsilon/2$ for all $p \in S$ and therefore $\delta'_S(\varphi, \psi) \leqslant \epsilon/2 < \epsilon$. Hence $\langle S, W' \rangle$ is contained in the ϵ-uniform neighborhood under δ'_S, i.e., $\hat{\mathfrak{w}}_S \leqslant \bigwedge \{\hat{\mathfrak{w}}_{\delta', S} : \delta' \in \Delta_b(\mathfrak{w}')\}$. By 33.5 we have

$$\mathfrak{w}' = \bigwedge \{\mathfrak{w}_{\delta'} : \delta' \in \Delta(\mathfrak{w}')\}.$$

[5] The trivial case where R is a one-point set is an exception: For then $\hat{A}$ consists of only one map and $\mathfrak{S}$ need not necessarily be a covering of R.

If $\delta'_1, \ldots, \delta'_n$ are a finite number of pseudometrics in $\Delta(\mathfrak{w}')$, then

$$\delta'_0(p', q') = \max\{\delta'_1(p', q'), \ldots, \delta'_n(p', q')\}$$

defines another pseudometric δ'_0. For the ϵ-uniform neighborhood of δ'_0 we have

$$W_\epsilon^{\delta'_0} = W_\epsilon^{\delta'_1} \cap \ldots \cap W_\epsilon^{\delta'_n} \in \mathfrak{w}',$$

whence $\delta'_0 \in \Delta(\mathfrak{w}')$. Conversely, given any $W' \in \mathfrak{w}$ we can find a $\delta'_0 \in \Delta(\mathfrak{w}')$ and an $\epsilon > 0$ such that $W_\epsilon^{\delta'_0} \subseteq W'$. By $\delta'(p', q') = \min\{\epsilon, \delta'_0(p', q')\}$ we define a new pseudometric on R'. Since $W_\eta^{\delta'} = W_\eta^{\delta'_0}$ for $\eta \leqslant \epsilon$, we have $\delta' \in \Delta_b(\mathfrak{w}')$. For each couple of maps (φ, ψ) in the ϵ-uniform neighborhood of δ'_S we have $\delta'(\varphi p, \psi p) < \epsilon$ and hence $(\varphi p, \psi p) \in W_\epsilon^{\delta'} = W_\epsilon^{\delta'_0} \subseteq W'$ for all $p \in S$. This ϵ-uniform neighborhood is therefore contained in $\langle S, W' \rangle$, whence

$$\bigwedge \{\hat{\mathfrak{w}}_{\delta', S} : \delta' \in \Delta_b(\mathfrak{w}')\} \leqslant \hat{\mathfrak{w}}_S . \blacksquare$$

If the set R bears a topology τ then special relationships must hold. We can then characterize topologically the system $\mathfrak{S}$ of sets which determine the uniform structure $\hat{\mathfrak{w}}_\mathfrak{S}$ on $\hat{A}$. In this case it is not so much the set $\hat{A}$ of all mappings but rather the subset $\hat{C}$ of all continuous mappings from (R, τ) into $(R', \tau_{\mathfrak{w}'})$ which is really interesting.

34.9 *If $\mathfrak{S}$ is an open covering of (R, τ) then $\hat{C}$ is a closed set under the topology on $\hat{A}$ induced by the uniform structure $\hat{\mathfrak{w}}_\mathfrak{S}$, i.e., all uniformly convergent filters of continuous maps converge on the sets of an open covering of (R, τ) to continuous maps.*

$\hat{C}$ is also closed whenever (R, τ) is a locally completely compact space and $\mathfrak{S}$ is the system of all completely compact subsets of (R, τ).

Proof: Let the topology induced by $\hat{\mathfrak{w}}_\mathfrak{S}$ be denoted by $\hat{\tau}$ and let φ be a map in $\hat{A}$ such that $\hat{\mathfrak{b}} = \hat{\tau}\varphi \wedge \overline{\hat{C}} \neq \mathfrak{o}$. Given any point $p \in R$ and any neighborhood $V' \in \tau_{\mathfrak{w}'}\varphi p$ we can then find symmetric uniform neighborhoods $W', W'_1 \in \mathfrak{w}'$ with $W'[\varphi p] \subseteq V'$ and $W'_1 \circ W'_1 \circ W'_1 \leqslant W'$. For our first contention, let S be an (open) set belonging to $\mathfrak{S}$ with $p \in S$, while for our second contention let S be a completely compact neighborhood belonging to τp. Since $\hat{\mathfrak{b}}$ converges uniformly on S to φ, there must exist a $\hat{B} \in \hat{\mathfrak{b}}$ such that $(\varphi \times \psi) S \subseteq W'_1$ for all $\psi \in B$. If φ is a fixed map in $\hat{B}$, then $(\varphi p, \psi p) \in W'_1$ and $(\varphi q, \psi q) \in W'_1$ for all $q \in S$. Because ψ is continuous there exists a neighborhood $U \in \tau p$ with $U \subseteq S$ and $\psi U \subseteq W'_1[\psi p]$. Consequently $(\psi p, \psi q) \in W'_1$; hence $(\varphi p, \varphi q) \in W'_1 \circ W'_1 \circ W'_1 \subseteq W'$ and $\varphi q \in W'[\varphi p] \subseteq V'$ for all $q \in U$. Thus $\varphi U \subseteq V'$ and φ is continuous. $\blacksquare$

The following special case has important applications: Let τ again be a topology on R and let $\mathfrak{R}$ be the system of all completely compact subsets of (R, τ).

Then the structure of uniform convergence $\hat{\mathfrak{w}}_{\mathfrak{K}}$ induces on the completely compact subsets of R in $\hat{A}$ a topology $\hat{\tau}$. On the other hand the topology $\mathfrak{w}'$ of R' induces a topology $\tau_{\mathfrak{w}'}$ in $\hat{A}$ which in turn determines the topology $\hat{\tau}_{\mathfrak{K}}$ of compact convergence (cf. p. 158).

34.10 *On the set $\hat{C}$ of all continuous maps from (R, τ) into $(R', \tau_{\mathfrak{w}'})$ we see that the topology $\hat{\tau}$ of uniform convergence restricted to the completely compact subsets of (R, τ) agrees thereon with the topology $\hat{\tau}_{\mathfrak{K}}$ of compact convergence, i.e., $(\hat{\tau})_{\hat{C}} = (\hat{\tau}_{\mathfrak{K}})_{\hat{C}}$. More generally, $\hat{\tau}\varphi = \hat{\tau}_{\mathfrak{K}}\varphi$ for all $\varphi \in \hat{C}$.*

Proof: Let φ be a fixed map in $\hat{C}$. Let K be a completely compact subset of (R, τ) and V' an open subset of $(R', \tau_{\mathfrak{w}'})$ such that $\varphi K \subseteq V'$. To prove that $\hat{\tau}\varphi \leqslant \hat{\tau}_{\mathfrak{K}}\varphi$ we must show that there exists a uniform neighborhood $W_0' \in \mathfrak{w}'$ with $\langle K, W_0' \rangle [\varphi] \subseteq \langle K, V' \rangle$, where $\langle K, W_0' \rangle [\varphi]$ is identical to the set of all $\psi \in \hat{A}$ such that $(\varphi \times \psi) K \subseteq W_0'$ and $\langle K', V' \rangle$ is the set of all $\psi \in \hat{A}$ with $\psi K \subseteq V'$. By the continuity of φ, if K is completely compact so is φK (19.1). To each $x \in K$ we can associate a $W_x' \in \mathfrak{w}'$, for which $W_x'[\varphi x]$ is open and $(W_x' \circ W_x') [\varphi x]$ is contained in V'. Then there exist a finite number of points $x_1, ..., x_r \in K$ such that $\varphi K \subseteq W_{x_1}'[\varphi x_1] \cup ... \cup W_{x_r}'[\varphi x]$. If we put

$$W_0' = W_{x_1}' \cap ... \cap W_{x_r}'$$

then we see that $\psi \in \langle K, W_0' \rangle [\varphi]$ implies $(\varphi p, \psi p) \in W_0'$ for all $p \in K$. To each $p \in K$ we can associate an x_p with $(\varphi x_p, \varphi p) \in W_{x_p}'$; thus

$$(\varphi x_p, \psi p) \in W_{x_p}' \circ W_0' \subseteq W_{x_p}' \circ W_{x_p}'$$

and therefore $\psi p \in (W_{x_p}' \circ W_{x_p}') [\varphi x_p] \subseteq V'$. We finally have $\psi K \subseteq V'$ and $\langle K, W_0' \rangle [\varphi] \subseteq \langle K, V' \rangle$.

To establish the converse we must show that $\hat{\tau}_{\mathfrak{K}}\varphi \leqslant \hat{\tau}\varphi$. Let us therefore suppose we have been given a completely compact subset K of (R, τ) and a uniform neighborhood $W' \in \mathfrak{w}'$. We must find completely compact subsets $K_1, ..., K_r$ of (R, τ) and open subsets $V_1', ..., V_r'$ of $(R', \tau_{\mathfrak{w}'})$ such that $\varphi K_\rho \subseteq V_\rho'$ for $\rho = 1, ..., r$ and $\langle K_1, V_1' \rangle \cap ... \cap \langle K_r, V_r' \rangle \leqslant \langle K, W' \rangle [\varphi]$. With W' we can associate, by 33.8, an open and symmetric uniform neighborhood W_0' in $(R', \tau_{\mathfrak{w}'}) \times (R', \tau_{\mathfrak{w}'})$ such that $\bar{W}_0' \circ \bar{W}_0' \circ \bar{W}_0' \subseteq W'$. Since φK is completely compact there exist points $x_1, ..., x_r \in K$ with $\varphi K \subseteq W_0'[\varphi x_1] \cup ... \cup W_0'[\varphi x_r]$. For $\rho = 1, ..., r$ we then see that $K_\rho = K \cap \varphi^{-1}\bar{W}_0'[\varphi x_\rho]$ is a completely compact subset of (R, τ) and $V_\rho' = (W_0' \circ W_0') (\varphi x_\rho)$ is an open subset of $(R', \tau_{\mathfrak{w}'})$. We have $\varphi K_\rho \subseteq \bar{W}_0'[\varphi x_\rho] \subseteq (W_0' \circ W_0') [\varphi x_\rho] = V_\rho'$ and $K = K_1 \cup ... \cup K_r$. From

$$\psi \in \langle K_1, V_1' \rangle \cap ... \cap \langle K_r, V_r' \rangle$$

and $p \in K$ it follows that $\psi p \in V'_\rho = (W'_0 \circ W'_0)[\varphi x_\rho]$, hence $(\varphi x_\rho, \psi p) \in W'_0 \circ W'_0$ for an appropriate ρ with $\rho \in K_\rho$. Moreover, we have $\varphi p \in \bar{W}'_0[\varphi x_\rho]$ and, since W'_0 is symmetrical, $(\varphi p, \varphi x_\rho) \in \bar{W}'_0$. Finally, we have

$$(\varphi p, \psi p) \in \bar{W}'_0 \circ W'_0 \circ W'_0 \subseteq W',$$

hence $(\varphi \times \psi) K \subseteq W'$ and therefore $\psi \in \langle K, W' \rangle [\varphi]$. ∎

A subset $\hat{\mathfrak{M}}$ of the set of all maps from R into the real line is said to be *increasingly filtrated*, whenever, given any two maps $\varphi_1, \varphi_2 \in \hat{\mathfrak{M}}$ we can find a $\varphi_3 \in \hat{\mathfrak{M}}$ such that $\varphi_3 p \geqslant \max(\varphi_1 p, \varphi_2 p\}$ for all $p \in R$. If we put

$$\hat{\mathfrak{M}}_\varphi = \{\varphi' : \varphi' \in \hat{\mathfrak{M}}, \varphi' p \geqq \varphi p \quad \text{for all } p \in R\},$$

then $\{\hat{\mathfrak{M}}_\varphi : \varphi \in \hat{\mathfrak{M}}\}$ is a filter basis generating a filter $\hat{\mathfrak{a}}(\hat{\mathfrak{M}})$.

In the following theorem we suppose that the real line bears the usual uniform structure while $\hat{C}$ will denote the set of all continuous maps of a space (R, τ) into the real line.

34.11 (Dini's Theorem) *Let $\hat{\mathfrak{M}}$ be an increasingly filtrated subset of $\hat{C}$ and suppose that the filter $\hat{\mathfrak{d}}$ of $\hat{C}$ satisfies the condition $\mathfrak{o} \leqslant \hat{\mathfrak{d}} \leqslant \hat{\mathfrak{a}}(\hat{\mathfrak{M}})$. If $\hat{\mathfrak{d}}$ converges pointwise to a continuous mapping $\psi \in \hat{C}$, then, on completely compact subsets of (R, τ), $\hat{\mathfrak{d}}$ and $\hat{\mathfrak{a}}(\hat{\mathfrak{M}})$ converge* uniformly *to ψ.*

Proof: It is true that $\varphi p \leqslant \psi p$ for all $\varphi \in \hat{\mathfrak{M}}$ and all $p \in R$: For otherwise we would have $\psi p < \varphi p \leqslant \varphi' p$ for an appropriate $\varphi \in \hat{\mathfrak{M}}$, an appropriate $p \in R$, and all $\varphi' \in \hat{\mathfrak{M}}_\varphi$, which, since $\hat{\mathfrak{d}} \leqslant \hat{\mathfrak{a}}(\hat{\mathfrak{M}})$, would contradict the pointwise convergence of $\hat{\mathfrak{d}}$ to ψ. Furthermore, given $p \in R$ and $\epsilon > 0$, there exists a $\hat{D} \in \hat{\mathfrak{d}}$ with $\epsilon_p \hat{D} \subseteq U_\epsilon(\psi p)$ (cf. p. 156). From $\varphi \in \hat{D}$ it then follows that $\epsilon_p \hat{\mathfrak{M}}_\varphi \subseteq U_\epsilon(\psi p)$ and hence $\hat{\mathfrak{a}}(\hat{\mathfrak{M}})$ converges pointwise to ψ.

In order to complete the proof of our theorem we need to show: Given a completely compact subset K of R and an $\epsilon > 0$ there exists a $\varphi_0 \in \hat{\mathfrak{M}}$ such that $(\varphi \times \psi) K \subseteq W_\epsilon$ for all $\varphi \in \hat{\mathfrak{M}}_{\varphi_0}$. With this in mind let us associate with each $\varphi \in \hat{\mathfrak{M}}$ the subset of $K \times \mathbf{R}$

$$B_\varphi = \{(p, x) : p \in K, \varphi p \leqq x \leqq \psi p\}.$$

Since φ and ψ are continuous in $K \times \mathbf{R}$, the set B_φ is closed. The collection of sets $\{B_\varphi : \varphi \in \hat{\mathfrak{M}}\}$ is a filter basis and therefore generates a closed filter $\mathfrak{b}$. From the pointwise convergence of $\hat{\mathfrak{a}}(\hat{\mathfrak{M}})$ toward ψ we see that

$$\cap \{B_\varphi : \varphi \in \hat{\mathfrak{M}}\} = \{(p, \psi p) : p \in K\}.$$

For a fixed $\varphi' \in \hat{\mathfrak{M}}$ the set $A = \{(p, x) : p \in K, \varphi' p \leqslant x \leqslant \psi p - \epsilon\}$ is completely compact and we have $\cap \{B_\varphi \cap A : \varphi \in \hat{\mathfrak{M}}\} = \emptyset$. By 11.19, we therefore also have $\mathfrak{b} \wedge \overline{A} = \mathfrak{o}$. Thus there exists a $\varphi_0 \in \hat{\mathfrak{M}}$ such that $\psi p - \epsilon < \varphi_0 p \leqslant \psi p$

for all $p \in K$. From $\varphi \in \hat{\mathfrak{M}}_{\varphi_0}$ it then follows that $\varphi_0 p \leqslant \varphi p \leqslant \psi p$; hence $| \varphi p - \psi p | < \epsilon$ for all $p \in K$ and finally $(\varphi \times \psi) K \subseteq W_\epsilon$. ∎

Definition 34c: *Let* (R, τ) *be a topological and* $(R', \mathfrak{w}')$ *be a uniform space. Then a set* $\hat{\mathfrak{M}}$ *of mappings* $\varphi : R \to R'$ *is said to be* equicontinuous *if, given any* $W' \in \mathfrak{w}'$ *and any* $p \in R$ *we can find a neighborhood* $U \in \tau p$ *such that* $\varphi U \subseteq W'[\varphi p]$ *for all* $\varphi \in \hat{\mathfrak{M}}$.

If the set of maps of $\hat{\mathfrak{M}}$ form an equicontinuous set then manifestly each of the member maps is continuous. In the following theorem, let us suppose that the uniform structure $\mathfrak{w}'$ of R' induces a T_1-topology, let $\hat{A}$ once again represent the set of all maps from R into R' and let $\hat{\tau}$ be the topology of uniform convergence on the completely compact subsets of (R, τ).

34.12 (Ascoli's Theorem) *Let* $\hat{\mathfrak{M}}$ *be a subset of equicontinuous maps in* $\hat{A}$ *and for each* $p \in R$ *let the set* $\epsilon_p M = \{\varphi p : \varphi \in \hat{\mathfrak{M}}\}$ *be completely compact in* $(R', \mathfrak{w}')$. *Then* $\hat{\mathfrak{M}}$ *itself is completely compact in* $(\hat{A}, \hat{\tau})$.

Proof: Let $\hat{\mathfrak{v}}$ be an ultrafilter in $\hat{\mathfrak{M}}$. For each $p \in R$ the image ultrafilter $\epsilon_p \mathfrak{v}$ converges to a unique point $p' \in R'$. Hence $\psi p = p'$ defines a map $\psi \in \hat{A}$. In order to prove that $\hat{\mathfrak{v}} \leqslant \hat{\tau}\psi$ it suffices to establish: Given any completely compact subset K of (R, τ) and any $W' \in \mathfrak{w}'$, then there must exist a $\hat{V} \in \hat{\mathfrak{v}}$ such that $(\varphi \times \psi) K \subseteq W'$ for all $\varphi \in \hat{V}$.

By 33.1 there exist symmetric uniform neighborhoods W_0', $W_1' \in \mathfrak{w}'$ with $W_0' \circ W_0' \subseteq W'$ and $W_1' \circ W_1' \subseteq W_0'$. By 33.8 we can therefore assume W_0' to be closed. Since $\hat{\mathfrak{M}}$ is an equicontinuous family, there also exists, for any fixed $p \in R$, an open neighborhood $U_p \in \tau p$ with $\varphi U_p \subseteq W_1'[\varphi p]$ for all $\varphi \in \hat{\mathfrak{M}}$. In addition, there exists a $V_p \in \hat{\mathfrak{v}}$ such that $\varphi p \in W_1'[\psi p]$ for all $\varphi \in \hat{V}_p$. From $\varphi \in \hat{V}_p$ and $q \in U_p$ it therefore follows that $(\varphi q, \psi p) \in W_1' \circ W_1' \subseteq W_0'$ and also $\epsilon_q \hat{\mathfrak{v}} \leqslant \overline{W_0'[\psi p]}$. Since W_0' is closed, we also have $(\psi q, \psi p) \in W_0'$ and

$$(\varphi q, \psi q) \in W_0' \circ W_0' \subseteq W'.$$

Hence $(\varphi \times \psi) U_p \subseteq W'$. Since $\{U_p : p \in K\}$ is an open covering of K, there exists a finite number of points $p_1, ..., p_n \in K$ with $K \subseteq U_{p_1} \cup ... \cup U_{p_n}$. If we put $\hat{V} = \hat{V}_{p_1} \cap ... \cap \hat{V}_{p_n}$, then for each $\varphi \in \hat{V}$, we see that

$$(\varphi \times \psi) K \subseteq (\varphi \times \psi) U_{p_1} \cup ... \cup (\varphi \times \psi) U_{p_n} \subseteq W'. \blacksquare$$

Notes and Exercises

34.A With appropriate hypotheses on (R, τ) we can establish the converse to 34.12 (we use the notations of 34.12).

Exercise: Let (R, τ) be a regular and l.c.c. space. If $\hat{\mathfrak{M}}$ is a completely compact family of continuous maps in $(\hat{A}, \hat{\tau})$, then $\hat{\mathfrak{M}}$ is also a equicontinuous family and for each $p \in R$ the family $\epsilon_p \hat{\mathfrak{M}}$ is completely compact in $(R', \mathfrak{w}')$.

34.B Let $(R', \mathfrak{w}')$ be a uniform space and let $\hat{A}$ be the set of all maps from R into R'.

Exercise: Determine the coarsest uniform structure $\hat{\mathfrak{w}}$ on $\hat{A}$ with respect to which all maps $\epsilon_p : \hat{A} \to R'(p \in R$; cf. Section 23) are uniformly continuous. The topology induced by $\hat{\mathfrak{w}}$ is the topology of pointwise convergence with respect to the topology induced on R' by $\mathfrak{w}'$.

35. Extension and Completion of Uniform Spaces

The extension of uniform structures from subspaces to larger spaces can, in the main, be handled as in Section 32. The key problem is how to properly carry over the tools and concepts of Section 32 to uniform spaces.

In a pseudometric space (R, δ) we called $\mathfrak{a}$ a δ-Cauchy filter whenever $\delta(\mathfrak{a}) = 0$ (30b). This property can also be described using the uniform structure $\mathfrak{w}_\delta$ induced by $\delta : \delta(\mathfrak{a}) = 0$ is equivalent to the fact that, given any $\epsilon > 0$ we can find an $A \in \mathfrak{a}$ such that $\delta(A) < \epsilon$, and hence such that $A \times A \subseteq W_\epsilon^\delta$. But again, this last condition is equivalent to $\mathfrak{a} \times \mathfrak{a} \leqslant \mathfrak{w}_\delta$. It is in this form that the concept of Cauchy filter can be immediately carried over to uniform spaces.

Definition 35a: *In a uniform space $(R, \mathfrak{w})$ we call a filter $\mathfrak{a}$ a* Cauchy-filter *(more precisely: a $\mathfrak{w}$-Cauchy-filter) whenever $\mathfrak{a} \times \mathfrak{a} \leqslant \mathfrak{w}$.*

By what precedes we can see that the old and new definitions of a Cauchy filter coincide in pseudometric spaces: The filter $\mathfrak{a}$ is a δ-Cauchy filter if and only if it is a $\mathfrak{w}_\delta$-Cauchy filter.

35.1 *A filter $\mathfrak{a}$ is a $\mathfrak{w}$-Cauchy filter if and only if for every pseudometric $\delta \in \Delta(\mathfrak{w})$ it is a δ-Cauchy filter.*

Proof: Since $\mathfrak{w} = \bigwedge \{\mathfrak{w}_\delta : \delta \in \Delta(\mathfrak{w})\}$ (33.5), the conditions $\mathfrak{a} \times \mathfrak{a} \leqslant \mathfrak{w}$ and $\mathfrak{a} \times \mathfrak{a} \leqslant \mathfrak{w}_\delta$ are equivalent for all $\delta \in \Delta(\mathfrak{w})$. ∎

35.2 *Let φ be a uniformly continuous map from $(R, \mathfrak{w})$ into $(R', \mathfrak{w}')$. Then if $\mathfrak{a}$ is a $\mathfrak{w}$-Cauchy filter, $\varphi\mathfrak{a}$ must be a $\mathfrak{w}'$-Cauchy filter.*

Proof: From $\mathfrak{a} \times \mathfrak{a} \leqslant \mathfrak{w}$ it follows that

$$\varphi\mathfrak{a} \times \varphi\mathfrak{a} = (\varphi \times \varphi)(\mathfrak{a} \times \mathfrak{a}) \leqslant (\varphi \times \varphi)\,\mathfrak{w} \leqslant \mathfrak{w}'. \blacksquare$$

The following theorem is a word-for-word transposition of theorem 30.4, from the case of pseudometric spaces to that of uniform spaces.

35.3 *In a uniform space* $(R, \mathfrak{w})$ *every convergent filter is a Cauchy filter. Conversely, if* $\mathfrak{a}$ *is a* $\mathfrak{w}$*-Cauchy filter and* M *a completely compact (k-compact) subset of* $(R, \tau_{\mathfrak{w}})$, *then* $\mathfrak{a} \wedge \overset{\frown}{M} \neq \mathfrak{o}$ *(and* $\mathrm{Rk}(\mathfrak{a}) \leqslant k$*) implies the convergence of the filter* $\mathfrak{a}$.

Proof: If $\mathfrak{a}$ converges to p in $(R, \mathfrak{w})$, then $\mathfrak{a}$ converges, definitely, for all $\delta \in \Delta(\mathfrak{w})$ to the point p in (R, δ). By 30.4, $\mathfrak{a}$ is therefore a δ-Cauchy filter for all $\delta \in \Delta(\mathfrak{w})$ and hence, by 35.1, also a $\mathfrak{w}$-Cauchy filter. From

$$\mathfrak{a} \wedge \overset{\frown}{M} \neq \mathfrak{o} \text{ (and } \mathrm{Rk}\,(\mathfrak{a}) \leqslant k)$$

we deduce the existence of an adherent point p to $\mathfrak{a}$. If $\mathfrak{a}$ is a $\mathfrak{w}$-Cauchy filter and hence a δ-Cauchy filter for all $\delta \in \Delta(\mathfrak{w})$ we see by 30.4 that $\mathfrak{a}$ is δ-convergent to p and by 33.5 that $\mathfrak{a}$ is $\mathfrak{w}$-convergent to p. ∎

In a pseudometric space (R, δ) we called $\mathfrak{a}$ a round filter whenever the filter $\mathfrak{r}(\mathfrak{a})$ generated by the filter basis $\{U_\epsilon^\delta(A) : A \in \mathfrak{a}, \epsilon > 0\}$ was identical to $\mathfrak{a}$ itself (32a). To be more precise we shall now call such filters δ-round filters and denote them by $\mathfrak{r}_\delta(\mathfrak{a})$ rather than $\mathfrak{a}$.

If A is a subset of R and W a subset of $R \times R$, we shall denote the set $\cup \{W[p] : p \in A\}$ by $W[A]$. With this notation in a pseudometric space (R, δ), we have $U_\epsilon^\delta(A) = W_\epsilon^\delta[A]$.

Definition 35b: *Let* $\mathfrak{a}$ *be a filter in a uniform space* $(R, \mathfrak{w})$. *We denote by* $\mathfrak{r}_\mathfrak{w}(\mathfrak{a})$ *the filter generated by the filter basis* $\{W[A] : W \in \mathfrak{w}, A \in \mathfrak{a}\}$. *Whenever* $\mathfrak{r}_\mathfrak{w}(\mathfrak{a}) = \mathfrak{a}$ *we shall call* $\mathfrak{a}$ *a* $\mathfrak{w}$-round *filter.*

In pseudometric spaces we manifestly have $\mathfrak{r}_\delta(\mathfrak{a}) = \mathfrak{r}_{\mathfrak{w}_\delta}(\mathfrak{a})$, the concepts of "$\delta$-round" and "$\mathfrak{w}_\delta$-round" becoming identical. We also have:

35.4
$$\mathfrak{r}_\mathfrak{w}(\mathfrak{a}) = \bigwedge \{\mathfrak{r}_\delta(\mathfrak{a}) : \delta \in \Delta(\mathfrak{w})\}.$$

Proof: Since $\mathfrak{w} \leqslant \mathfrak{w}_\delta$ for $\delta \in \Delta(\mathfrak{w})$, given any $\epsilon > 0$ we can find a $W \in \mathfrak{w}$ with $W \subseteq W_\epsilon^\delta$. We have $W[A] \subseteq W_\epsilon^\delta[A]$ and hence $\mathfrak{r}_\mathfrak{w}(\mathfrak{a}) \leqslant \mathfrak{r}_\delta(\mathfrak{a})$. Conversely, by 33.5 and 31.2, given a $W \in \mathfrak{w}$ there exists an $\epsilon > 0$ and a $\delta \in \Delta(\mathfrak{w})$ with $W_\epsilon^\delta \subseteq W$. Consequently $W_\epsilon^\delta[A] \subseteq W[A]$ and $\bigwedge \{\mathfrak{r}_\delta(\mathfrak{a}) : \delta \in \Delta(\mathfrak{w})\} \leqslant \mathfrak{r}_\mathfrak{w}(\mathfrak{a})$. ∎

Theorem 32.2 can also be transposed word-for-word, giving:

35.5 (1) *Every* $\mathfrak{w}$*-round filter of* R *is open in* $(R, \mathfrak{w})$.

(2) *The filter* $\mathfrak{r}_\mathfrak{w}(\mathfrak{a})$ *is* $\mathfrak{w}$*-round.*

(3) *The neighborhoods-filter* $\tau_\mathfrak{w} p$ *are* $\mathfrak{w}$*-round.*

(4) $\mathfrak{r}_\mathfrak{w}(\mathfrak{a})$ *is a Cauchy filter if and only if* $\mathfrak{a}$ *is a Cauchy filter.*

(5) *If* $\mathfrak{a}$ *is a* $\mathfrak{w}$*-round filter and* $\mathfrak{b}$ *a* $\mathfrak{w}$*-Cauchy filter, from* $\mathfrak{a} \wedge \mathfrak{b} \neq \mathfrak{o}$ *we deduce* $\mathfrak{b} \leqslant \mathfrak{a}$.

Proof

(1) Let $\mathfrak{a}$ be a $\mathfrak{w}$-round filter. By 32.2, $\mathfrak{r}_\delta(\mathfrak{a})$ is an open filter in (R, δ) for

each $\delta \in \Delta(\mathfrak{w})$, and since $\mathfrak{w} \leqslant \mathfrak{w}_\delta$, $\mathfrak{r}_\delta(\mathfrak{a})$ is all the more so open in $(R, \mathfrak{w})$. Hence $\mathfrak{a} \doteq \mathfrak{r}_{\mathfrak{w}}(\mathfrak{a}) = \bigwedge \{\mathfrak{r}_\delta(\mathfrak{a}) ; \delta \in \Delta(\mathfrak{w})\}$ is an open filter in $(R, \mathfrak{w})$ (35.4; 5.4).

(2) We have $W[W[A]] = (W \circ W)[A]$ and hence $\mathfrak{r}_{\mathfrak{w}}(\mathfrak{r}_{\mathfrak{w}}(\mathfrak{a})) = \mathfrak{r}_{\mathfrak{w}}(\mathfrak{a})$.

(3) By 32.2 we have $\tau_\delta p = \mathfrak{r}_\delta(\ulcorner p \urcorner)$ for each $\delta \in \Delta(\mathfrak{w})$ and hence

$$\tau_{\mathfrak{w}} p = \wedge \{\tau_\delta p : \delta \in \Delta(\mathfrak{w})\} = \wedge \{\mathfrak{r}_\delta(\ulcorner p \urcorner) : \delta \in \Delta(\mathfrak{w})\} = \mathfrak{r}_{\mathfrak{w}}(\ulcorner p \urcorner).$$

Our contention then follows from (2).

(4) If $\mathfrak{a}$ is a $\mathfrak{w}$-Cauchy filter then it certainly is a δ-Cauchy filter for each $\delta \in \Delta(\mathfrak{w})$. By 32.2, $\mathfrak{r}_\delta(\mathfrak{a})$ is then a δ-Cauchy filter. Since $\mathfrak{r}_{\mathfrak{w}}(\mathfrak{a}) \leqslant \mathfrak{r}_\delta(\mathfrak{a})$, $\mathfrak{r}_{\mathfrak{w}}(\mathfrak{a})$ must be a δ-Cauchy filter for every $\delta \in \Delta(\mathfrak{w})$ and hence also a $\mathfrak{w}$-Cauchy filter (35.1). The converse follows directly from $\mathfrak{a} \leqslant \mathfrak{r}_{\mathfrak{w}}(\mathfrak{a})$.

(5) From $\mathfrak{a} \wedge \mathfrak{b} \neq \mathfrak{o}$ it follows that $\mathfrak{r}_\delta(\mathfrak{a}) \wedge \mathfrak{b} \neq \mathfrak{o}$. Since $\mathfrak{b}$ is a δ-Cauchy filter for every $\delta \in \Delta(\mathfrak{w})$ (35.1), by 32.2 we see that $\mathfrak{b} \leqslant \mathfrak{r}_\delta(\mathfrak{a})$ and hence

$$\mathfrak{b} \leqslant \bigwedge \{\mathfrak{r}_\delta(\mathfrak{a}) : \delta \in \Delta(\mathfrak{w})\} = \mathfrak{r}_{\mathfrak{w}}(\mathfrak{a}) = \mathfrak{a}. \blacksquare$$

Definition 35c: *Let* $(R, \mathfrak{w})$ *be a uniform space and* $(\hat{R}, \hat{\tau})$ *an extension of* $(R, \tau_{\mathfrak{w}})$. *We say that a uniform structure* $\hat{\mathfrak{w}}$ *of* $\hat{R}$ *is a* continuation of $\mathfrak{w}$ to $(\hat{R}, \hat{\tau})$ *if* $\tau = \tau\hat{\mathfrak{w}}$ *and if* $\mathfrak{w}$ *is the trace of* $\hat{\mathfrak{w}}$ *on* $R \times R$.

In order to better describe continuations of uniform structures we shall use the following notational abbreviations, analogous to those used in Section 25:

If $(\hat{R}, \hat{\tau})$ is an extension of $(R, \tau_{\mathfrak{w}})$ we associate with each subset W of $R \times R$ a subset $\hat{s}(W)$ of $\hat{R} \times \hat{R}$ defined by

$$\hat{s}(W) = \{(\hat{p}, \hat{q}) : \hat{p}, \hat{q} \in \hat{R}, \operatorname{Tr}(\hat{\tau}\hat{p}) \times \operatorname{Tr}(\hat{\tau}\hat{q}) \leqslant \ulcorner W \urcorner\}$$

35.6 *Let* $\hat{\mathfrak{w}}$ *be a continuation of the uniform structure* $\mathfrak{w}$ *of* R *to the extension* $(\hat{R}, \hat{\tau})$ *of* $(R, \tau_{\mathfrak{w}})$. *Then* $\{\hat{s}(W) : W \in \mathfrak{w}\}$ *is a basis of* $\hat{\mathfrak{w}}$.

Proof: By 33.8, $\hat{\mathfrak{w}}$ is an open and closed filter in $(\hat{R}, \hat{\tau}) \times (\hat{R}, \hat{\tau})$. To each $W \in \mathfrak{w}$ we can therefore associate an open set $\hat{W} \in \hat{\mathfrak{w}}$ such that $\hat{W} \cap (R \times R) \subseteq W$. It follows from $(\hat{p}, \hat{q}) \in \hat{W}$ that $\operatorname{Tr}(\hat{\tau}\hat{p}) \times \operatorname{Tr}(\hat{\tau}\hat{q}) \leqslant \ulcorner W \urcorner$, and hence $\hat{W} \subseteq \hat{s}(W)$. Conversely, let $\hat{W}$ be a closed set in $\hat{\mathfrak{w}}$ with trace $W = \hat{W} \cap (R \times R)$. From $(\hat{p}, \hat{q}) \in \hat{s}(W)$, and hence from $\operatorname{Tr}(\hat{\tau}\hat{p}) \times \operatorname{Tr}(\hat{\tau}\hat{q}) \leqslant \ulcorner W \urcorner$, it follows that $(\hat{\tau}\hat{p} \times \hat{\tau}\hat{q}) \wedge \ulcorner \hat{W} \urcorner \neq \mathfrak{o}$ and $(\hat{p}, \hat{q}) \in \hat{W}$. Hence $\hat{s}(W) \subseteq \hat{W}$ and $\{\hat{s}(W) : W \in \mathfrak{w}\}$ must be a basis of $\hat{\mathfrak{w}}$. $\blacksquare$

35.7 *In the uniform space* $(R, \mathfrak{w})$ *let* T^* *be a trace system and* $(\hat{R}, \hat{\tau})$ *the* T^**-extension of* $(R, \tau_{\mathfrak{w}})$.

(1) $\mathfrak{w}$ *possesses a continuation to* $(\hat{R}, \hat{\tau})$ *if and only if* T^* *is completely composed of* $\mathfrak{w}$*-round Cauchy filters. Then the continuation* $\hat{\mathfrak{w}}$ *is uniquely determined by* $\mathfrak{w}$.

(2) *Every pseudometric* $\delta \in \Delta(\mathfrak{w})$ *likewise possesses an extension* $\hat{\delta}$ *to* $\hat{R}$ *with* $\tau_{\hat{\delta}} \geqslant \hat{\tau}$, *and we have* $\Delta(\hat{\mathfrak{w}}) = \{\hat{\delta} : \delta \in \Delta(\mathfrak{w})\}$.

(3) *With each extension* $(\tilde{R}, \tilde{\mathfrak{w}})$ *of* $(R, \mathfrak{w})$ *having the trace-system* T^* *we can associate a uniformly continuous map of* $(\tilde{R}, \tilde{\mathfrak{w}})$ *onto* $(\hat{R}, \hat{\mathfrak{w}})$, *satisfying the trace condition and equal to the identity on* R. *This property uniquely characterizes* $(\hat{R}, \hat{\mathfrak{w}})$ *up to uniform isomorphism.*

(4) *If* $\mathfrak{w}$ *induces a* T_1*-topology, so does* $\hat{\mathfrak{w}}$.

Proof: If $\mathfrak{w}$ admits of a continuation $\hat{\mathfrak{w}}$ to $(\hat{R}, \hat{\tau})$, then, by 35.5, $\hat{\tau}\hat{p} = \tau_{\hat{\mathfrak{w}}}\hat{p}$ is a round Cauchy filter in $(\hat{R}, \hat{\mathfrak{w}})$ and hence $\operatorname{Tr}(\hat{\tau}\hat{p})$ is a round Cauchy filter in $(R, \mathfrak{w})$. Conversely, suppose T^* is entirely composed of $\mathfrak{w}$-round Cauchy filters. By 32.3 every pseudometric $\delta \in \Delta(\mathfrak{w})$ must then possess a unique extension $\hat{\delta}$ to $\hat{R}$ with $\tau_{\hat{\delta}} \geqslant \hat{\tau}$. By

$$\hat{\mathfrak{w}} = \bigwedge \{\mathfrak{w}_{\hat{\delta}} : \delta \in \Delta(\mathfrak{w})\}$$

we can define a uniform structure $\hat{\mathfrak{w}}$ on $\hat{R}$, whose trace in $R \times R$ is manifestly $\mathfrak{w}$ itself. By 33.3, we see that

$$\tau_{\hat{\mathfrak{w}}} = \bigwedge \{\tau_{\hat{\delta}} : \delta \in \Delta(\mathfrak{w})\} \geqslant \hat{\tau}.$$

This implies $\operatorname{Tr}(\tau_{\hat{\mathfrak{w}}}\hat{p}) \geqslant \operatorname{Tr}(\hat{\tau}\hat{p})$. But since $\operatorname{Tr}(\tau_{\hat{\mathfrak{w}}}p)$ is a $\mathfrak{w}$-Cauchy filter and $\operatorname{Tr}(\hat{\tau}\hat{p})$ is by hypothesis a $\mathfrak{w}$-round filter, we see by 35.5 (5) that $\operatorname{Tr}(\tau_{\hat{\mathfrak{w}}}\hat{p}) = \operatorname{Tr}(\hat{\tau}\hat{p})$. By 25.1, $\hat{\tau}$ is the coarsest topology on $\hat{R}$ with trace system T^*. Hence $\tau_{\hat{\mathfrak{w}}} = \hat{\tau}$, and $\hat{\mathfrak{w}}$ is a continuation of $\mathfrak{w}$ to $(\hat{R}, \hat{\tau})$. Since each $\hat{\delta} \in \Delta(\hat{\mathfrak{w}})$ must be an extension to $\hat{R}$ of a $\delta \in \Delta(\mathfrak{w})$, we see immediately from the very definition of $\hat{\mathfrak{w}}$ that $\Delta(\hat{\mathfrak{w}}) = \{\hat{\delta} : \delta \in \Delta(\mathfrak{w})\}$.

If the uniform space $(\tilde{R}, \tilde{w})$ is an extension of $(R, \mathfrak{w})$ with trace system T^*, then by 25.1 there exists a continuous map φ from $(\tilde{R}, \tau_{\tilde{\mathfrak{w}}})$ onto $(\hat{R}, \hat{\tau})$, obeying the trace condition $\operatorname{Tr}(\tau_{\tilde{\mathfrak{w}}}\tilde{p}) = \operatorname{Tr}(\hat{\tau}\varphi\tilde{p})$ and reducing to the identity on R. We then have

$$\begin{aligned}(\varphi \times \varphi)\, s(W) &= \{(\varphi\tilde{p}, \varphi\tilde{q}) : \tilde{p}, \tilde{q} \in \tilde{R}, \operatorname{Tr}(\tau_{\tilde{\mathfrak{w}}}\tilde{p}) \times \operatorname{Tr}(\tau_{\tilde{\mathfrak{w}}}\tilde{q}) \leqslant \overset{\sqcap}{W}\} \\ &= \{(\varphi\tilde{p}, \varphi\tilde{q}) : \tilde{p}, \tilde{q} \in \tilde{R}, \operatorname{Tr}(\hat{\tau}\varphi\tilde{p}) \times \operatorname{Tr}(\hat{\tau}\varphi\tilde{q}) \leqslant \overset{\sqcap}{W}\} \subseteq \hat{s}(W).\end{aligned}$$

By 35.6 we see that $(\varphi \times \varphi)\, \tilde{\mathfrak{w}} \leqslant \hat{\mathfrak{w}}$; i.e., φ is a uniformly continuous map of $(\tilde{R}, \tilde{\mathfrak{w}})$ onto $(\hat{R}, \hat{\mathfrak{w}})$. By the same arguments as in 25.1 it follows that $(\hat{R}, \hat{\mathfrak{w}})$ is uniquely determined up to uniform isomorphism by the above property.

Since $\operatorname{Tr}(\hat{\tau}\hat{p})$ and $\operatorname{Tr}(\hat{\tau}\hat{q})$ are round Cauchy filters, by 35.5 (5) we deduce from $\operatorname{Tr}(\hat{\tau}\hat{p}) \wedge \operatorname{Tr}(\hat{\tau}\hat{q}) \neq \mathfrak{o}$ that $\operatorname{Tr}(\hat{\tau}\hat{p}) \leqslant \operatorname{Tr}(\hat{\tau}\hat{q})$ and also $\operatorname{Tr}(\hat{\tau}\hat{q}) \leqslant \operatorname{Tr}(\hat{\tau}\hat{p})$, whence $\operatorname{Tr}(\hat{\tau}\hat{p}) = \operatorname{Tr}(\hat{\tau}\hat{q})$. By 25.3, $(\hat{R}, \hat{\tau})$ is therefore a T_1-space. ∎

Definition 35d: *A uniform space is said to be* complete *if all its Cauchy filters converge within it.*

Every completely compact uniform space is complete. As in Section 32, we can also define the completion of a uniform space, as follows: In the uniform space $(R, \mathfrak{w})$ let now T^* denote the collection of all round Cauchy filters. By 35.5, T^* is entirely composed of open filters and it contains the neighborhoods filters $\tau_{\mathfrak{w}}p$ for all $p \in R$, and therefore T^* is a trace system. By 35.7, $\mathfrak{w}$ possesses a continuation $\hat{\mathfrak{w}}$ to the T^*-extension $(\hat{R}, \hat{\tau})$ of $(R, \tau_{\mathfrak{w}})$. The uniform space $(\hat{R}, \hat{\mathfrak{w}})$ is called the *completion* of the uniform space $(R, \mathfrak{w})$.

35.8 *The completion $(\hat{R}, \hat{\mathfrak{w}})$ of a uniform space $(R, \mathfrak{w})$ is complete.*

Proof: Let $\hat{\mathfrak{a}}$ be a Cauchy filter in $(\hat{R}, \hat{\mathfrak{w}})$. Then by 35.5, $\hat{\mathfrak{b}} = \mathfrak{r}_{\hat{\mathfrak{w}}}(\hat{\mathfrak{a}})$ is a round Cauchy filter and an open filter of $(\hat{R}, \hat{\mathfrak{w}})$. Hence its trace $\mathfrak{b} = \mathrm{Tr}\,(\hat{\mathfrak{b}})$ is distinct from $\mathfrak{o}$ and a $\mathfrak{w}$-round Cauchy filter, and therefore it is a filter belonging to the trace system of $(\hat{R}, \hat{\mathfrak{w}})$. Thus $\mathfrak{b}$ converges to a point $\hat{p} \in \hat{R}$. Since $\mathfrak{o} \neq \mathfrak{b} \leqslant \hat{\mathfrak{b}} \wedge \tau_{\hat{\mathfrak{w}}}\hat{p}$ we see, by 35.5, that $\hat{\mathfrak{b}} = \tau_{\hat{\mathfrak{w}}}\hat{p}$ and then $\hat{\mathfrak{a}} \leqslant \tau_{\hat{\mathfrak{w}}}\hat{p}$. Every Cauchy filter of $(\hat{R}, \hat{\mathfrak{w}})$ is therefore convergent. ∎

From 35.7 we can now deduce: The completion of a uniform T_1-space is again a T_1-space.

35.9 *Every closed subspace of a complete uniform space is itself complete. In a uniform T_1-space every complete subspace is closed.*

Proof: Let $(R, \mathfrak{w})$ be a uniform space and A a closed subset of $(R, \tau_{\mathfrak{w}})$. If $\mathfrak{a}$ is a $\mathfrak{w}$-Cauchy filter with $\mathfrak{a} \leqslant \ulcorner A \urcorner$, then it converges to a point $p \in R$. Since A is closed we must have $p \in A$. Conversely, let A be a complete subspace of the uniform T_1-space $(R, \mathfrak{w})$. From $p \in \bar{A}$ we deduce that $\mathfrak{a} = \tau_{\mathfrak{w}}p \wedge \ulcorner A \urcorner \neq \mathfrak{o}$. By 35.3, $\mathfrak{a}$ is a Cauchy filter, which, since A is complete, converges to a point $q \in A$. $(R, \mathfrak{w})$, being a uniform T_1-space, must therefore be completely regular and hence a T_2-space. Since $\mathfrak{o} < \mathfrak{a} \leqslant \tau_{\mathfrak{w}}p \wedge \tau_{\mathfrak{w}}q$ we therefore have $p = q$ and finally $p \in A$. ∎

35.10 *In the uniform product space $(R, \mathfrak{w}) = \times\,\{(R_\iota, \mathfrak{w}_\iota) : \iota \in I\}$ $\mathfrak{a}$ is a $\mathfrak{w}$-Cauchy filter if and only if every projection $\pi_\iota \mathfrak{a}$ is a $\mathfrak{w}_\iota$-Cauchy filter. $(R, \mathfrak{w})$ is complete if and only if all coordinate spaces $(R_\iota, \mathfrak{w}_\iota)$ are complete.*

Proof: Since $\mathfrak{w} = \bigwedge \{(\pi_\iota \times \pi_\iota)^{-1}\mathfrak{w}_\iota : \iota \in I\}$ we see by 22.1 that $\mathfrak{w}_\iota = (\pi_\iota \times \pi_\iota)\mathfrak{w}$ Therefore the conditions, $\mathfrak{a} \times \mathfrak{a} \leqslant \mathfrak{w}$ and $\pi_\iota \mathfrak{a} \times \pi_\iota \mathfrak{a} \leqslant \mathfrak{w}_\iota$, for all $\iota \in I$, must be equivalent.

Let all coordinate spaces $(R_\iota, \mathfrak{w}_\iota)$ be complete. If $\mathfrak{a}$ is a $\mathfrak{w}$-Cauchy filter, the projections $\pi_\iota \mathfrak{a}$, being $\mathfrak{w}_\iota$-Cauchy filters, must converge to points $p_\iota \in R_\iota$. Hence $\mathfrak{a}$ converges, by 22.3, to the point $\times\, p_\iota$ of R; i.e., $(R, \mathfrak{w})$ is complete. Conversely, suppose the product space is complete and for a $\kappa \in I$ let $\mathfrak{a}_\kappa$ be a $\mathfrak{w}_\kappa$-Cauchy filter. If p_ι is a point in R_ι for each $\iota \neq \kappa$, then

$$\mathfrak{a} = \pi_\kappa^{-1}\mathfrak{a}_\kappa \wedge \bigwedge \{\pi_\kappa^{-1}\ulcorner p_\iota \urcorner : \iota \neq \kappa\}$$

is a $\mathfrak{w}$-Cauchy filter which converges to a point $p \in R$. By 22.2 we have $\mathfrak{a}_\kappa = \pi_\kappa \mathfrak{a}$ and, by 22.3, $\mathfrak{a}_\kappa$ converges to $\pi_\kappa p$; i.e., $(R_\kappa, \mathfrak{w}_\kappa)$ is also complete. ∎

35.11 *Let $(R', \mathfrak{w}')$ be a complete uniform space, $(\tilde{R}, \tilde{\mathfrak{w}})$ a uniform space and $(R, \mathfrak{w})$ a dense subspace of $(\tilde{R}, \tilde{\mathfrak{w}})$. Then every uniformly continuous map $\varphi : (R, \mathfrak{w}) \to (R', \mathfrak{w}')$ admits of an extension $\tilde{\varphi}$ to $\tilde{R}$, where $\tilde{\varphi}$ is a uniformly continuous map from $(\tilde{R}, \tilde{\mathfrak{w}})$ into $(R', \mathfrak{w}')$. If $(R', \mathfrak{w}')$ is a T_1-space, this extension $\tilde{\varphi}$ is unique.*

Proof: For each $\hat{p} \in \tilde{R}$, $\mathrm{Tr}\,(\tau_{\tilde{\mathfrak{w}}}\hat{p})$ is a Cauchy filter in $(R, \mathfrak{w})$. By 35.2, $\varphi\,\mathrm{Tr}\,(\tau_{\tilde{\mathfrak{w}}}\hat{p})$ is a Cauchy filter in $(R', \mathfrak{w}')$ and it is convergent because of the completeness of $(R', \mathfrak{w}')$. Now $(R', \mathfrak{w}')$, being a uniform space, must be a T_3-space and hence, by 19.9, φ possesses a continuous extension $\tilde{\varphi}$ to $(\tilde{R}, \tau_{\tilde{\mathfrak{w}}})$. Since φ is uniformly continuous, given any $W' \in \mathfrak{w}'$ we can find a $W \in \mathfrak{w}$ such that $(\varphi \times \varphi)\, W \subseteq W'$. From $(\hat{p}, \tilde{q}) \in \tilde{s}(W)$, and hence from

$$\mathrm{Tr}\,(\tau_{\tilde{\mathfrak{w}}}\hat{p}) \times \mathrm{Tr}\,(\tau_{\tilde{\mathfrak{w}}}\tilde{q}) \leqslant \overleftrightarrow{W},$$

and because

$$(\varphi \times \varphi)\,(\mathrm{Tr}\,\tau_{\tilde{\mathfrak{w}}}\hat{p}) \times \mathrm{Tr}\,(\tau_{\tilde{\mathfrak{w}}}\tilde{q})) \leqslant \tilde{\varphi}\tau_{\tilde{\mathfrak{w}}}\hat{p} \times \tilde{\varphi}\tau_{\tilde{\mathfrak{w}}}\tilde{q} \leqslant \tau_{\tilde{\mathfrak{w}}'}\tilde{\varphi}\hat{p} \times \tau_{\tilde{\mathfrak{w}}'}\tilde{\varphi}\tilde{q}$$

we deduce that

$$(\tau_{\mathfrak{w}'}\tilde{\varphi}\hat{p} \times \tau_{\mathfrak{w}'}\tilde{\varphi}q) \wedge \overleftrightarrow{W'} \geqslant (\varphi \times \varphi)\,[\mathrm{Tr}\,(\tau_{\tilde{\mathfrak{w}}}\hat{p}) \times \mathrm{Tr}\,(\tau_{\tilde{\mathfrak{w}}}\tilde{q})) \wedge \overleftrightarrow{W}] > \mathfrak{o}.$$

Therefore $(\tilde{\varphi}\hat{p}, \tilde{\varphi}\tilde{q}) \in \overline{W'}$ and hence $(\tilde{\varphi} \times \tilde{\varphi})\tilde{s}(W) \subseteq \overline{W'}$. By 35.6 $\{\tilde{s}(W) : W \in \mathfrak{w}\}$ is a basis of $\tilde{\mathfrak{w}}$ and, by 33.8, $\mathfrak{w}'$ is a closed filter, which facts imply that $(\tilde{\varphi} \times \tilde{\varphi})\, \tilde{\mathfrak{w}} \leqslant \overline{\mathfrak{w}'} = \mathfrak{w}'$ and hence that $\tilde{\varphi}$ is uniformly continuous. If $(R', \mathfrak{w}')$ is a T_1-space then $(R', \mathfrak{w}')$ is a T_2-space and our contention on the uniqueness of the extension $\tilde{\varphi}$ then follows from 19.8. ∎

An analogous result holds for pseudometric spaces.

35.12 *Let (R', δ') be a complete pseudometric space, $(\tilde{R}, \tilde{\delta})$ a pseudometric space, and (R, δ) a dense subspace of $(\tilde{R}, \tilde{\delta})$. Then every metric-preserving map φ from (R, δ) into (R', δ') admits of an extension $\tilde{\varphi}$ to $\tilde{R}$, where $\tilde{\varphi}$ is a metric-preserving map of $(\tilde{R}, \tilde{\delta})$ into (R', δ'). If (R', δ') is a metric space, then $\tilde{\varphi}$ is unique.*

Proof: From $\delta(p, q) = \delta'(\varphi p, \varphi q)$ it follows immediately that φ is a uniformly continuous map from $(R, \mathfrak{w}_\delta)$ into $(R', \mathfrak{w}_{\delta'})$. By the previous theorem there must then exist an extension $\tilde{\varphi}$ of φ to $\tilde{R}$, $\tilde{\varphi}$ being a (uniformly) continuous map from $(\tilde{R}, \tilde{\delta})$ into (R', δ'). We have $\varphi\,\mathrm{Tr}\,(\tau_{\tilde{\delta}}\hat{p}) \leqslant \tilde{\varphi}\tau_{\tilde{\delta}}\hat{p} \leqslant \tau_{\delta'}\tilde{\varphi}\hat{p}$ and then, using 32.1,

$$\tilde{\delta}(\hat{p}, \tilde{q}) = \delta(\mathrm{Tr}\,(\tau_{\tilde{\delta}}p) \vee \mathrm{Tr}\,(\tau_{\tilde{\delta}}\tilde{q})) = \delta'(\varphi\,\mathrm{Tr}\,(\tau_{\tilde{\delta}}\hat{p}) \vee \varphi\,\mathrm{Tr}\,(\tau_{\tilde{\delta}}\tilde{q}))$$

$$= \delta'(\tau_{\delta'}\tilde{\varphi}\hat{p} \vee \tau_{\delta'}\tilde{\varphi}\tilde{q}) = \delta'(\tilde{\varphi}\hat{p}, \tilde{\varphi}\tilde{q}),$$

which shows that $\tilde{\varphi}$ is even metric-preserving. ∎

35.13 *Let $(R, \mathfrak{w})$ be a uniform T_1-space. Every completely uniform T_1-space $(\tilde{R}, \tilde{\mathfrak{w}})$ containing $(R, \mathfrak{w})$ as a dense subspace is uniformly isomorphic to the completion $(\hat{R}, \hat{\mathfrak{w}})$ of $(R, \mathfrak{w})$.*

Proof: If $(R, \mathfrak{w})$ is a T_1-space, then so is $(\hat{R}, \hat{\mathfrak{w}})$ (35.7). Since the identity of R is a uniformly continuous map from $(R, \mathfrak{w})$ into $(\hat{R}, \hat{\mathfrak{w}})$, and also into $(\tilde{R}, \tilde{\mathfrak{w}})$, by 35.11 this identity possesses two unique extensions $\varphi : (\tilde{R}, \tilde{\mathfrak{w}}) \rightarrow (\hat{R}, \hat{\mathfrak{w}})$ and $\psi : (\hat{R}, \hat{\mathfrak{w}}) \rightarrow (\tilde{R}, \tilde{\mathfrak{w}})$. Then $\varphi\psi$ is a uniformly continuous map of $(\hat{R}, \hat{\mathfrak{w}})$ into itself and also an extension to $\hat{R}$ of the identity map, on R. But the identity of $\hat{R}$ is also such an extension and since there can only be one such extension, it follows that $\varphi\psi$ must be the identity of $\hat{R}$. Likewise, $\psi\varphi$ is the identity of $\tilde{R}$. Hence φ is a 1-1 map from $\tilde{R}$ onto $\hat{R}$ with inverse equal to ψ. Since φ and ψ are both uniformly continuous, φ must even be a uniform isomorphism. ∎

We can characterize those uniform spaces whose completion is completely compact, just as we did in Section 32.

Definition 35e: *A uniform space $(R, \mathfrak{w})$ is said to be* precompact[6] *if, given any $W \in \mathfrak{w}$ we can find a finite number of points $x_1, \ldots, x_n \in R$ such that $R = W[x_1] \cup \ldots \cup W[x_n]$.*

In pseudometric spaces this definition of a precompact space becomes identical to the definition of precompact space given in Section 32 (32d).

35.14 *A uniform space $(R, \mathfrak{w})$ is precompact if and only if, for each $\delta \in \Delta(\mathfrak{w})$, the pseudometric space (R, δ) is precompact.*

Proof: Let $(R, \mathfrak{w})$ be precompact. Given $\epsilon > 0$ and $\delta \in \Delta(\mathfrak{w})$ there exists a $W \in \mathfrak{w}$ with $W \subseteq W_\epsilon^\delta$. In addition, there exists a finite number of points $x_1, \ldots, x_n \in R$ such that

$$R = W[x_1] \cup \ldots \cup W[x_n] \subseteq W_\epsilon^\delta[x_1] \cup \ldots \cup W_\epsilon^\delta[x_n]$$

$$= U_\epsilon^\delta(x_1) \cup \ldots \cup U_\epsilon^\delta(x_n).$$

Hence (R, δ) must be precompact. Conversely, using 31.2 and 33.5, we see that given any $W \in \mathfrak{w}$ we can find a $\delta \in \Delta(\mathfrak{w})$ and an $\epsilon > 0$ such that $W_\epsilon^\delta \subseteq W$. If (R, δ) is precompact, then $R = U_\epsilon^\delta(x_1) \cup \ldots \cup U_\epsilon^\delta(x_n) \leqslant W[x_1] \cup \ldots \cup W[x_n]$ for appropriate $x_1, \ldots, x_n \in R$. ∎

35.15 *The following statements are pairwise equivalent:*

(1) *The uniform space $(R, \mathfrak{w})$ is precompact.*

(2) *The completion $(\hat{R}, \hat{\mathfrak{w}})$ of $(R, \mathfrak{w})$ is precompact.*

(3) *The completion $(\hat{R}, \hat{\mathfrak{w}})$ of $(R, \mathfrak{w})$ is completely compact.*

(4) *Given any filter $\mathfrak{a} \neq \mathfrak{o}$ of R, we can find a $\mathfrak{w}$-Cauchy filter $\mathfrak{b} \neq \mathfrak{o}$ such that $\mathfrak{b} \leqslant \mathfrak{a}$.*

[6] See Definition 32d.

Conditions (2) *and* (3) *tell us that in complete uniform spaces the concepts of "precompact" and "completely compact" coincide.*

Proof

(1) → (2): Let $\hat{W}$ be a closed uniform neighborhood of $\hat{\mathfrak{w}}$. We can associate with the uniform neighborhood $W = \hat{W} \cap (R \times R)$ in $\mathfrak{w}$, points $x_1, ..., x_n \in R$ such that $R = W[x_1] \cup ... \cup W[x_n]$. If the closure symbol "—" denotes closures taken in $(\hat{R}, \hat{\mathfrak{w}})$, then $\hat{R} = \bar{R} = \overline{W[x_1]} \cup ... \cup \overline{W[x_n]} = \hat{W}[x_1] \cup ... \cup \hat{W}[x_n]$. Since by 33.8, $\hat{\mathfrak{w}}$ is a closed filter, this implies that $(\hat{R}, \hat{\mathfrak{w}})$ is precompact.

(2) → (3): Let $\hat{\mathfrak{v}}$ be an ultrafilter of $\hat{R}$. Given any $\hat{W} \in \hat{\mathfrak{w}}$ we can find a symmetric uniform neighborhood $\hat{W}_0 \in \hat{\mathfrak{w}}$ with $\hat{W}_0 \circ \hat{W}_0 \subseteq \hat{W}$ (33.1) and we can find points $\hat{x}_1, ..., \hat{x}_n \in \hat{R}$ with $\hat{R} = \hat{W}_0[\hat{x}_1] \cup ... \cup \hat{W}_0[\hat{x}_n]$. For at least one index ν we then have $\hat{\mathfrak{v}} \leqslant \ulcorner \hat{W}_0[\hat{x}_\nu] \urcorner$ and hence $\hat{\mathfrak{v}} \times \hat{\mathfrak{v}} \leqslant \ulcorner \hat{W}_0 \circ \hat{W}_0 \urcorner \leqslant \ulcorner \hat{W} \urcorner$. Consequently $\hat{\mathfrak{v}} \times \hat{\mathfrak{v}} \leqslant \hat{\mathfrak{w}}$. This means that $\hat{\mathfrak{v}}$ is a $\hat{\mathfrak{w}}$-Cauchy filter, which, since $(\hat{R}, \hat{\mathfrak{w}})$ is complete, must converge to a point of $\hat{R}$ (11.2).

(3) → (4): The filter $\mathfrak{a}$ of R possesses an adherent point $\hat{p}$ in $\hat{R}$. The filter $\mathfrak{b} = \hat{\mathfrak{a}} \wedge \tau_{\hat{\mathfrak{w}}}\hat{p}$, where $\hat{\mathfrak{a}}$ is the filter in $\hat{R}$ determined by $\mathfrak{a}$, is different from $\mathfrak{o}$ and a $\hat{\mathfrak{w}}$-Cauchy filter. Since $\mathfrak{b} \leqslant \mathfrak{a} \leqslant \ulcorner R \urcorner$, $\mathfrak{b}$ is even a $\mathfrak{w}$-Cauchy filter.

(4) → (1): Let W be a uniform neighborhood in $\mathfrak{w}$, and let $\mathfrak{E}$ be the system of all finite subsets of R. Then the collection $\{\complement W[E] : E \in \mathfrak{E}\}$ is a filter basis, generating a filter $\mathfrak{a}$ of R. From $\mathfrak{a} \neq \mathfrak{o}$ we deduce the existence of a $\mathfrak{w}$-Cauchy filter $\mathfrak{b}$ with $\mathfrak{o} < \mathfrak{b} \leqslant \mathfrak{a}$. In $\mathfrak{b}$ there exists a set B with $B \times B \subseteq W$. If x is therefore a fixed point in B, then $B \subseteq W[x]$, hence $\mathfrak{b} \leqslant \ulcorner W[x] \urcorner$. Since $\{x\} \in \mathfrak{E}$ we must also have $\mathfrak{b} \leqslant \mathfrak{a} \leqslant \ulcorner \complement W[x] \urcorner$, contradicting $\mathfrak{b} \neq \mathfrak{o}$. This implies $\mathfrak{a} = \mathfrak{o}$ and hence $R = W[E] = W[x_1] \cup ... \cup W[x_n]$ for an appropriate set $E = \{x_1, ..., x_n\}$ in $\mathfrak{E}$. ∎

We shall terminate by examining the completeness of spaces of maps. Let therefore $(R', \mathfrak{w}')$ be a uniform space, $\hat{A}$ the set of all maps from R into R', and $\mathfrak{S}$ a covering of R. By Section 34, the uniform structure $\hat{\mathfrak{w}}_{\mathfrak{S}}$ of uniform convergence on the sets of $\mathfrak{S}$ is well-defined in $\hat{A}$.

35.16 *A filter of $\hat{A}$ converges in $(\hat{A}, \hat{\mathfrak{w}}_{\mathfrak{S}})$ to a map φ if and only if it is a $\hat{\mathfrak{w}}_{\mathfrak{S}}$-Cauchy filter which converges pointwise to φ.*

If $(R', \mathfrak{w}')$ is complete, then $(\hat{A}, \hat{\mathfrak{w}}_{\mathfrak{S}})$ is also complete. If in addition $\mathfrak{S}$ is an open covering of (R, τ), then the subspace of all continuous maps from (R, τ) into $(R', \tau_{\mathfrak{w}'})$ is also complete. The same conclusion holds whenever (R, τ) is a locally completely compact space and $\mathfrak{S}$ is the system of all completely compact subsets of (R, τ).

Proof: If $\hat{\mathfrak{b}}$ converges in $(\hat{A}, \hat{\mathfrak{w}}_{\mathfrak{S}})$ to φ, then, by 35.3, $\hat{\mathfrak{b}}$ is a $\hat{\mathfrak{w}}_{\mathfrak{S}}$-Cauchy filter. If p is a point in R then there exists an $S \in \mathfrak{S}$ with $p \in S$. Given any $W' \in \mathfrak{w}'$

there exists a $\hat{B} \in \hat{\mathfrak{b}}$ with $(\varphi x, \psi x) \in W'$; hence $\psi x \in W'[\varphi x]$ for all $\psi \in \hat{B}$ and all $x \in S$. Since $p \in S$ we have $\psi p \in W'[\varphi p]$ for all $\psi \in \hat{B}$ and also $\epsilon_p \hat{\mathfrak{b}} \leqslant \mathfrak{w}'[\varphi p] = \tau_{\mathfrak{w}'}\varphi p$ (We recall that the map $\epsilon_p : \hat{A} \to R'$ was defined by $\epsilon_p \varphi = \varphi p$.) Hence $\hat{\mathfrak{b}}$ converges pointwise to φ. Conversely, let $\hat{\mathfrak{b}}$ be a $\hat{\mathfrak{w}}_{\mathfrak{S}}$-Cauchy filter, pointwise convergent to φ. Given $S \in \mathfrak{S}$ and $W' \in \mathfrak{w}'$ then there exists a $\hat{B} \in \hat{\mathfrak{b}}$ with $\hat{B} \times \hat{B} \subseteq \langle S, W' \rangle$. Hence $(\psi_1 x, \psi_2 x) \in W'$ for all $\psi_1, \psi_2 \in \hat{B}$ and $x \in S$. With a fixed $p \in S$ we can moreover associate, because of the pointwise convergence of $\hat{\mathfrak{b}}$ to φ, a $\hat{B}_0 \in \hat{\mathfrak{b}}$ with $(\varphi p, \chi p) \in W'$ for all $\chi \in \hat{B}_0$. Hence $\chi \in \hat{B}_0 \cap \hat{B}$ implies $(\varphi p, \chi p) \in W'$ and $(\chi p, \psi p) \in W'$, hence $(\varphi p, \psi p) \in W' \circ W'$ for all $\psi \in \hat{B}$ and $p \in S$. Consequently $\hat{B} \subseteq \langle S, W' \circ W' \rangle [\varphi]$ and $\hat{\mathfrak{b}} \leqslant \hat{\mathfrak{w}}_{\mathfrak{S}}(\varphi)$; i.e., $\hat{\mathfrak{b}}$ also converges uniformly on the sets of $\mathfrak{S}$ to the function φ.

From now on let us suppose that $(R', \mathfrak{w}')$ is complete. With any $p \in R$ we can associate an $S_p \in \mathfrak{S}$ where $p \in S_p$. If therefore $\hat{\mathfrak{a}}$ is a $\hat{\mathfrak{w}}_{\mathfrak{S}}$-Cauchy filter, then

$$\epsilon_p \hat{\mathfrak{a}} \times \epsilon_p \hat{\mathfrak{a}} = (\epsilon_p \times \epsilon_p)(\hat{\mathfrak{a}} \times \hat{\mathfrak{a}}) \leqslant (\epsilon_p \times \epsilon_p)\hat{\mathfrak{w}}_{\mathfrak{S}} \leqslant (\epsilon_p \times \epsilon_p)\hat{\mathfrak{w}}_{S_p} \leqslant \mathfrak{w}'.$$

Therefore $\epsilon_p \hat{\mathfrak{a}}$ is a $\mathfrak{w}'$-Cauchy filter, which, because of the completeness of $(R', \mathfrak{w}')$ converges to a point $p' \in R'$. We see that $\hat{\mathfrak{a}}$ converges pointwise to the map $\varphi \in \hat{A}$ defined by $\varphi p = p'$. By what has preceded, $\hat{\mathfrak{a}}$ also converges uniformly on the sets of $\mathfrak{S}$ to φ; i.e., $(\hat{A}, \hat{\mathfrak{w}}_{\mathfrak{S}})$ is also complete. If, in addition, $\mathfrak{S}$ is an open covering of (R, τ), then by 34.9 the set $\hat{C}$ of all continuous maps of (R, τ) into $(R', \tau_{\mathfrak{w}'})$ is closed in $(\hat{A}, \hat{\mathfrak{w}}_{\mathfrak{S}})$. The same results hold for our second contention. By 35.9 the subspace of $(\hat{A}, \hat{\mathfrak{w}}_{\mathfrak{S}})$ generated by $\hat{C}$ is likewise complete. ∎

Notes and Exercises

35A The construction of the completion of a uniform space is often carried out by means of classes of equivalent Cauchy filters. For example, in the usual construction of the reals by completing the set of rationals, each real number is represented by a class of equivalent Cauchy sequences. The concept of round filter enables us to avoid such classes. A round filter is precisely the sup-filter of all Fréchet filters belonging to equivalent Cauchy filters.

35B Let $(R, \mathfrak{w}) = \times \{(R_\iota, \mathfrak{w}_\iota) : \iota \in I\}$.

Exercise: If $\mathfrak{a}$ is a $\mathfrak{w}$-round filter, then every projection $\pi_\iota \mathfrak{a}$ is a $\mathfrak{w}_\iota$-round filter. Conversely, if for each $\iota \in I$, $\mathfrak{a}_\iota$ is a $\mathfrak{w}_\iota$-round filter, then $\mathfrak{a} = \bigwedge \pi_\iota^{-1}\mathfrak{a}_\iota$ is a $\mathfrak{w}$-round filter. The completion of $(R, \mathfrak{w})$ is uniformly isomorphic to the product of the completions of the spaces $(R_\iota, \mathfrak{w}_\iota)$.

35C Let $(\hat{R}, \hat{\tau})$ be the Stone-Čech compactification of a complete and regular space (R, τ). By 33.9 there exists one and only one uniform structure $\hat{\mathfrak{w}}$ on R

which induces the topology $\hat{\tau}$. On the subspace R, $\hat{\mathfrak{w}}$ induces a uniform structure $\mathfrak{w}$ with $\tau_{\mathfrak{w}} = \tau$.

Exercise: $(\hat{R}, \hat{\mathfrak{w}})$ is the completion of $(R, \mathfrak{w})$. Furthermore, $\mathfrak{w}$ is the coarsest uniform structure on R under which all continuous maps from (R, τ) into $[0, 1]$ are uniformly continuous maps from $(R, \mathfrak{w})$. In general, however, $\mathfrak{w}$ is not the finest uniform structure on R which induces the topology τ.

35D *Exercise:* The subspace of $(R, \mathfrak{w})$ determined by a subset M is precompact if and only if, given any $W \in \mathfrak{w}$ we can find points $x_1, ..., x_n \in R$ such that $M \subseteq W[x_1] \cup ... \cup W[x_n]$.

The product space $\times \{(R_\iota, \mathfrak{w}_\iota) : \iota \in I\}$ is precompact if and only if each of its coordinate spaces $(R_\iota, \mathfrak{w}_\iota)$ is precompact.

VII

Topological Groups. Applications

Many of the spaces which appear in mathematics and its applications posess, besides their topological structure, an equally important algebraic structure. For instance, the real line $\mathbf{R}$ is also a field, while $\mathbf{R}^n$ is a vector space over the field of reals. When a set carries both a topological and an algebraic structure, we can expect nontrivial connections between these structures only if they are "compatible" with each other. The "compatibility condition" below demands that all the algebraic operations be continuous. If the algebraic structure is that of a group, this procedure leads us to the notion of topological group.

In this chapter we shall only give the elementary, fundamental facts of the theory of topological groups and allied structures. We assume that the reader is familiar with the basic concepts of modern algebra.

36. Topological Groups

In this section G will always denote a (not necessarily abelian) group. The group operation will be written multiplicatively. The product of the elements $a, b \in G$ will be denoted by $a \cdot b$; a^{-1} will denote the inverse element to a while e will denote the neutral element (the multiplicative 1) of the group. If A and B are subsets of G, let

$$A^{-1} = \{a^{-1} : a \in A\} \qquad \text{and} \qquad A \cdot B = \{a \cdot b : a \in A, b \in B\}.$$

If $\mathfrak{a}$ and $\mathfrak{b}$ are filters in G, then the systems

$$(A^{-1} : A \in \mathfrak{a}\} \qquad \text{and} \qquad \{A \cdot B : A \in \mathfrak{a}, B \in \mathfrak{b}\}$$

are filter bases, generating filters which we shall designate by $\mathfrak{a}^{-1}$ and $\mathfrak{a} \cdot \mathfrak{b}$, respectively. The following relations then hold:

$$\mathfrak{a}_1 \leqslant \mathfrak{a}_2 \text{ and } \mathfrak{b}_1 \leqslant \mathfrak{b}_2 \text{ imply } \mathfrak{a}_1^{-1} \leqslant \mathfrak{a}_2^{-1} \text{ and } \mathfrak{a}_1 \cdot \mathfrak{b}_1 \leqslant \mathfrak{a}_2 \cdot b_2$$

$$(\bigvee \mathfrak{a}_\iota)^{-1} = \bigvee \mathfrak{a}_\iota^{-1} \quad \text{and} \quad (\bigwedge \mathfrak{a}_\iota)^{-1} = \bigwedge \mathfrak{a}_\iota^{-1}.$$

$$(\bigvee \mathfrak{a}_\iota) \cdot (\bigvee \mathfrak{b}_\kappa) = \bigvee (\mathfrak{a}_\iota \cdot \mathfrak{b}_\kappa) \quad \text{and} \quad (\bigwedge \mathfrak{a}_\iota) \cdot (\bigwedge \mathfrak{b}_\kappa) \leqslant \bigwedge (\mathfrak{a}_\iota \cdot \mathfrak{b}_\kappa).$$

The passage from an element to its inverse is really a map from G onto G, while the passage from two elements to their product is a map from $G \times G$ onto G. If, in addition, G carries a topology τ, the aformentioned compatibility conditions demand that the formation of inverses and products be continuous maps onto (G, τ), from the domains (G, τ) and $(G, \tau) \times (G, \tau)$, respectively.

Definition 36a: *A topology τ on G is called a* group topology *or said to be* compatible *with the group structure of G, if the formation of products and inverses are both continuous maps. In more symbolic language, this definition is equivalent to the following two conditions:*

$$(\mathbf{G}_1) \quad (\tau a) \cdot (\tau b) \leqq \tau(a \cdot b) \quad \text{and} \quad (\mathbf{G}_2) \quad (\tau a)^{-1} \leqq \tau a^{-1}$$

for all $a, b \in G$.

The couple (G, τ) is called a topological group.

The following two conditions are equivalent to $(\mathbf{G}_1)$ and $(\mathbf{G}_2)$:

$$(G') \quad (\tau a) \cdot (\tau b)^{-1} \leqslant \tau(a \cdot b^{-1}), \quad (G'') \quad (\tau a)^{-1} (\tau b) \leqslant \tau(a^{-1} \cdot b)$$

for all $a, b \in G$.

Condition (G') says that the map $(a, b) \rightarrow a \cdot b^{-1}$ is continuous, while (G'') says that the map $(a, b) \rightarrow a^{-1}b$ is continuous. From $(\mathbf{G}_2)$ we deduce that $(\tau a^{-1})^{-1} \leqslant \tau a$; hence $\tau a^{-1} \leqslant (\tau a)^{-1}$ and finally $(\tau a)^{-1} = \tau a^{-1}$. This means that the taking of inverses is a homeomorphism of the topological group (G, τ) onto itself. For a fixed $a \in G$, the translations defined by $\theta_a^l x = a \cdot x$ and $\theta_a^r x = x \cdot a$ are known as the *left and right translations*, respectively, of (G, τ) onto itself. They are continuous maps: For we have

$$\vartheta_a^l \tau x = \dot{a} \cdot \tau x \leqq (\tau a) \cdot (\tau x) \leqq \tau(a \cdot x) = \tau(\vartheta_a^l x)$$

and, similarly, $\theta_a^r \tau x = \tau x \cdot \dot{a} \leqslant \tau(\theta_a^r x)$. The respective inverse maps are $(\theta_a^l)^{-1} = \theta_{a^{-1}}^l$ and $(\theta_a^r)^{-1} = \theta_{a^{-1}}^r$. Since these inverse maps are also translations, they must be topological maps of (G, τ) onto itself. The same result holds for each *inner automorphism* $x \rightarrow a^{-1} \cdot x \cdot a = \theta_{a^{-1}}^l \cdot \theta_a^r$ of G.

Since translations in a topological group (G, τ) are topological mappings, we have $\ulcorner a \urcorner \cdot \tau e = \tau a = \tau e \cdot \ulcorner a \urcorner$ for each $a \in G$. Hence a topological group is uniquely determined as soon as we know the neighborhoods filter τe of the neutral element. We shall henceforth denote this special neighborhoods filter by $\mathfrak{u}_\tau$. Then $\ulcorner a \urcorner \cdot \mathfrak{u}_\tau = \tau a = \mathfrak{u}_\tau \cdot \ulcorner a \urcorner$.

36.1 *In a topological group* (G, τ) *the neighborhoods filter* $\mathfrak{u}_\tau$ *of the neutral element possesses the following properties:*

(a) $\ulcorner e \urcorner \leqslant \mathfrak{u}_\tau$.

(b) $\mathfrak{u}_\tau \cdot \mathfrak{u}_\tau \leqslant \mathfrak{u}_\tau$.

(c) $\mathfrak{u}_\tau^{-1} \leqslant \mathfrak{u}_\tau$.

(d) $\ulcorner a \urcorner^{-1} \cdot \mathfrak{u}_\tau \cdot \ulcorner a \urcorner \leqslant \mathfrak{u}_\tau$.

Conversely, if $\mathfrak{u}$ *is a filter in the group G satisfying properties* (a)–(d), *then there exists a unique topology* τ *on G such that* $\mathfrak{u} = \tau e$. *For this topology* $\tau a = \ulcorner a \urcorner \cdot \mathfrak{u} = \mathfrak{u} \cdot \ulcorner a \urcorner$ *and in conditions* (b)–(d) *the equality sign even holds.*

Proof: If τ is a group topology, conditions (a)–(d) follow directly from the continuity of the group operation. Since the formation of inverses and of inner automorphisms are topological maps, the equality sign even holds in (c) and (d). Since $\tau e = \ulcorner e \urcorner \cdot \tau e \leqslant (\tau e) \cdot (\tau e)$ we also have equality in (b).

Conversely, let us suppose that the filter $\mathfrak{u}$ of the group G satisfies properties (a)–(d). By (d) we have $\mathfrak{u} \cdot \ulcorner a \urcorner \leqslant \ulcorner a \urcorner \cdot \mathfrak{u}$ and also (replacing a by a^{-1}) $\ulcorner a \urcorner \cdot \mathfrak{u} \leqslant \mathfrak{u} \cdot \ulcorner a \urcorner$; hence $\ulcorner a \urcorner \cdot \mathfrak{u} = \mathfrak{u} \cdot \ulcorner a \urcorner$ for all $a \in G$. If we put $\tau a = \ulcorner a \urcorner \cdot \mathfrak{u}$, then $\ulcorner e \urcorner \leqslant \tau e$ by (a). An arbitrary set in τa is of the form $a \cdot U$ for $U \in \mathfrak{u}$. By (b) there exists a $U' \in \mathfrak{u}$ such that $U' \cdot U' \leqslant U$. From $b \in a \cdot U'$ we then deduce $b \cdot U' \leqslant a \cdot U' \cdot U' \leqslant a \cdot U$. Hence τ is a topology on G. By (b) we further have

$$(\tau a) \cdot (\tau b) = \ulcorner a \urcorner \cdot \mathfrak{u} \cdot \ulcorner b \urcorner \cdot \mathfrak{u} = \ulcorner a \urcorner \cdot \ulcorner b \urcorner \cdot \mathfrak{u} \cdot \mathfrak{u} \leqq \ulcorner a \cdot b \urcorner \cdot \mathfrak{u} = \tau(a \cdot b)$$

and also, from (c),

$$(\tau a)^{-1} = (\ulcorner a \urcorner \cdot \mathfrak{u})^{-1} = \mathfrak{u}^{-1} \cdot \ulcorner a \urcorner^{-1} \leqq \mathfrak{u} \cdot \ulcorner a \urcorner^{-1} = \tau a^{-1}.$$

Hence τ is a group topology. It is the only group topology on G such that $\tau e = \mathfrak{u}$, for in any topological group we must always have $\tau a = \ulcorner a \urcorner \cdot \tau e = \tau e \cdot \ulcorner a \urcorner$. ∎

Examples

36.I The discrete and trivial topologies on any group are group topologies.

36.II The set of real numbers, under addition as the group operation and bearing the usual topology of the line, is an example of a topological group. The set of real numbers distinct from zero, under the usual topology with ordinary multiplication now being the group operation, also form a topological group.

36.III Let G be the affine group on the real line; i.e., the group of all maps $\varphi_{a,b} : \mathbf{R} \to \mathbf{R}$ defined by $\varphi_{a,b}x = ax + b$, $a \neq 0$ with the group operation being the usual composition of maps. The sets

$$U_\epsilon = \{\varphi_{a,b} : |a - 1| < \epsilon, |b| < a\epsilon\} \qquad (\epsilon > 0)$$

form a filter basis, generating a filter $\mathfrak{u}$ which determines the group topology on G.

36.2 *A group topology τ is a T_1-topology if and only if $\cap \{U : U \in \mathfrak{u}_\tau\} = \{e\}$.*

Proof: Since $\cap \{V : V \in \tau a\} = \cap \{a \cdot U : U \in \mathfrak{u}_\tau\} = a \cdot \cap \{U : U \in \mathfrak{u}_\tau\}$, the condition of this theorem is equivalent to $\cap \{V : V \in \tau a\} = \{a\}$ for all $a \in G$ (8.1). ∎

For every subset M of the group G we define the following subsets of $G \times G$:

$$W_M^l = \{(x, y) : x, y \in G, x^{-1} \cdot y \in M\} \text{ and } W_M^r = \{(x, y) : x, y \in G, x \cdot y^{-1} \in M\}$$

If τ is a group topology, then the systems of sets $\{W_U^l : U \in \mathfrak{u}_\tau\}$ and $\{W_U^r : U \in \mathfrak{u}_\tau\}$ constitute filter bases, generating in $G \times G$ the filters $\mathfrak{w}_\tau^l$ and $\mathfrak{w}_\tau^r$, respectively.

36.3 *$\mathfrak{w}_\tau^l$ and $\mathfrak{w}_\tau^r$ are uniform structures on G, each of which induces the topology τ.*

Proof: Since $e \in U$ we have $(a, a) \in W_U^l$ for all $U \in \mathfrak{u}_\tau$ and hence $\overset{\sqcap}{D(G)} \leqslant \mathfrak{w}_\tau^l$. By 36.1, given any $U \in \mathfrak{u}_\tau$ we can find a $V \in \mathfrak{u}_\tau$ with $V \cdot V \subseteq U$. From $(a, b) \in W_V^l$ and $(b, c) \in W_V^l$ we deduce that $a^{-1} \cdot c = (a^{-1} \cdot b)(b^{-1} \cdot c) \in V \cdot V \subseteq U$, hence $(a, c) \in W_U^l$. Thus $W_V^l \circ W_V^l \subseteq W_U^l$ and $\mathfrak{w}_\tau \cdot \mathfrak{w}_\tau^l \leqslant \mathfrak{w}_\tau^l$. Lastly, $(W_U^l)^{-1} = W_{U^{-1}}^l$, whence, since $\mathfrak{u}_\tau^{-1} = \mathfrak{u}_\tau$, we see that $(\mathfrak{w}_\tau^l)^{-1} = \mathfrak{w}_\tau^l$. This means that $\mathfrak{w}_\tau^l$ is a uniform structure on G. From $W_U^l[A] = \{b : a^{-1} \cdot b \in U\} = a \cdot U$, it follows that $\mathfrak{w}_\tau^l[a] = \overset{\sqcap}{a} \cdot \mathfrak{u}_\tau = \tau a$, which tells us that $\mathfrak{w}_\tau^l$ also induces the topology τ. The contention on $\mathfrak{w}_\tau^r$ follows by similar arguments. ∎

The vicinities filters $\mathfrak{w}_\tau^l$ and $\mathfrak{w}_\tau^r$ are known as the *left uniform* and *right uniform structures*, respectively, of the topological group (G, τ). By 33.3, $\mathfrak{w}_\tau = \mathfrak{w}_\tau^l \wedge \mathfrak{w}_\tau^r$ is also a uniform structure on G, inducing the topology τ. This structure is known as the (*two-sided*) *uniform structure* on (G, τ). The previously established theorem (33.7) and the existence of this (two-sided) uniform structure, taken together, imply:

36.4 *Every topological group is a T_{3a}-space; every T_1-topological group is completely regular and thus, all the more so, a T_2-space.*

36.5 *Let M be an arbitrary and U an open subset of the topological group (G, τ). Then the sets $M \cdot U$ and $U \cdot M$ are both open.*

Proof: Since translations are topological maps and if U is open so are $a \cdot U$ and $U \cdot a$. Hence $M \cdot U = \cup \{a \cdot U : a \in M\}$ and $U \cdot M = \cup \{U \cdot a : a \in M\}$ are open. ∎

If H is a subgroup of G and τ a group topology on G, then we see that the relative topology τ_H is evidently a group topology on H. Hence every subgroup of a topological group is itself a topological group.

36.6 *Let H be a subgroup of the topological group (G, τ).*

(1) *H is open if and only if $\overset{\circ}{H} \neq \emptyset$.*

(2) *If H is open, then H must also be closed.*

(3) *$\bar{H}$ is a subgroup of G. If H is a normal subgroup, so is $\bar{H}$.*

(4) *There are only these two possibilities: Either H is closed, or else $\bar{H} \cap \complement H$ is dense in $\bar{H}$.*

Proof

(1) If H is open, then $e \in H = \overset{\circ}{H}$, and hence $\overset{\circ}{H} \neq \emptyset$. Conversely, suppose $a \in \overset{\circ}{H}$. If $b \in H$ then $b \cdot a^{-1}H = H$. Since the left-translation generated by $b \cdot a^{-1}$ is a homeomorphism, we have

$$b = b \cdot a^{-1} \cdot a \in b \cdot a^{-1}(\overset{\circ}{H}) = \overset{\circ}{\widehat{b \cdot a^{-1} \cdot H}} = \overset{\circ}{H}$$

for all $b \in H$ and hence $H \subseteq \overset{\circ}{H}$.

(2) If H is open, then $a \cdot H$ is also open for each $a \in G$. Hence

$$H = \complement \bigcup \{a \cdot H : a \notin H\}$$

is closed.

(3) The map $\varphi : G \times G \to G$ defined by $\varphi(x, y) = x \cdot y^{-1}$ is a continuous map from $(G, \tau) \times (G, \tau)$ onto (G, τ), satisfying $\varphi(H \times H) = H$. From $a, b \in \bar{H}$ it therefore follows that $a \cdot b^{-1} \in \varphi(\bar{H} \times \bar{H}) = \varphi(\overline{H \times H}) \subseteq \bar{H}$ (22.3; 17.1); i.e., $\bar{H}$ is also a subgroup. Since the inner automorphism defined by x is also a topological map, we also have $x^{-1} \cdot \bar{H} \cdot x = \overline{x^{-1} \cdot H \cdot x}$, and hence, if H is a normal subgroup, so is $\bar{H}$.

(4) By (3), $\bar{H}$ is itself a topological group. If $\bar{H} \cap \complement H$ is not dense in $\bar{H}$, then the interior of H in $\bar{H}$ is nonempty. By (2), H is an open subgroup of $\bar{H}$ while, by (3), H is a closed subgroup of $\bar{H}$, whence $H = \bar{H}$. ∎

Let M be a nonempty subset of the group G. By the subgroup of G generated by M we mean the smallest subgroup of G containing M. A subgroup H of the topological group (G, τ) is said to be *discrete* if the relative topology τ_H is discrete.

36.7 *The subgroup generated by a nonempty open subset of a topological group (G, τ) is simultaneously open and closed. If (G, τ) is connected, then G can be generated by any neighborhood of the neutral element. Every discrete normal subgroup of a connected topological group is contained in the center of the group* (*the center of a group G is the set of all elements $z \in G$ such that $z \cdot a = a \cdot z$ for all $a \in G$*).

Proof: The subgroup generated by a nonempty set has a nonempty interior and hence, by 36.6, this subgroup must be simultaneously open and closed. In particular, every subgroup generated by a neighborhood of the neutral element is both open and closed—if (G, τ) is connected, this subgroup must therefore equal all of G. Lastly, let H be a discrete normal subgroup of the connected group (G, τ). Being discrete, H must be totally disconnected and hence all the components of H are one-point subsets. For a fixed $h \in H$ the map $\varphi : G \to H$ defined by $\varphi x = x^{-1} \cdot h \cdot x$ is continuous. By 19.3, φG is a connected subset of H and hence φG is a one-point set. Since $\varphi e = h$ we see that $h \in \varphi G$ and therefore $\varphi G = \{h\}$. We therefore have $x^{-1} \cdot h \cdot x = h$ for all $x \in G$, which means that h lies in the center of G. ∎

36.8 *The component G_0 containing the neutral element of a topological group (G, τ) is a closed normal subgroup. The component containing a is the translate $a \cdot G_0$.*

Proof: Since the left-translation defined by the element a is a topological map, $a \cdot G_0$ must be the component containing a. From $a \in G_0$, we see that $e \in G_0 \cap a^{-1} \cdot G_0$. Since G_0 and $a^{-1} \cdot G_0$ are connected, we see by 14.5 that $G_0 \cup a^{-1} \cdot G_0$ is also connected. By the maximality of G_0 we therefore have $a^{-1} \cdot G_0 \subseteq G_0$. From $a, b \in G_0$ we therefore deduce that $a^{-1} \cdot b \in G_0$, and hence G_0 must be a subgroup of G. The inner automorphism defined by any element $x \in G$ is also a topological map. Since $e \in G_0 \cap x^{-1} \cdot G_0 \cdot x$, we also see that $x^{-1} \cdot G_0 \cdot x \subseteq G_0$, i.e., G_0 is even a normal subgroup. By (14.8) every component of a space is closed and hence G_0 must itself be closed. ∎

36.9 *Every locally completely compact topological group is paracompact.*

Proof: Let U be a completely compact neighborhood belonging to τe. By $(\mathbf{T}_3)$ and (36.4) we can assume that U is closed. Then $V = U \cap U^{-1}$ is also closed and hence a completely compact neighborhood of e. Suppose

$$V^n = V \times \ldots \times V \qquad \text{and} \qquad V^{(n)} = V \cdot \ldots \cdot V$$

(n factors in both cases). The map φ from V^n onto $V^{(n)}$ defined by

$$\varphi(x_1, \ldots, x_n) = x_1 \cdot \ldots \cdot x_n$$

is then continuous. Since V^n is completely compact (22.6), $V^{(n)}$ must also be completely compact (19.1). The subgroup H generated by V is both open and closed (36.6) and we have $H = \cup \{V^{(n)} : n \in \mathbf{N}\}$. Being a subgroup of (G, τ), H is locally completely compact (12.5) and countable at infinity. The sets

$a \cdot H(a \in G)$ are then also of this nature. Suppose now that Q is a complete system of representatives of the cosets of G modulo H. Then

$$(G, \tau) = \Sigma\{(a \cdot H, \tau_{a.H} : a \in Q\},$$

and our contention follows from 24.7. ∎

Let $\{G_\iota : \iota \in I\}$ be a collection of groups. In $\times\ G_\iota$ we can define a multiplication, $(\times\ a_\iota) \cdot (\times\ b_\iota) = \times\ (a_\iota \cdot b_\iota)$, with respect to which $\times\ G_\iota$ is a group. We have $(\times\ a_\iota)^{-1} = \times\ a_\iota^{-1}$. If, in addition, τ_ι is a group topology on G_ι for every $\iota \in I$, then the product space $\times\ \{(G_\iota, \tau_\iota) : \iota \in I\}$ is also a topological group: Since $\pi_\kappa((\times\ a_\iota) \cdot (\times\ b_\iota)) = \pi_\kappa \times (a_\iota b_\iota) = a_\kappa \cdot b_\kappa$ and since

$$\pi_\kappa(\times\ a_\iota)^{-1} = \pi_\kappa \times a_\iota^{-1} = a_\kappa^{-1}$$

we see by 22.9, that the taking of products and inverses are continuous maps on $\times\ (G_\iota, \tau_\iota)$.

With every subgroup H of G we can uniquely associate the equivalence relation on G defined by

$$\tilde{H} = \{(a, b) : a, b \in G,\ a^{-1} \cdot b \in H\}.$$

If (G, τ) is a topological group, the quotient space $(G, \tau)/\tilde{H}$ is known as the *homogeneous space* determined by the subgroup H.

36.10 *The canonical map κ from (G, τ) onto $(G, \tau)/\tilde{H}$ is continuous and open.*

For each left-translation $\theta_a^l(a \in G)$, the map $\theta_a' = \kappa\theta_a^l\kappa^{-1}$ is a topological map of $(G, \tau)/\tilde{H}$ onto itself.

The homeomorphisms group of $(G, \tau)/\tilde{H}$ is transitive. (Homogeneous space, cf. p. 118.)

$(G, \tau)/\tilde{H}$ is a T_1-space if and only if H is closed subgroup.

If H is an open subgroup, then $(G, \tau)/\tilde{H}$ is discrete.

Proof: By 21.7, κ is a continuous map. Let us assume that M is an open subset of (G, τ). The image $M' = \kappa M$ satisfies $\kappa^{-1}M' = M \cdot H$. By 36.5, $\kappa^{-1}M'$ is open in $(G, \tau)/\tilde{H}$ and, by 21.1, M' is open in $(G, \tau)/\tilde{H}$. Hence κ is also an open map.

From $\kappa x = \kappa y$ it follows that $(\theta_a^l x)^{-1} \cdot (\theta_a^l y) = (a \cdot x)^{-1} \cdot (a \cdot y) = x^{-1}y \in H$ and hence $\kappa\theta_a^l x = \kappa\theta_a^l y$. Therefore $\theta_a' = \kappa\theta_a^l\kappa^{-1}$ is a map of $G/\tilde{H}$ into itself; and even a map onto $G/\tilde{H}$, because θ_a^l is a map of G onto itself. Since $\theta_{a^{-1}}^l = \kappa\theta_a^l$, we see that $\theta_a'\kappa$ is continuous and hence, by 27.1, θ_a' is also continuous. Now $\theta_{a^{-1}}^l$ is the inverse to θ_a^l and hence $\theta_{a^{-1}}'\theta_a' = \kappa\theta_{a^{-1}}^l\kappa^{-1}\kappa\theta_a^l\kappa^{-1}$ is the identity on $G/\tilde{H}$. Therefore θ_a' is a 1-1 map with continuous inverse $\theta_{a^{-1}}'$; i.e., θ_a' is a homeomorphism.

For arbitrary $x, y \in G$ and $a = yx^{-1}$ we have $\theta_a'(\kappa x) = \kappa y$. The homeomorphisms group of $(G, \tau)/\tilde{H}$ is therefore transitive.

By 21.7, $(G, \tau)/\tilde{H}$ is a T_1-space if and only if the equivalence classes $a \cdot H (a \in G)$ are closed. This, however, can only happen when H itself is closed.

If H is open then all the classes $a \cdot H$ are open and hence all points of $(G, \tau)/H$ are open. The quotient topology must in such a case be discrete. ∎

Whenever H is a normal subgroup of G, modern algebra tells us that the quotient set $G/\tilde{H}$ can be given a group structure. The resulting group is denoted by G/H and called the factor group of G modulo H. In this situation we denote the quotient space of the topological space (G, τ) modulo the equivalence relation $\tilde{H}$ by the symbol $(G, \tau)/H$. $(G, \tau)/H$ is itself a topological group. In this case the canonical map onto $(G, \tau)/H$ is a continuous and open (group) homomorphism.

36.11 *In a topological group (G, τ), the subgroup $\{\bar{e}\}$ is normal. The factor group $(G, \tau)/\{\bar{e}\}$ is a T_1-space (and hence completely regular).*

Proof: Since $\{e\}$ is a normal subgroup of G, by 36.6 we see that $\{\bar{e}\}$ must be a normal subgroup also. The other contentions follow from 36.10 and 36.4. ∎

36.12 *Let (G, τ) and (G', τ') be two topological groups, and let φ be a homomorphism*[1] *of G into G'. Then $\varphi : (G, \tau) \to (G', \tau')$ if and only if $\varphi \mathfrak{u}_\tau \leqslant \mathfrak{u}_{\tau'}$.*

Proof: Since φ is a homomorphism, we have $\varphi \tau a = \varphi(\overset{\sqcap}{a} \cdot \mathfrak{u}_\tau) = (\varphi \overset{\sqcap}{a}) \cdot (\varphi \mathfrak{u}_\tau)$. From $\varphi \mathfrak{u}_\tau \leqslant \mathfrak{u}_{\tau'}$ it therefore follows that $\varphi \tau a \leqslant (\varphi \overset{\sqcap}{a}) \cdot \mathfrak{u}_{\tau'} = \tau' \varphi a$ for all $a \in G$; i.e., φ is continuous. Conversely, if φ is continuous, then

$$\varphi \mathfrak{u}_\tau = \varphi \tau e \leqslant \tau' \varphi e = \mathfrak{u}_{\tau'}. \blacksquare$$

Two topological groups (G, τ) and (G', τ') are said to be *isomorphic* if their exists an (algebraic) isomorphism of G onto G' such that this group isomorphism is also a homeomorphism of the topological spaces (G, τ) and (G', τ').

36.13 *Let φ be a continuous homomorphism of (G, τ) onto (G', τ'), with kernel $K = \varphi^{-1}e'$, where e' is the neutral element in G'. And let κ denote the canonical map from G onto G/K. Then $\psi = \varphi\kappa^{-1}$ is a continuous 1-1 map of $(G, \tau)/K$ onto (G', τ'). If φ is also open, then the topological groups $(G, \tau)/K$ and (G', τ') are isomorphic.*

Proof: We know for sure that ψ is an (algebraic) isomorphism of G/K onto G'. The equivalence $\tilde{K}$ associated with the kernel K of φ is identical to the equivalence $\tilde{\varphi}$ induced by φ(cf. p. 7). From 21.8 we therefore see that ψ is a homomorphism of $(G, \tau)/K$ onto (G', τ_φ). Since τ_φ is the finest topology on

[1] N. Bourbaki uses the following terminology: A homomorphism of G into G' is called a representation. A continuous homomorphism φ of (G, τ) into (G', τ') which sends open subsets of G onto open subsets of G', is simply called a homomorphism (of topological groups).

G' with respect to which φ is continuous, we must have $\tau_q \leqslant \tau'$. Hence ψ is also a continuous map of $(G, \tau)/K$ onto (G', τ'). If φ is, in addition, an open map, then $\tau' = \tau_q$ (21.6) and ψ must even be a homeomorphism. ∎

Notes and Exercises

36A *Exercise:* The projection maps on the product group $\times (G_\iota, \tau_\iota)$ are continuous and open homomorphisms.

36B *Exercise:* Let H and K be normal subgroups of the group G with $K \subseteq H \subseteq G$. Then the topological groups

$$(G, \tau)/H \qquad \text{and} \qquad (G, \tau)/K/H/K$$

are isomorphic.

36C Let H be a subgroup and K a normal subgroup of the topological group (G, τ). Let κ denote the canonical map of G onto G/K and let τ' be the topology induced by τ on the subgroup $H \cdot K$.

Exercise: The (topological) subgroup κH of $(G, \tau)/K$ and the topological group $(H \cdot K, \tau')/K$ are isomorphic.

On the contrary, the topological groups $(H \cdot K, \tau')/K$ and $(H, \tau_H)/H \cap K$ are generally not isomorphic.

Exercise: Let (G, τ) be the additive group of all real numbers, under the usual topology (36.II), let K be the subgroup of all positive and negative integers (0 included), and let H be the subgroup of all integral multiples of some fixed irrational number. (Since G is abelian, all these subgroups are normal.)

1. $(H, \tau_H)/H \cap K$ is a discrete group.
2. The group $H + K$ is dense in (G, τ). ($H + K$ is, of course, the additive analog of $H \cdot K$.)
3. The quotient group $(H + K, \tau')/K$ is not discrete and therefore not isomorphic to $(H, \tau_H)/H \cap K$ [8, IV, p. 14].

36D *Exercise:* The left- and right-uniform structures on the topological group of 36.III are identical.

36E *Exercise:* For each $\iota \in I$, let τ_ι be a group topology on the group G. Then $\tau = \bigwedge \{\tau_\iota : \iota \in I\}$ is a group topology on G.

The sup-topology of even a finite number of group topologies need not, in general, be a group topology: Let B be the additive group of all integers. For a fixed prime number p let U_n be the set of all integers divisible by p^n for some n and let $\mathfrak{u}_p$ be the filter in G generated by the filter basis $\{U_n : n \in \mathbf{N}\}$.

Exercise

1. We can define a T_1-group topology on G by $\tau_p 0 = \mathfrak{u}_p$.
2. For distinct primes p, q the sup-topology $\tau_p \vee \tau_q$ is not a group topology.
3. In general the set of group topologies on an arbitrary group G forms a complete lattice, but this lattice is, in general, not a sublattice of the lattice of all topologies on G viewed as a set.

37. Extensions and Completions of Topological Groups

A topological group $(\hat{G}, \hat{\tau})$ is said to be an *extension* of the topological group (G, τ) if, as a topological space $(\hat{G}, \hat{\tau})$ is an extension of (G, τ) (25a), and if, in addition, G is a subgroup of $\hat{G}$. Since in topological groups we have at our disposal uniform structures (36.3), we can use the results of Section 35 in our study of extensions of such topological groups.

Since in this section we shall always be studying topological groups (G, τ) with uniform structures $\mathfrak{w}_\tau^l$, $\mathfrak{w}_\tau^r$, and $\mathfrak{w}_\tau$, we shall employ the following notational simplifications: Instead of $\mathfrak{w}_\tau^l$-Cauchy filters we shall simply speak of l-Cauchy filters, and likewise for r-Cauchy filters. The $\mathfrak{w}_\tau$-Cauchy filters will simply be called Cauchy filters. The forming of round filters $\mathfrak{r}(\mathfrak{a})$ with respect to $\mathfrak{w}_\tau^l$, $\mathfrak{w}_\tau^r$, or $\mathfrak{w}_\tau$ will accordingly be denoted by $\mathfrak{r}_l(\mathfrak{a})$, $\mathfrak{r}_r(\mathfrak{a})$, and $\mathfrak{r}(\mathfrak{a})$, respectively, and we shall simply speak of l-round, r-round, and round filters, respectively. In abelian groups these distinctions are, of course, superfluous, for in this case the three uniform structures in question become identical.

37.1 *$\mathfrak{a}$ is an l-Cauchy filter (r-Cauchy filter) in (G, τ) if and only if $\mathfrak{a}^{-1} \cdot \mathfrak{a} \leqslant \mathfrak{u}_\tau$ ($\mathfrak{a} \cdot \mathfrak{a}^{-1} \leqslant \mathfrak{u}_\tau$). If both these inequalities hold we then have an equality which characterizes Cauchy filters.*

Proof: $\mathfrak{a}$ is an l-Cauchy filter if and only if $\mathfrak{a} \times \mathfrak{a} \leqslant \mathfrak{w}_\tau^l$ (35a), hence if and only if, given any $U \in \mathfrak{u}_\tau$ we can find an $A \in \mathfrak{a}$ such that $x^{-1} \cdot y \in U$ for all $x, y \in A$. This is equivalent to $A^{-1} \cdot A \subseteq U$, which proves the contention involving $\mathfrak{a}^{-1} \cdot \mathfrak{a} \leqslant \mathfrak{u}_\tau$. By similar reasoning we can prove the contention on r-Cauchy filters.[2] Since $\mathfrak{w}_\tau = \mathfrak{w}_\tau^l \wedge \mathfrak{w}_\tau^r$, the last contention follows immediately. ∎

In general an l-Cauchy filter need not be an r-Cauchy filter (cf. 37A).

[2] Here, and in similar situations, a detailed proof of a "right-sides contention" is really superflous, as the following duality-principle shows: Let (G, τ) be a topological group. We can create in G a new "multiplication," denoted by $*$ and defined by $a * b = b \cdot a$. With respect to this new multiplication G is still a group and τ still a group topology. Hence we can associate with every (topological) group a second (topological) group, each arising from the other by interchanging "left" and "right" (*groupes opposés*). As in Section 2, we can enunciate a duality principle: If in a theorem on (topological groups we interchange the concepts "left" and "right," we obtain a new, still valid theorem.

37.2 *For any arbitrary filter* $\mathfrak{a}$ *in* (G, τ) *we have*

$$\mathfrak{r}_l(\mathfrak{a}) = \mathfrak{a} \cdot \mathfrak{u}_\tau, \qquad \mathfrak{r}_r(\mathfrak{a}) = \mathfrak{u}_\tau \cdot \mathfrak{a}, \qquad \mathfrak{r}(\mathfrak{a}) = \mathfrak{a} \cdot \mathfrak{u}_\tau \wedge \mathfrak{u}_\tau \cdot \mathfrak{a}.$$

Proof: The sets $W_U^l[A]$, where $U \in \mathfrak{u}_\tau$ and $A \in \mathfrak{a}$, constitute a basis of $\mathfrak{r}_l(\mathfrak{a})$ (35b). We have $W_U^l[a] = \{x : a^{-1} \cdot x \in U\} = a \cdot U$, hence

$$W_U^l[A] = \cup \{W_U^l[a] : a \in A\} = A \cdot U$$

and hence $\mathfrak{r}_l(\mathfrak{a}) = \mathfrak{a} \cdot \mathfrak{u}_\tau$. The other contentions follow similarly. ∎

37.3 *Let* $\mathfrak{b}$ *be an l-Cauchy filter (r-Cauchy filter) of a topological group* (G, τ). *Then* $\mathfrak{u}_\tau \cdot \mathfrak{b} \leqslant \mathfrak{b} \cdot \mathfrak{u}_\tau$ $(\mathfrak{b} \cdot \mathfrak{u}_\tau \leqslant \mathfrak{u}_\tau \cdot \mathfrak{b})$.

If $\mathfrak{a}$ *and* $\mathfrak{b}$ *are both l-Cauchy filters (r-Cauchy filters) or Cauchy filters), then their product is also of the same nature.*

Proof: Let B be a set belonging to the l-Cauchy filter $\mathfrak{b}$ and U a neighborhood in $\mathfrak{u}_\tau$. Since $\mathfrak{u}_\tau \cdot \mathfrak{u}_\tau \leqslant \mathfrak{u}_\tau$ (36.1) there must exist a $U' \in \mathfrak{u}_\tau$ with $U' \cdot U' \subseteq U$. By 37.1, there even exists a $B' \in \mathfrak{b}$ with $B'^{-1} \cdot B' \subseteq U'$. For a fixed $b \in B \cap B'$ we then have $B' \subseteq b \cdot U'$, and since $\dot{b}^{-1} \cdot \mathfrak{u}_\tau \cdot \dot{b} \leqslant \mathfrak{u}_\tau$ (36.1) there exists a $U'' \in \mathfrak{u}_\tau$ such that $U'' \cdot b \subseteq b \cdot U'$. For the sets $B'' = B \cap B'$ of $\mathfrak{b}$ we then have

$$U'' \cdot B'' \subseteq U'' \cdot b \cdot U' \subseteq b \cdot U' \cdot U' \subseteq b \cdot U \subseteq B \cdot U,$$

and consequently $\mathfrak{u}_\tau \cdot \mathfrak{b} \leqslant \mathfrak{b} \cdot \mathfrak{u}_\tau$. Our contention on r-Cauchy filters follows similarly.

Let us now suppose that $\mathfrak{a}$ and $\mathfrak{b}$ are two l-Cauchy filters; i.e., suppose $\mathfrak{a}^{-1} \cdot \mathfrak{a} \leqslant \mathfrak{u}_\tau$ and $\mathfrak{b}^{-1} \cdot \mathfrak{b} \leqslant \mathfrak{u}_\tau$ (37.1). By what we have just proved, we must have

$$(\mathfrak{a} \cdot \mathfrak{b})^{-1} \cdot (\mathfrak{a} \cdot \mathfrak{b}) = \mathfrak{b}^{-1} \cdot \mathfrak{a}^{-1} \cdot \mathfrak{a} \cdot \mathfrak{b} \leqq \mathfrak{b}^{-1} \cdot \mathfrak{u}_\tau \cdot \mathfrak{b} \leqq \mathfrak{b}^{-1} \cdot \mathfrak{b} \cdot \mathfrak{u}_\tau \leqq \mathfrak{u}_\tau \cdot \mathfrak{u}_\tau \leqq \mathfrak{u}_\tau.$$

Hence $\mathfrak{a} \cdot \mathfrak{b}$ is also an l-Cauchy filter. The contention on r-Cauchy filters follows similarly. If $\mathfrak{a}$ and $\mathfrak{b}$ are both Cauchy filters then $\mathfrak{a} \cdot \mathfrak{b}$ is both an l- and an r-Cauchy. filter and hence also a Cauchy filter. ∎

If $\mathfrak{a}$ is an l-Cauchy filter, then $\mathfrak{a}^{-1}(\mathfrak{a}^{-1})^{-1} = \mathfrak{a}^{-1} \cdot \mathfrak{a} \leqslant \mathfrak{u}_\tau$, and hence $\mathfrak{a}^{-1}$ is an r-Cauchy filter. But, in general, this $\mathfrak{a}^{-1}$ need not be an l-Cauchy filter: Similarly, if $\mathfrak{a}$ is an r-Cauchy filter, $\mathfrak{a}^{-1}$ must be an l-Cauchy filter. These partial results imply:

37.4 *If* $\mathfrak{a}$ *is a Cauchy filter, then so is* $\mathfrak{a}^{-1}$.

37.5 *In a topological group* (G, τ) *the Cauchy filter* $\mathfrak{a}$ *is round if and only if* $\mathfrak{a} = \mathfrak{a} \cdot \mathfrak{u}_\tau$ $(\mathfrak{a} = \mathfrak{u}_\tau \cdot \mathfrak{a})$.

Proof: $\mathfrak{a}$ is also an l- and r-Cauchy filter and hence, by 37.3, $\mathfrak{a} \cdot \mathfrak{u}_\tau = \mathfrak{u}_\tau \cdot \mathfrak{a}$. By 37.2 we have $\mathfrak{r}(\mathfrak{a}) = \mathfrak{a} \cdot \mathfrak{u}_\tau \wedge \mathfrak{u}_\tau \cdot \mathfrak{a} = \mathfrak{a} \cdot \mathfrak{u}_\tau$, from which our contention follows immediately. ∎

The group operation or multiplication of G induces in the set of all filters of G another associative multiplication. In general, however, the set of all filters of G cannot be endowed with a group structure using this multiplication —unless we limit ourselves to only the round Cauchy filters.

37.6 *With respect to induced multiplication the set $RC(G, \tau)$ of all round Cauchy filters of a topological group (G, τ) is itself a group.*

Proof: Let $\mathfrak{a}$ and $\mathfrak{b}$ be round Cauchy filters. Then the filters $\mathfrak{a}^{-1}$ and $\mathfrak{a} \cdot \mathfrak{b}$ are certainly Cauchy filters (37.4; 37.3). Since $\mathfrak{u}_\tau^{-1} = \mathfrak{u}_\tau$ we have

$$\mathfrak{a}^{-1} \cdot \mathfrak{u}_\tau = \mathfrak{a}^{-1} \cdot \mathfrak{u}_{\tau^{-1}} = (\mathfrak{u}_\tau \cdot \mathfrak{a})^{-1} = \mathfrak{a}^{-1} \quad (37.5)$$

and $\mathfrak{a}^{-1}$ is also a round filter. Likewise, since $(\mathfrak{a} \cdot \mathfrak{b}) \cdot \mathfrak{u}_\tau = \mathfrak{a} \cdot (\mathfrak{b} \cdot \mathfrak{u}_\tau) = \mathfrak{a} \cdot \mathfrak{b}$ (37.5), the filter $\mathfrak{a} \cdot \mathfrak{b}$ must be round. The neighborhoods filter $\mathfrak{u}_\tau$ is also a round Cauchy filter (35.5). For round filters $\mathfrak{a}$ we have $\mathfrak{u}_\tau \cdot \mathfrak{a} = \mathfrak{a} = \mathfrak{a} \cdot \mathfrak{u}_\tau$ (37.5). This means that $\mathfrak{u}_\tau$ is the neutral element for the multiplication of $RC(G, \tau)$. For all $\mathfrak{a} \in RC(G, \tau)$ we finally see that $\mathfrak{a}^{-1} \cdot \mathfrak{a} \leqslant \mathfrak{u}_\tau$ and $\mathfrak{a} \cdot \mathfrak{a}^{-1} \leqslant \mathfrak{u}_\tau$ (37.1). But, by 35.5 and the fact that the filters $\mathfrak{a}^{-1} \cdot \mathfrak{a}$ and $\mathfrak{a} \cdot \mathfrak{a}^{-1}$ are round, the above two inequalities must really be equalities. ∎

The set $\{\tau a : a \in G\}$ of all neighborhoods filters of a topological group (G, τ) is evidently a subgroup of $RC(G, \tau)$. If τ is a T_1-topology (and therefore also a T_2-topology), the above subgroup is (algebraically) isomorphic to G itself. Let the topological group $(\hat{G}, \hat{\tau})$ be an extension of (G, τ). Then the trace map is an (algebraic) isomorphism of the group of all neighborhoods filters of $(\hat{G}, \hat{\tau})$ into the group $RC(G, \tau)$, having as its image the trace system T^* of $(\hat{G}, \hat{\tau})$. Hence if the topological group $(\hat{G}, \hat{\tau})$ is an extension of the topological group (G, τ), its trace system must be a subgroup of $RC(G, \tau)$. Under certain conditions, this result can be reversed.

37.7 *Let (G, τ) be a T_1-topological group, T^* a trace-system on (G, τ), and the topological space $(\hat{G}, \hat{\tau})$ a T^*-extension of (G, τ).*

Then we can define a group structure in $\hat{G}$ so that $(\hat{G}, \hat{\tau})$ is a topological group and an extension of (G, τ) if and only if T^* is a subgroup of $RC(G, \tau)$.

Proof: The above remarks imply that our condition is necessary. As for sufficiency, let T^* be a subgroup of $RC(G, \tau)$. By 35.7 the uniform structure $\mathfrak{w}_\tau$ on (G, τ) admits of a unique continuation to $(\hat{G}, \hat{\tau})$, and $\hat{\tau}$ is also a T_1-topology. The trace-map $\sigma : \hat{G} \to T^*$ is a 1-1 map. Since T^* is by hypothesis a group, we can define a group structure in $\hat{G}$ by $\hat{a} \cdot \hat{b} = \sigma^{-1}(\mathrm{Tr}\,(\hat{\tau}\hat{a})) \cdot (\mathrm{Tr}\,(\hat{\tau}\hat{b}))$. Since G is isomorphic to the subgroup of T^* determined by the neighborhoods filter $\tau a (a \in G)$, we see that G is also a subgroup of $\hat{G}$. We have as an immediate consequence of the very definition of the group structure on $\hat{G}$, that the filter $\hat{\mathfrak{u}} = \hat{\tau} e$ possesses properties (*a*)–(d) of 36.1. Hence $\hat{\tau}'\hat{a} = \ulcorner\hat{a}\urcorner \cdot \hat{\mathfrak{u}}$ defines a group topology $\hat{\tau}'$ on G. The uniform structure associated with this topology is also a continuation of $\mathfrak{w}_\tau$ to $\hat{G}$. But, by 35.7, $\mathfrak{w}_\tau$ can be continued in only one way and hence these two uniform structures on G must be identical. Therefore $\hat{\tau}' = \hat{\tau}$ and $(\hat{G}, \hat{\tau})$ is a topological group. ∎

A topological group (G, τ) is said to be *complete* (l- or r-complete) if every one of its Cauchy filters (every one of its l- or r-Cauchy filters, respectively) converges within it.

37.8 *Every l-complete topological group is also r-complete and complete.*

Proof: Let the topological group (G, τ) be l-complete and let $\mathfrak{a}$ be an r-Cauchy filter. Then $\mathfrak{a}^{-1}$ is an l-Cauchy filter. Consequently $\mathfrak{a}^{-1} \leqslant \tau x$ and therefore $\mathfrak{a} \leqslant (\tau x)^{-1} \leqslant \tau x^{-1}$ for a certain $x \in G$. Hence (G, τ) is also r-complete. Since $\mathfrak{w}_\tau \leqslant \mathfrak{w}_\tau^l$, every Cauchy filter is surely an l-Cauchy filter, and we see that (G, τ) must be complete. ∎

By 35.8 and 37.7, whenever (G, τ) is a T_1-topological group then the completion of the uniform space $(G, \mathfrak{w}_\tau)$ is also a topological group and an extension of (G, τ). It is called the *completion of the topological group* (G, τ). We can also consider the completions of the uniform spaces $(G, \mathfrak{w}_\tau^l)$ and $(G, \mathfrak{w}_\tau^r)$, called the l-completion and r-completion, respectively, of (G, τ). In general we cannot extend the group structure of G to these one-sided completions.

37.9 *The following statements about the T_1-topological group (G, τ) are pairwise equivalent:*

(1) *The group structure of G can be extended in a continuous fashion to the l-completion $(\hat{G}, \hat{\tau})$ of (G, τ); i.e., the l-completion is an l-complete topological group and an extension of (G, τ).*

(2) *If $\mathfrak{a}$ is an l-Cauchy filter of (G, τ) then so is $\mathfrak{a}^{-1}$.*

(3) *Every l-Cauchy filter of (G, τ) is also an r-Cauchy filter (and hence a Cauchy-filter).*

(4) *The l-completion of (G, τ) is identical to the completion of (G, τ).*

Proof

(1) → (2): Let $\mathfrak{a}$ be an l-Cauchy filter of (G, τ), then $\mathfrak{a}^{-1}$ is an r-Cauchy filter. Since $(\hat{G}, \hat{\tau})$ is, by 37.8, also r-complete, $\mathfrak{a}^{-1}$ must converge in $(\hat{G}, \hat{\tau})$ and is therefore also an l-Cauchy filter.

(2) → (3): If $\mathfrak{a}$, and hence also $\mathfrak{a}^{-1}$, is an l-Cauchy filter, then $\mathfrak{a} = (\mathfrak{a}^{-1})^{-1}$ must be an r-Cauchy filter.

(3) → (1): Every Cauchy filter of (G, τ) is all the more so an l-Cauchy filter. If, conversely, every l-Cauchy filter is an r-Cauchy filter (and hence a Cauchy filter), then the l-Cauchy filters are identical to the Cauchy filters and the l-completion of (G, τ) is, as a topological space, identical to the completion

of (G, τ). But the completion of (G, τ) is a topological group and an extension of (G, τ) and thus our contention has been proved. ∎

37.10 *Every locally completely compact group (G, τ) is l-complete (hence also r-complete and therefore complete)* (37.8).

Proof: Let U be a completely compact neighborhood in $\mathfrak{u}_\tau$. Since left-translations are topological maps, for each $a \in G$, $a \cdot U$ must also be completely compact. Suppose that $\mathfrak{a}$ is an l-Cauchy filter of (G, τ). Then there exists an $A \in \mathfrak{a}$ with $A^{-1} \cdot A \subseteq U$. For fixed $a \in A$ we have $A \subseteq a \cdot U$, hence $\mathfrak{a} \leqslant \overset{\frown}{a \cdot U}$. Since $a \cdot U$ is completely compact, the filter $\mathfrak{a}$ must be convergent (35.3). ∎

A topological group (G, τ) is said to be l-precompact (r-precompact, precompact) if it is precompact with respect to the uniform structure $\mathfrak{w}_\tau^l(\mathfrak{w}_\tau^r, \mathfrak{w}_\tau)$. This is equivalent to the condition that, given any $U \in \mathfrak{u}_\tau$ we can find a finite number of elements $a_1, ..., a_n \in G$ such that

$$G \subseteq a_1 \cdot U \cup ... \cup a_n \cdot U \qquad (l\text{-precompact}),$$
$$G \subseteq U \cdot a_1 \cup ... \cup U \cdot a_n \qquad (r\text{-precompact}),$$
$$G \subseteq (a_1 \cdot U \cap U \cdot a_1) \cup ... \cup (a_n \cdot U \cap U \cdot a_n) \qquad (\text{precompact}).$$

37.11 *In topological groups the concepts of l-precompact, r-precompact, and precompact all coincide.*

Proof: We shall use 35.15(4). Since every Cauchy filter of the topological group (G, τ) is also an l- and r-Cauchy filter, precompactness must imply both l- and r-precompactness. Conversely, suppose (G, τ) is l-precompact and let $\mathfrak{a} > 0$ be a filter of G. Then $\mathfrak{a}^{-1} > \mathfrak{o}$ also, and there exists an l-Cauchy filter $\mathfrak{b}$ such that $\mathfrak{o} < \mathfrak{b} \leqslant \mathfrak{a}^{-1}$. Hence $\mathfrak{o} < \mathfrak{b}^{-1} \leqslant \mathfrak{a}$. Since $\mathfrak{b}^{-1}$ is an r-Cauchy filter, we have shown that (G, τ) is r-precompact. Moreover, since $\mathfrak{b}^{-1} > \mathfrak{o}$, there must exist an l-Cauchy filter $\mathfrak{c}$ such that $\mathfrak{o} < \mathfrak{c} \leqslant \mathfrak{b}^{-1} \leqslant \mathfrak{a}$. Since $\mathfrak{c} \leqslant \mathfrak{b}^{-1}$, $\mathfrak{c}$ is also an r-Cauchy filter and hence even a Cauchy filter. Therefore (G, τ) is also precompact. ∎

A subset M of a topological group (G, τ) is said to be precompact if the subspace of the uniform space $(G, \mathfrak{w}_\tau)$ determined by M is precompact.

37.12 *The completion $(\hat{G}, \hat{\tau})$ of a topological group (G, τ) is locally completely compact if and only if $\mathfrak{u}_\tau$ contains a precompact neighborhood.*

Proof: Since translations are topological maps, $(\hat{G}, \hat{\tau})$ is locally completely compact if and only if the neutral element e possesses a completely compact neighborhood $\hat{U}$. But a neighborhood $\hat{U} \in \mathfrak{u}_{\hat{\tau}}$ is completely compact if and only if the neighborhood $\hat{U} \cap G \in \mathfrak{u}_\tau$ is precompact (35.15). ∎

Notes and Exercises

37A Let G be the set of all topological maps of the real unit interval $[0, 1]$ onto itself. Consider now the set of all bounded maps of φ of $[0, 1]$ into R. We can put a norm on this set of maps by

$$\| \varphi \| = \sup \{ | \varphi x | : x \in [0, 1] \}.$$

This norm induces in G the topology of uniform convergence (6E). With respect to the multiplication of maps G is a group.

Exercise

1. (G, τ) is a complete T_1-topological group.
2. Suppose

$$\varphi_n x = \begin{cases} \left(2 - \dfrac{1}{n}\right) x & \text{for} \quad 0 \leqslant x \leqslant \frac{1}{2} \\ \dfrac{1}{n} x + 1 - \dfrac{1}{n} & \qquad \frac{1}{2} < x \leqslant 1 \end{cases}$$

The maps φ_n belong to G. The sequence $(\varphi_n)_{n \in \mathbf{N}}$ converges uniformly to a map which, however, does not belong to G.

3. The Fréchet filter of the sequence $(\varphi_n)_{n \in \mathbf{N}}$ is an r-Cauchy filter of (G, τ), but not an l-Cauchy filter. Hence the group (G, τ) is neither l- nor r-complete [24, p. 212].

37B *Exercise:* Let (G, τ) be a topological group. Then the group $RC(G, \tau)$ is (algebraically) isomorphic to the factor group $G/\overline{\{e\}}$.

37C Let (G, τ) and (G', τ') be two topological groups and φ a continuous homomorphism from (G, τ) into (G', τ').

Exercise

1. φ is a uniformly continuous map of $(G, \mathfrak{w}_\tau^l)$ into $(G', \mathfrak{w}_{\tau'}^l)$, of $(G, \mathfrak{w}_\tau^r)$ into $(G', \mathfrak{w}_{\tau'}^r)$, and of $(G, \mathfrak{w}_\tau)$ into $(G', \mathfrak{w}_{\tau'})$.
2. If τ' is a T_1-topology and (G', τ') is complete, then φ admits of a unique continuous extension $\hat{\varphi}$ to the completion of (G, τ). Then $\hat{\varphi}$ is even a homomorphism.
3. Let the complete T_1-topological group $(\tilde{G}, \tilde{\tau})$ be an extension of the topological group (G, τ). Then there exists an (algebraic and topological) isomorphism of $(\tilde{G}, \tilde{\tau})$ onto the completion of (G, τ) which reduces to the identity on G. Hence the completion of a T_1-topological group is uniquely determined up to isomorphism.

38. Topological Rings and Fields

Analogously to the definition of a topological group, we can define a *topological ring* (A, τ) to be an object which is a ring A in the sense of algebra and also a topological space, these two structures being linked in compatibility by the condition that the ring operations of addition, subtraction and multiplication be continuous mappings. The topology τ on A is then known as the *ring topology*. A topological ring is a topological group, with respective to its additive structure. The following axioms characterize a ring topology τ:

$(\mathbf{R}_1)$ $\quad (\tau a) + (\tau b) \leqslant \tau(a + b).$

$(\mathbf{R}_2)$ $\quad -(\tau a) \leqslant \tau(-a).$

$(\mathbf{R}_3)$ $\quad (\tau a) \cdot (\tau b) \leqslant \tau(a \cdot b).$

Axioms $(\mathbf{R}_1)$ and $(\mathbf{R}_2)$ are nothing but the group axioms $(\mathbf{G}_1)$ and $(\mathbf{G}_2)$, written additively, however (36a). Axiom $(\mathbf{R}_3)$ ensures the continuity of multiplication. In a ring, the neutral element with respect to addition is the zero element 0. The ring topology can be completely and uniquely stipulated by giving the neighborhoods filter $\tau 0$ of the zero element, which filter we shall again denote by $\mathfrak{u}_\tau$. We have $\tau a = \ulcorner a \urcorner + \mathfrak{u}_\tau$.

38.1 *The neighborhoods filter* $\mathfrak{u}_\tau$ *of the zero element of a topological ring* (A, τ) *possesses the following properties:*

(a) $\quad \ulcorner 0 \urcorner \leqslant \mathfrak{u}_\tau.$

(b) $\quad \mathfrak{u}_\tau + \mathfrak{u}_\tau \leqslant \mathfrak{u}_\tau.$

(c) $\quad -\mathfrak{u}_\tau \leqslant \mathfrak{u}_\tau.$

(d) $\quad \mathfrak{u}_\tau \cdot \mathfrak{u}_\tau \leqslant \mathfrak{u}_\tau.$

(e) $\quad \ulcorner a \urcorner \cdot \mathfrak{u}_\tau \leqslant \mathfrak{u}_\tau$ *and* $\mathfrak{u}_\tau \cdot \ulcorner a \urcorner \leqslant \mathfrak{u}_\tau$ *for all* $a \in A$.

Conversely, if $\mathfrak{u}$ *is a filter of the ring* A *verifying properties* (a)-(e) *above, then there exists one and only one ring topology* τ *on* A *such that* $\tau 0 = \mathfrak{u}$.

Proof: Since the topological ring (A, τ) is also an additive topological group, by 36.1 we see that (a)-(c) must hold (of course, we must think of addition rather than multiplication as being the group operation!). Since $0 \cdot 0 = 0$, condition (d) follows from $(\mathbf{R}_3)$. Also, using $(\mathbf{R}_3)$ again, since $\ulcorner a \urcorner \leqslant \tau a$ we have $\ulcorner a \urcorner \cdot \mathfrak{u}_\tau \leqslant (\tau a) \cdot (\tau 0) \leqslant \tau(a \cdot 0) = \mathfrak{u}_\tau$ and, similarly, $\mathfrak{u}_\tau \cdot \ulcorner a \urcorner \leqslant \mathfrak{u}_\tau$.

Conversely, suppose the filter $\mathfrak{u}$ satisfies conditions (a)–(e). By 36.1, $\tau a = \ulcorner a \urcorner + \mathfrak{u}$ then defines a group topology on the additive group A. (Property (d) of 36.1 is automatically satisfied, since addition is commutative.) By (b), (d), and (e) we also have

$$\begin{aligned}(\tau a) \cdot (\tau b) = (\ulcorner a \urcorner + \mathfrak{u}) \cdot (\ulcorner b \urcorner + \mathfrak{u}) &\leqslant \ulcorner a \urcorner \cdot \ulcorner b \urcorner + \ulcorner a \urcorner \cdot \mathfrak{u} + \mathfrak{u} \cdot \ulcorner b \urcorner + \mathfrak{u} \cdot \mathfrak{u} \\ &\leqslant \ulcorner a \cdot b \urcorner + \mathfrak{u} + \mathfrak{u} + \mathfrak{u} \leqslant \ulcorner a \cdot b \urcorner + \mathfrak{u} = \tau(a \cdot b).\end{aligned}$$

Thus τ is also a ring topology. The statement on unicity was already a consequence of 36.1. ∎

Since every topological ring is a topological group with respect to its addition, the theorems we have already proved for topological groups must hold for topological rings (viewed as groups) as well. *The residue class ring of a topological ring* (A, τ) *modulo a two-sided ideal* J *of* A *is again a topological ring:* For $(A, \tau)/J$ is certainly a topological group and the canonical map κ is a continuous and open homomorphism (36.10; κ is even a ring homomorphism). If τ' is the quotient topology, then

$$\tau'(\kappa a) \cdot \tau'(\kappa b) = (\kappa\tau a) \cdot (\kappa\tau b) = \kappa((\tau a) \cdot (\tau b)) \leqslant \kappa\tau(a \cdot b) = \tau'\kappa(a \cdot b)$$
$$= \tau'((\kappa a) \cdot (\kappa b)).$$

Hence multiplication in the quotient space is continuous.

A topological ring (A, τ) is a T_1-space if and only if (A, τ), viewed as an additive group, is a T_1-topological group— which can only happen if

$$\cap \{U : U \in \mathfrak{u}_\tau\} = \{0\} \tag{36.2}$$

38.2 *In a topological ring* (A, τ) *the set* $\overline{\{0\}}$ *is a two-sided ideal, and* $(A, \tau)/\overline{\{0\}}$ *is a* T_1*-topological ring.*

Proof: $(A, \tau)/\overline{\{0\}}$ is a T_1-topological group. (36.11). For each $a \in A$ the map $\varphi_a : A \to A$ defined by $\varphi_a x = a \cdot x$ is continuous. Hence

$$a \cdot \overline{\{0\}} = \varphi_a \overline{\{0\}} \subseteq \overline{\varphi_a \{0\}} = \overline{\{0\}}.$$

Similarly, we can show that $\overline{\{0\}} \cdot a \subseteq \overline{\{0\}}$. The ideal $\overline{\{0\}}$ is therefore two-sided and $(A, \tau)/\overline{\{0\}}$ is a T_1-topological ring. ∎

By the *uniform structure* $\mathfrak{w}_\tau$ of a topological ring (A, τ) we mean the uniform structure of its additive group. Since this group is abelian we do not need to distinguish between left-uniform and right-uniform structures. We have ($W_U = W_U^l = W_U^r$; cf. p. 242)

$$W_U = \{(a, b) : a - b \in U\} \qquad (U \in \mathfrak{u}_\tau).$$

A filter $\mathfrak{c}$ is a Cauchy filter if and only if $\mathfrak{c} - \mathfrak{c} \leqslant \mathfrak{u}_\tau$ (37.1). The condition $\mathfrak{c} = \mathfrak{c} + \mathfrak{u}_\tau$ characterizes $\mathfrak{c}$ as a round filter (37.2). A filter $\mathfrak{b}$ of (A, τ) is said to be *left-* or *right-bounded* whenever the conditions $\mathfrak{u}_\tau \cdot \mathfrak{b} \leqslant \mathfrak{u}_\tau$ or $\mathfrak{b} \cdot \mathfrak{u}_\tau \leqslant \mathfrak{u}_\tau$ are respectively satisfied. A filter which is both left- and right-bounded is said to be *bounded.* (Since boundedness pertains to the multiplication structure of the group rather than the additive abelian structure, we must, of course, distinguish between left- and right-boundedness.)

38.3 *Every Cauchy filter in a topological ring (A, τ) is bounded. If $\mathfrak{b}$ and $\mathfrak{c}$ are Cauchy filters, then $\mathfrak{b} \cdot \mathfrak{c}$ is a Cauchy filter.*

Proof: Given $U \in \mathfrak{u}_\tau$ we can find neighborhoods $U', U'' \in \mathfrak{u}_\tau$ such that $U' + U' \subseteq U$ and $U'' \cdot U'' \subseteq U'$ (38.1). Suppose that $\mathfrak{c}$ is a Cauchy filter; then $\mathfrak{c} - \mathfrak{c} \leqslant \mathfrak{u}_\tau$. We can then find a $C \in \mathfrak{c}$ with $C - C \subseteq U''$, and for a fixed $c \in C$ we have $C = c + U''$. With c and U' we can moreover associate a $U''' \in \mathfrak{u}_\tau$ such that $c \cdot U''' \subseteq U'$ and $U''' \cdot c \subseteq U'$. For the neighborhood $U_0 = U'' \cap U'''$ belonging to $\mathfrak{u}_\tau$, we have

$$C \cdot U_0 \subseteq U''' + U'' \cdot U'' \subseteq U' + U' \subseteq U$$

and, similarly, $U_0 \cdot C \subseteq U$. Hence $\mathfrak{c} \cdot \mathfrak{u}_\tau \leqslant \mathfrak{u}_\tau$ and $\mathfrak{u}_\tau \cdot \mathfrak{c} \leqslant \mathfrak{u}_\tau$; i.e., $\mathfrak{c}$ is bounded.

If $\mathfrak{b}$ and $\mathfrak{c}$ are Cauchy filters, it follows that

$$\mathfrak{b} \cdot \mathfrak{c} - \mathfrak{b} \cdot \mathfrak{c} \leqslant \mathfrak{b} \cdot (\mathfrak{c} - \mathfrak{c}) + (\mathfrak{b} - \mathfrak{b}) \cdot \mathfrak{c} \leqslant \mathfrak{b} \cdot \mathfrak{u}_\tau + \mathfrak{u}_\tau \cdot \mathfrak{c} \leqslant \mathfrak{u}_\tau + \mathfrak{u}_\tau \leqslant \mathfrak{u}_\tau .$$

which proves that $\mathfrak{b} \cdot \mathfrak{c}$ is also a Cauchy filter. ∎

The completion $(\hat{A}, \hat{\tau})$ of a T_1-topological ring (A, τ) is always, by the above paragraphs, a topological group. By (38.3), (19.8), and (19.9) the multiplication operation can also be extended continuously to $(\hat{A}, \hat{\tau})$ and this multiplication satisfies all the ring-axioms. Hence the completion of a T_1-topological ring is itself a topological ring.

38.4 *In a topological ring (A, τ) the component A_0 containing the zero element of the ring is a closed two-sided ideal. A (two-sided) simple*[3] *topological ring is connected or totally disconnected.*

Proof: By 36.8, A_0 is a closed subgroup of the additive group (A, τ). For any $a \in A$ the map $\varphi_a : A \to A$ defined by $\varphi_a x = a \cdot x$ is continuous. Since A_0 is connected, so is $\varphi_a A_0$ (19.3). And since $0 \in A_0 \cap \varphi_a A_0$ we see that $A_0 \cup \varphi_a A_0$ is also connected (14.5). By the maximality of A_0 we have $\varphi \cdot A_0 = \varphi_a A_0 = A_0$. Similarly, $A_0 \cdot a \subseteq A_0$ and A_0 must be a two-sided ideal. If A is a simple ring, then $A_0 = \{0\}$ or $A_0 = A$. If the second of these cases holds then (A, τ) is connected. If the first case holds $\{a\} = a + A_0$ is the component containing the element a. Hence each component is a one-point set and (A, τ) is totally disconnected. ∎

Every (not necessarily commutative) field is both a ring and, with respect to its additive structure, an (abelian group). Moreover, the set $K^{\times} = K \cap \complement 0$ of all elements of K distinct from zero, is a group with respect to multiplication (called the multiplicative group of the field). The inverse filter $\mathfrak{a}^{-1}$ to the filter $\mathfrak{a}$ of K is well-defined if and only $\mathfrak{a} \leqslant \overline{K^{\times}}$, or equivalently, if and only if $\mathfrak{a} \wedge \overline{0} = \mathfrak{o}$.

A topology τ on the field K is said to be a *field topology* and (K, τ) a *topological*

[3] A ring A is said to be (two-sided) *simple* if the zero-ideal and the whole ring A are the only two-sided ideals of A. For the purposes of the above theorem it is sufficient to assume that the zero-ideal and A itself are the only two-sided *closed* ideals in A.

field if the field operations are continuous maps: i.e., if (K, τ) is a topological ring and if the taking of multiplicative inverses is a continuous map of the subspace $(K^{\times}, \tau_{K^*})$ onto itself. By 38.2, $\{\ddot{0}\} = \cap\{U : U \in \tau 0\}$ must then be a two-sided ideal in K, and hence equal to all of K, or to merely $\{0\}$. In the first case τ must be the trivial topology while in the second case τ must be a T_1-topology and therefore completely regular (36.4). We shall henceforth not count the trivial topology among possible field topologies, which convention allows us to enunciate: *Every field topology is completely regular.*

Whenever (K, τ) is a topological field, then (K, τ) is a topological group with respect to addition, while $(K^{\times}, \tau_{K^{\times}})$ is a topological group with respect to multiplication. If, as before, we let $\mathfrak{u}_\tau$ denote the neighborhoods filter of 0, then $\tau a = \ulcorner a \urcorner \cdot (\tau 1) = (\tau 1) \cdot \ulcorner a \urcorner = \ulcorner a \urcorner \cdot (\ulcorner 1 \urcorner + \mathfrak{u}_\tau) = (\ulcorner 1 \urcorner + \mathfrak{u}_\tau)\,\ulcorner a \urcorner$ (1 represents the unit element of the field). By $(\mathbf{T}_1)$, we then have $\tau 1 \wedge \ulcorner 0 \urcorner = \mathfrak{o}$. The filter $(\tau 1)^{-1}$ is therefore well-defined, and by the continuity of the inverse map, we see that $(\tau 1)^{-1} \leqslant \tau 1$. Hence the neighborhoods filter $\mathfrak{u}_\tau$ possesses, besides properties (a)–(e) of 38.1, the additional property

(f) $(\ulcorner 1 \urcorner + \mathfrak{u}_\tau)^{-1} \leqq \ulcorner 1 \urcorner + \mathfrak{u}_\tau$.

Conversely, if $\mathfrak{u}$ is a filter in the field K satisfying conditions (a)-(f), then there exists one and only one field topology on K such that $\tau 0 = \mathfrak{u}$.

38.5 *A topological field is either connected or totally disconnected.*

Proof: Since fields are simple rings, this theorem follows from 38.4. ∎

By the *uniform structure* of a topological field (κ, τ) we shall mean the uniform structure $\mathfrak{w}_\tau$ of the additive topological group (K, τ). Whenever we speak of Cauchy filters, of completions, etc., these concepts will be understood to be defined with respect to the above uniform structure. We can, however, consider other uniform structures: Since $(K^{\times}, \tau_{K^{\times}})$ is a multiplicative topological group, it determines on its own a left-uniform, a right-uniform, and a uniform structure. The uniform structure of $K^{\times}$ is called the *multiplicative uniform structure* of the topological field (K, τ). This uniform structure is generally different from that induced on the group $K^{\times}$ by $\mathfrak{w}_\tau$. Cauchy filters with respective to the multiplicative uniform structure will be called *multiplicative Cauchy filters.*

38.6 *In a topological field* (K, τ) *the filter* $\mathfrak{a}$ *of* $K^{\times}$ *is a multiplicative Cauchy filter if and only if* $\mathfrak{a}$ *is a Cauchy filter and* $\mathfrak{a}^{-1}$ *a bounded filter.*

Proof: Let the filter $\mathfrak{a}$ of $K^{\times}$ be a multiplicative Cauchy filter. By 37.1 we have $\mathfrak{a}^{-1} \cdot \mathfrak{a} \leqslant \tau 1 = \ulcorner 1 \urcorner + \mathfrak{u}_\tau$, and $\mathfrak{a} \cdot \mathfrak{a}^{-1} \leqslant \ulcorner 1 \urcorner + \mathfrak{u}_\tau$. Since $(\tau 0) \cdot (\tau 1) \leqslant \tau 0$, given $U \in \mathfrak{u}_\tau$ we can find a $V \in \mathfrak{u}_\tau$ such that $V \cdot (1 + V) \subseteq U$. In addition there exists an $A \in \mathfrak{a}$ with $A^{-1} \cdot A \subseteq 1 + V$, and for fixed $a \in A$ we therefore have $A \subseteq a \cdot (1 + V)$. And lastly, since $(\tau 0) \cdot \ulcorner a \urcorner \leqslant (\tau 0) \cdot (\tau a) \leqslant \tau 0$ there must exist a $W \in \mathfrak{u}_\tau$ with $W \cdot a \subseteq V$. Consequently

$$W \cdot A \subseteq W \cdot a \cdot (1 + V) \subseteq V \cdot (1 + V) \subseteq U,$$

hence $\mathfrak{u}_\tau \cdot \mathfrak{a} \leqslant \mathfrak{u}_\tau$. Since $a_1 - a_2 = (a_1 \cdot a_2^{-1} - 1)\, a_2$, the above relation implies that

$$\mathfrak{a} - \mathfrak{a} \leqslant (\mathfrak{a} \cdot \mathfrak{a}^{-1} - \overset{\ulcorner\urcorner}{1}) \cdot \mathfrak{a} \leqslant \mathfrak{u}_\tau \cdot \mathfrak{a} \leqslant \mathfrak{u}_\tau;$$

i.e., $\mathfrak{a}$ is also a Cauchy filter. By 37.4, $\mathfrak{a}^{-1}$ must be a multiplicative Cauchy filter and, by what we have just proved, also a Cauchy filter. Finally, by 38.3, $\mathfrak{a}^{-1}$ is a bounded filter.

Conversely, let $\mathfrak{a}$ be a Cauchy filter in $K^\times$ and let the filter $\mathfrak{a}^{-1}$ be bounded. Then

$$\mathfrak{a}^{-1} \cdot \mathfrak{a} - \overset{\ulcorner\urcorner}{1} \leqq \mathfrak{a}^{-1} \cdot (\mathfrak{a} - \mathfrak{a}) \leqq \mathfrak{a}^{-1} \cdot \mathfrak{u}_\tau \leqq \mathfrak{u}_\tau,$$

hence $\mathfrak{a}^{-1} \cdot \mathfrak{a} \leqslant \tau 1$ and likewise $\mathfrak{a} \cdot \mathfrak{a}^{-1} \leqslant \tau 1$. Hence $\mathfrak{a}$ is therefore also a multiplicative Cauchy filter. ∎

The completion $(\hat{K}, \hat{\tau})$ of a topological field (K, τ) is surely a topological ring, by the results previously proved for topological rings. But $(\hat{K}, \hat{\tau})$ is, in general, not a topological field: In fact the ring $\hat{K}$ can even possess zero divisors! A necessary and sufficient condition for $(\hat{K}, \hat{\tau})$ to be a topological field is evidently that the operation of taking inverses admit of a continuous extension to $(\hat{K}, \hat{\tau})$. And this can only occur if, for every convergent Cauchy filter which does not converge to 0, the filter $\mathfrak{a}^{-1}$ is a Cauchy filter. (Since $\tau 0$ is a round filter, $\tau 0 \wedge \mathfrak{a} \neq \mathfrak{o}$ implies $\mathfrak{a} \leqslant \tau 0$ (35.5). Hence if $\mathfrak{a}$ does not converge to 0, we must have $\tau 0 \wedge \mathfrak{a} = \mathfrak{o}$, which guarantees the existence of $\mathfrak{a}^{-1}$.)

38.7 *The completion* $(\hat{K}, \hat{\tau})$ *of a topological field* (K, τ) *is a topological field if and only if, for every Cauchy filter* $\mathfrak{a}$ *not convergent to* 0, *the filter* $\mathfrak{a}^{-1}$ *is bounded.*

Proof: Since $a_1^{-1} - a_2^{-1} = a_1^{-1}(a_2 - a_1)\, a_2^{-1}$ we see that

$$\mathfrak{a}^{-1} - \mathfrak{a}^{-1} \leqslant \mathfrak{a}^{-1}(\mathfrak{a} - \mathfrak{a}) \cdot \mathfrak{a}^{-1}.$$

If $\mathfrak{a}$ is a Cauchy filter then $\mathfrak{a}^{-1} - \mathfrak{a}^{-1} \leqslant \mathfrak{a}^{-1} \cdot \mathfrak{u}_\tau \cdot \mathfrak{a}^{-1}$. If, in addition, $\mathfrak{a}^{-1}$ is bounded, we then have $\mathfrak{a}^{-1} - \mathfrak{a}^{-1} \leqslant \mathfrak{u}_\tau \cdot \mathfrak{a}^{-1} \leqslant \mathfrak{u}_\tau$, and hence $\mathfrak{a}^{-1}$ is a Cauchy filter. The converse is immediate, since every Cauchy filter must be bounded (38.3). ∎

Using both 38.6 and 38.7, we see that:

38.8 *If the completion* $(\hat{K}, \hat{\tau})$ *of a topological field* (K, τ) *is itself a topological field, then the multiplicative group of* K *is also complete with respect to its multiplicative uniform structure.*

The condition of 38.7 is, in any case, satisfied in the topological field (K, τ) if the topology τ satisfies the following property: For every neighborhood $U \in \mathfrak{u}_\tau$, the set $(\complement U^{-1})$ is bounded; i.e., $\mathfrak{u}_\tau \cdot \overset{\ulcorner\urcorner}{(\complement U)^{-1}} \leqslant \mathfrak{u}_\tau$ and $\overset{\ulcorner\urcorner}{(\complement U)^{-1}} \cdot \mathfrak{u}_\tau \leqslant \mathfrak{u}_\tau$. Such topologies are known as *valuation-topologies or v-topologies*. The most important topological fields are the fields with valuation (38b) and they are *v*-topological fields.

Notes and Exercises

38A A topological field (K, τ) is said to be *locally bounded* whenever $\mathfrak{u}_\tau$ contains a bounded neighborhood U (i.e., $\mathfrak{u}_\tau \cdot \lceil U \rceil \leqslant \mathfrak{u}_\tau$ and $\lceil U \rceil \cdot \mathfrak{u}_\tau \leqslant \mathfrak{u}_\tau$). An element $t \neq 0$ of a topological field is said to be *analytically nilpotent* if the sequence $(t^n)_{n \in \mathbf{N}}$ converges to 0.

Exercise

1. Every $\mathfrak{v}$-topological field is locally bounded.

2. Every locally completely compact and locally bounded topological field containing an analytically nilpotent element is $\mathfrak{v}$-topological.

3. A connected and locally bounded topological field always contains an analytically nilpotent element.

4. A locally completely compact, connected topological field is locally bounded if and only if satisfies the first axiom of countability.

38B By a (multiplicative) *valuation* of the field K we mean a map $\beta : K \to \mathbf{R}$ possessing the following properties:

(a) $\beta a \geqslant 0$ for all $a \in K$.

(b) If $\beta a = 0$, then $a = 0$.

(c) $\beta(a \cdot b) = (\beta a) \cdot (\beta b)$.

(d) $\beta(a + b) \leqslant (\beta a) + (\beta b)$.

For a given valuation β of K let $U_\epsilon = \{a : a \in K, \beta a < \epsilon\}$. Then $\{U_\epsilon : \epsilon > 0\}$ is a filter basis generating the filter $\mathfrak{u}$.

Exercise

1. By $\tau a = \lceil a \rceil + \mathfrak{u}$ we can define a field-topology τ_β on K. It is called the topology induced by the valuation.

2. Every topology induced by a valuation is a $\mathfrak{v}$-topology.

By (2), the completion $(\hat{K}, \hat{\tau}_\beta)$ of a topological field (K, τ_β) is itself a topological field, which we can call the completion of the given field with valuation. We can extend the valuation β, from K to $\hat{K}$, in such a way that it induces on K exactly the topology $\hat{\tau}_\beta$. We also have the converse: *Given any commutative v-topological field* (K, τ) *containing an analytically nilpotent element* (*38A*), *we can find a valuation* β *on* K *such that* $\tau_\beta = \tau$.

The usual absolute value is a valuation of the field $\mathbf{Q}$ of rational numbers. We can find other valuations on $\mathbf{Q}$ as follows: Let p be a prime number. Every $a \in \mathbf{Q}$ can, of course, be expressed as a quotient of two integers m and n ($a = m/n$). Let e_m and e_n denote the highest exponents of p appearing in the prime factor decomposition of m and n, respectively. Then $\beta_p a = p^{e_n - e_m}$ defines a valuation β_p on $\mathbf{Q}$, known as the *p-adic valuation.*

Exercise

1. The p-adic valuation on $\mathbf{Q}$ satisfies the strong inequality

$$\beta_p(a+b) \leqslant \max\{\beta_p a, \beta_p b\}.$$

2. The field $\mathbf{Q}$ is precompact with respect to the topology induced by a p-adic valuation.

Two valuations which induce the same topology are said to be *topologically equivalent.* Up to topological equivalence the usual absolute value and the p-adic valuations are the only valuations possible on $\mathbf{Q}$. The completion of $\mathbf{Q}$ under the usual absolute value is the field of reals. The completion of $\mathbf{Q}$ under a p-adic valuation is a locally completely compact, totally disconnected field, known as the *field of p-adic numbers.*

38C Let U_q denote the set of all rationals satisfying the following property: If $a = m/n$ is a representation of a as quotient of relatively prime integers, then m is exactly divisible by 6^q. Furthermore, let $0 \in U_q$. Then $\{U_q : q \in \mathbf{N}\}$ is a filter basis, generating a filter $\mathfrak{u}$.

Exercise

1. We can define a field-topology on $\mathbf{Q}$ by $\tau a = \ulcorner\!\urcorner a + \mathfrak{u}$.

2. τ is the inf-topology of the topologies induced by the 2-valuation and the 3-valuation.

3. The completion of $(\mathbf{Q}, \tau)$ is a ring which contains divisors of zero.

39. Topological Vector Spaces. The Stone-Weierstrass Approximation Theorem

As we all know, a vector space V over the field K is an additive (Abelian) group V having K as (left-) domain of operators. Let (K, τ) be a topological field and V a vector-space over K. Then we say that a topology τ' on V is a *vector-space topology* and we call (V, τ') a *topological vector space* over the topological field (K, τ) if (V, τ') is a topological group with respect to vector addition and if the map $(a, x) \to a \cdot x$ of $K \times V$ into V is continuous from $(K, \tau) \times (V, \tau')$ into (V, τ').

Besides the properties satisfied by a group topology, a vector-space topology τ' also verifies the condition

(V) $\quad (\tau a) \cdot (\tau' x) \leqslant \tau'(a \cdot x)$ *for all* $a \in K$ *and* $x \in V$.

39.1 *The neighborhoods filter* $\mathfrak{u}_{\tau'} = \tau'\mathbf{0}$ *of the null vector* $\mathbf{0}$ *of a topological vector space* (V, τ') *over a topological field* (K, τ) *possesses the following properties:*

(a) $\lceil \mathbf{0} \rceil \leqslant \mathfrak{u}_{\tau'}$.

(b) $\mathfrak{u}_{\tau'} + \mathfrak{u}_{\tau'} \leqslant \mathfrak{u}_{\tau'}$.

(c) $(\tau 0) \cdot \mathfrak{u}_{\tau'} \leqslant \mathfrak{u}_{\tau'}$.

(d) $\lceil a \rceil \cdot \mathfrak{u}_{\tau'} \leqslant \mathfrak{u}_{\tau'}$ *for all* $a \in K$.

(e) $(\tau 0) \cdot \lceil x \rceil \leqslant \mathfrak{u}_{\tau'}$ *for all* $x \in V$.

Conversely, if $\mathfrak{u}$ *is a filter of* V *verifying* (a)-(e), *then there exists one and only one topology* τ' *on* V *such that* $\tau'\mathbf{0} = \mathfrak{u}$, *and such that* (V, τ') *is a topological vector space over the topological field* (K, τ).

Proof: (a) and (b) are the additive analogues of the corresponding properties proved in 36.1. The relations (c), (d), and (e) follow from **(V)** because

$$0 \cdot \mathbf{0} = a \cdot \mathbf{0} = 0 \cdot x = \mathbf{0}.$$

Conversely, suppose the filter $\mathfrak{v}$ of V possesses properties (a)-(e). Since $-\mathfrak{u} = (-1)\,\mathfrak{u}$ it follows from (d) that $-\mathfrak{u} \leqslant \mathfrak{u}$. Hence the filter $\mathfrak{u}$ possesses the first three properties of 36.1. Since 36.1 (d) is automatically fulfilled by the commutativity of vector addition, we can define a group topology on V by $\tau'x = \lceil x \rceil + \mathfrak{u}$. Using (c), (d), and (e) we have

$$\begin{aligned}(\tau a) \cdot (\tau' x) &= (\lceil a \rceil + \tau 0) \cdot (\lceil x \rceil + \mathfrak{u}) \leqq \lceil a \rceil \cdot \lceil x \rceil + (\tau 0) \cdot \lceil x \rceil + \lceil a \rceil \cdot \mathfrak{u} + (\tau 0) \cdot \mathfrak{u} \\ &\leqq \lceil a \cdot x \rceil + \mathfrak{u} + \mathfrak{u} + \mathfrak{u} \leqq \lceil a \cdot x \rceil + \mathfrak{u} = \tau'(a \cdot x).\end{aligned}$$

Hence (V, τ') is a topological vector space over (K, τ). The unicity contention follows from the validity of 36.1. ∎

Just as in 36.10 we can prove:

39.2 *If* L *is a linear subspace of a topological vector space* (V, τ') *over the topological field* (K, τ), *then the quotient space* $(V, \tau')/L$ *is also a topological vector space over* (K, τ). *The canonical map is continuous and open.*
$(V, \tau')/L$ *is* T_1*-topological if and only if the subspace* L *is closed.*

$\overline{\{\mathbf{0}\}}$ *is always a closed subspace and* $(V, \tau')/\overline{\{\mathbf{0}\}}$ *is a* T_1*-topological vector space.*

By the *uniform structure* $\mathfrak{w}_{\tau'}$ of a topological vector space (V, τ') we again mean the uniform structure of the additive topological group (V, τ'). A filter $\mathfrak{c}'$ of V is a Cauchy filter if and only if $\mathfrak{c}' - \mathfrak{c}' \leqslant \mathfrak{u}_{\tau'}$. A filter $\mathfrak{b}'$ of V is said to be *bounded* if $(\tau 0) \cdot \mathfrak{b}' \leqslant \mathfrak{u}_{\tau'}$.

39.3 *Let (V, τ') be a topological vector space over the topological field (K, τ). Every Cauchy filter (V, τ') is bounded. For each Cauchy filter $\mathfrak{a}$ of (K, τ) we have $\mathfrak{a} \cdot \mathfrak{u}_{\tau'} \leqslant \mathfrak{u}_{\tau'}$. If $\mathfrak{a}$ is a Cauchy filter of (K, τ) and $\mathfrak{c}'$ a Cauchy filter of (V, τ'), then $\mathfrak{a} \cdot \mathfrak{c}'$ is a Cauchy filter of (V, τ'). If τ is not the discrete topology of K, then we have $\mathfrak{a} \cdot \mathfrak{u}_{\tau'} \leqslant \mathfrak{u}_{\tau'}$ for every bounded filter $\mathfrak{a}$ of (K, τ) (whether $\mathfrak{a}$ be Cauchy or not).*

Proof: Given any $U' \in \mathfrak{u}_{\tau'}$, we can find neighborhoods $U_1', U_2' \in \mathfrak{u}_{\tau'}$ and $U \in \tau 0$ such that $U_1' + U_1' \subseteq U'$ and $U \cdot U_2' \subseteq U_1'$. If $\mathfrak{c}'$ is a Cauchy filter of (V, τ) (or if $\mathfrak{a}$ is a Cauchy filter of (K, τ) then there exists a $C' \in \mathfrak{c}'$ (an $A \in \mathfrak{a}$) such that $C' - C' \subseteq U_2'(A - A \subseteq U)$. For fixed elements $c' \in C'$, $a \in A$ we have $C' \subseteq c' + U_2'$ and $A \subseteq a + U$. Since $(\tau 0) \cdot \ulcorner c' \urcorner \leqslant \mathfrak{u}_{\tau'} (\ulcorner a \urcorner \cdot \mathfrak{u}_{\tau'} \leqslant \mathfrak{u}_{\tau'})$ there exist neighborhoods $U_1 \in \tau 0$ and $U_3' \in \mathfrak{u}_{\tau'}$ such that $U_1 \cdot c' \subseteq U_1'$ and $a \cdot U_3' \subseteq U_1'$. For the neighborhoods $U_2 = U \cap U_1 \in \tau 0 (U_4' = U_2' \cap U_3' \in \mathfrak{u}_{\tau'})$ we have

$$U_2 \cdot C' \subseteq U_1 \cdot c' + U \cdot U_2' \subseteq U_1' + U_1' \subseteq U',$$

$$A \cdot U_4' \subseteq a \cdot U_3' + U \cdot U_2' \subseteq U_1' + U_1' \subseteq U'.$$

Hence $(\tau 0) \cdot \mathfrak{c}' \leqslant \mathfrak{u}_{\tau'}$ and $\mathfrak{a} \cdot \mathfrak{u}_{\tau'} \leqslant \mathfrak{u}_{\tau'}$. Moreover,

$$\mathfrak{a} \cdot \mathfrak{c}' - \mathfrak{a} \cdot \mathfrak{c}' \leqq \mathfrak{a} \cdot (\mathfrak{c}' - \mathfrak{c}') + (\mathfrak{a} - \mathfrak{a}) \cdot \mathfrak{c}' \leqq \mathfrak{a} \cdot \mathfrak{u}_{\tau'} + (\tau 0) \cdot \mathfrak{c}' \leqq \mathfrak{u}_{\tau'} + \mathfrak{u}_{\tau'} \leqslant \mathfrak{u}_{\tau'}.$$

Hence $\mathfrak{a} \cdot \mathfrak{c}'$ is also a Cauchy filter of (V, τ').

Finally, suppose τ is not the discrete topology. Then every neighborhood $U \in \tau 0$ contains an element $U \neq 0$. By 39.1 (c), given any $U' \in \mathfrak{u}_{\tau'}$, we can find a $U_1' \in \mathfrak{u}_{\tau'}$ such that $u^{-1} \cdot U_1' \subseteq U'$, and hence $U_1' \subseteq u \cdot U' \subseteq U \cdot U'$. This implies that $\mathfrak{u}_{\tau'} \leqslant (\tau 0) \cdot \mathfrak{u}_{\tau'}$. If, therefore, $\mathfrak{a}$ is a bounded filter on (K, τ), we must have

$$\mathfrak{a} \cdot \mathfrak{u}_{\tau'} \leqq \mathfrak{a} \cdot (\tau 0) \cdot \mathfrak{u}_{\tau'} \leqq (\tau 0) \cdot \mathfrak{u}_{\tau'} \leqq \mathfrak{u}_{\tau'}. \blacksquare$$

The completion $(\hat{V}, \hat{\tau})$ of a T_1-topological vector space over (K, τ) is, by Section 37, an additive topological group. By 19.8, 19.9, and 39.3, we can extend to $(\hat{V}, \hat{\tau}')$ the multiplication of vectors of V by elements of K in a continuous way and so that $(\hat{V}, \hat{\tau}')$ becomes a topological vector space over (K, τ). More generally, this multiplication can even be extended continuously to $(\hat{K}, \hat{\tau}) \times (\hat{V}, \hat{\tau}')$, where $(\hat{K}, \hat{\tau})$ is the completion of (K, τ). Since this extension is unique (19.8) it must be linear; i.e., the topological vector space $(\hat{V}, \hat{\tau}')$ admits of the topological ring $(\hat{K}, \hat{\tau})$ as a continuous ring of operators upon it. If $(\hat{K}, \hat{\tau})$ is a topological field then $(\hat{V}, \hat{\tau}')$ is even a topological vector space over $(\hat{K}, \hat{\tau})$. We have demonstrated

39.4 *The completion $(\hat{V}, \hat{\tau}')$ of a T_1-topological vector space (V, τ') over (K, τ) is a topological vector space over (K, τ). If the completion $(\hat{K}, \hat{\tau})$ of (K, τ) is a topological field, then $(\hat{V}, \hat{\tau})$ is also a topological vector space over $(\hat{K}, \hat{\tau})$.*

Whenever (K, τ) is a topological field, then we can make the product space $(K, \tau)^n = (K, \tau) \times \dots \times (K, \tau)$ (n-factors, $n \in \mathbf{N}$) into a topological vector space over (K, τ) by defining addition and multiplication by elements of K in the usual coordinate-by-coordinate fashion. If (K, τ) is a complete, nondiscrete v-topological field, the following theorem tells us that the above procedure gives us all finite dimensional T_1-topological vector spaces over (K, τ).

39.5 *Let (K, τ) be a nondiscrete, complete $\mathfrak{v}$-topological field and (V, τ') an n-dimensional T_1-topological vector space over (K, τ). Then (V, τ') is algebraically and topologically isomorphic to $(K, \tau)^n$; i.e., there exists a homeomorphism of (V, τ) onto $(K_1\tau)^n$ which is also a linear mapping.*

Proof: Let $z_1, \dots, z_n$ be a K-basis of V. Then every vector $x \in V$ can be represented uniquely in the form $x = x_1 z_1 + \dots + x_n z_n$ where $x_1, \dots, x_n \in K$. Consider the map $\varphi_\nu : V \to K$ defined by $\varphi_\nu x = x_\nu$. As we know, $\varphi = \varphi_1 \times \dots \times \varphi_n$ is a 1-1 linear map from V onto K^n and hence an algebraic isomorphism. In order to show that φ is also a topological map of (V, τ') onto $(K, \tau)^n$ it suffices to show that, a filter $\mathfrak{a}'$ on V converges to $\mathbf{0}$ if and only if all coordinate filters $\varphi_\nu \mathfrak{a}'$ converge to 0 in (K, τ): For we then have

$$\varphi \mathfrak{u}_{\tau'} = \varphi_1 \mathfrak{u}_{\tau'} \times \dots \times \varphi_n \mathfrak{u}_{\tau'} = \tau 0 \times \dots \times \tau 0$$

and, since φ is linear,

$$\varphi \tau x = \varphi(\ulcorner x \urcorner + \mathfrak{u}_{\tau'}) = \ulcorner \varphi x \urcorner + \varphi \mathfrak{u}_{\tau'} = \ulcorner \varphi x \urcorner + (\tau 0 \times \dots \times \tau 0).$$

But the filter on the right in this equality is none other than the neighborhoods filter of φx in $(K, \tau)^n$.

From $\varphi_\nu \mathfrak{a}' \leqslant \tau 0$ for $\nu = 1, \dots, n$ we deduce that

$$\mathfrak{a}' \leqq (\varphi_1 \mathfrak{a}') \cdot \ulcorner z_1 \urcorner + \dots + (\varphi_n \mathfrak{a}') \cdot \ulcorner z_n \urcorner \leqq (\tau 0) \cdot \ulcorner z_1 \urcorner + \dots + (\tau 0) \cdot \ulcorner z_n \urcorner$$
$$\leqq \mathfrak{u}_{\tau'} + \dots + \mathfrak{u}_{\tau'} \leqq \mathfrak{u}_{\tau'}.$$

We must now prove the converse, that $\varphi_\nu \mathfrak{u}_{\tau'} \leqslant \tau 0$ for $\nu = 1, \dots, n$. The proof can be carried out by induction on the dimension n. For $n = 1$ we have $\mathfrak{u}_{\tau'} = (\varphi_1 \mathfrak{u}_{\tau'}) \cdot \ulcorner z_1 \urcorner$. If $\varphi_1 \mathfrak{u}_{\tau'} \leqslant \tau 0$ were false, then we could find a $U \in \tau 0$ and an ultrafilter $\mathfrak{v} \leqslant \varphi_1 \mathfrak{u}_{\tau'}$, such that $\mathfrak{v} \leqslant \ulcorner \complement U \urcorner$. By hypothesis, τ is a v-topology and hence $\mathfrak{v}^{-1}$ is bounded. By 39.3, this would imply that

$$\ulcorner z_1 \urcorner \leqq \mathfrak{v}^{-1} \cdot (\varphi_1 \mathfrak{u}_{\tau'}) \cdot \ulcorner z_1 \urcorner = \mathfrak{v}^{-1} \cdot \mathfrak{u}_{\tau'} \leqq \mathfrak{u}_{\tau'},$$

and by $(\mathbf{T}_1)$, we would finally have $z_1 = \mathbf{0}$, which is a contradiction. In general we must have

$$\mathfrak{u}_{\tau'} \leqq (\varphi_1 \mathfrak{u}_{\tau'}) \cdot \overline{z_1} + \ldots + (\varphi_n \mathfrak{u}_{\tau'}) \cdot \overline{z_n}.$$

Suppose $\varphi_n \mathfrak{u}_{\tau'} \leqslant \tau 0$ were false. Then, again, we could find a $U \in \tau 0$ and an ultrafilter $\mathfrak{v} \leqslant \varphi_n \mathfrak{u}_{\tau'}$, such that $\mathfrak{v} \leqslant \overline{\complement U}$, and we would have

$$\mathfrak{u}_{\tau'} - \overline{z_n} \leqq \mathfrak{v}^{-1} \cdot (\varphi_1 \mathfrak{u}_{\tau'}) \cdot \overline{z_1} + \ldots + \mathfrak{v}^{-1} \cdot (\varphi_{n-1} \mathfrak{u}_{\tau'}) \cdot \overline{z_{n-1}}.$$

The filter to the left in this relation converges in (V, τ') to z_n and is therefore a Cauchy filter. But by the right side of this relation we see that $\mathfrak{u}_{\tau'} - \overline{z_n}$ is a filter in the $(n-1)$-dimensional space V_{n-1} spanned by the basis-vectors $z_1, \ldots, z_{n-1}$. But by our inductive hypothesis, this space is algebraically and topologically isomorphic to $(K, \tau)^{n-1}$, and hence complete (35.10). This means that $\mathfrak{u}_{\tau'} - \overline{z_n}$ also converges to a vector in V_{n-1}. But the validity of $(\mathbf{T}_1)$ implies that of $(\mathbf{T}_2)$, and hence we have the contradiction $z_n \in V_{n-1}$. ∎

The most important examples of vector spaces occur when V is a vector space over the field of real or complex numbers, topologized in the usual way. Every topology defined by a norm of V (cf. 6V) is then automatically a vector-space topology; i.e., under its norm topology every normed vector space is a topological vector space over the real or complex numbers. Complete normal vector spaces (over the real or complex numbers) are called *Banach Spaces*.

A ring A is said to be an (associative) *algebra* over the commutative field K whenever A is also a vector space over K such that

$$a \cdot (x \cdot y) = (a \cdot x) \cdot y = x \cdot (a \cdot y)$$

for all $a \in K$ and all $x, y \in A$ (these relations imply that K must be *commutative*). By a topological algebra (A, τ') over the commutative topological field (K, τ) we shall mean an algebra A whose topology τ' is simultaneously a ring topology and a vector space topology with respect to (K, τ). *A Banach-algebra* is a topological algebra over the real or complex numbers, whose topology is induced by a norm which makes the algebra into a Banach space.

Let (R, τ) be a topological space and $\hat{C} = \hat{C}(R, \tau)$ be the set of all continuous maps of (R, τ) into the reals. In addition, let $\hat{\tau}$ denote on $\hat{C}$ the topology of uniform convergence on the completely compact subsets of (R, τ), or equivalently, the topology of compact convergence (34.10). If we define the usual addition and multiplication of function in $\hat{C}$ by

$$(\varphi + \psi)\, p = (\varphi p) + (\psi p), \qquad (\varphi \cdot \psi)\, p = (\varphi p) \cdot (\psi p),$$

then $\hat{C}$ becomes a commutative ring. If with each real number a we associate the constant map $\hat{a} \in \hat{C}$ defined by $\hat{a}p = a$, then we establish an isomorphism of the field of real numbers into the ring $\hat{C}$. The map $\hat{\imath}$ serves as the unit element of $\hat{C}$. Lastly, if for all real numbers a and all maps $\varphi \in \hat{C}$ we define the map $a \cdot \varphi$ by $\hat{a} \cdot \varphi$, then $\hat{C}$ becomes an algebra over the real numbers. Using 35.16 we see that $(\hat{C}, \hat{\tau})$ is a complete topological algebra over the field of real numbers.

If the space (R, τ) is completely compact, a topology $\hat{\tau}$ is induced by the norm on $\hat{C}$,

$$\| \varphi \| = \sup \{| \varphi p | : p \in R\}$$

and $(\hat{C}, \hat{\tau})$ turns out to be a Banach algebra. In general, to each completely compact subset K of R we can associate a seminorm on C,

$$\| \varphi \|_K = \sup \{| \varphi p | : p \in K\}$$

and $\hat{\tau}$ is the inf-topology of all the topologies induced by such seminorms.

If φ and ψ are maps of R into the real line, we can define new maps from R into the reals by

$$| \varphi | p = | \varphi p |, \qquad (\max \{\varphi, \psi\}) p = \max \{\varphi p, \psi p\},$$

and

$$(\min \{\varphi, \psi\}) p = \min \{\varphi p, \psi p\}$$

These maps are respectively denoted by $| \varphi |$, $\max \{\varphi, \psi\}$, and $\min \{\varphi, \psi\}$. If $\varphi \in \hat{C}$, then $| \varphi | \in \hat{C}$, because the taking of absolute values is a continuous map of the real line into itself. If φ and ψ are elements of $\hat{C}$, so are $\max \{\varphi, \psi\}$ and $\min \{\varphi, \psi\}$ (17.5).

39.6 (Stone's Approximation Theorem) *Let $\hat{L}$ be a linear subspace of the topological vector space $(\hat{C}, \hat{\tau})$ satisfying the following conditions*

(1) If $\psi \in \hat{L}$, then $| \psi | \in \hat{L}$.

(2) *For any two distinct points $p, q \in R$ and any two real numbers a, b we can find a $\psi \in \hat{L}$ such that $\psi p = a$ and $\psi q = b$.*

Then $\hat{L}$ is dense in $(\hat{C}, \hat{\tau})$. Property (2) *can be replaced by the following two properties:*

(2a) $\hat{1} \in \hat{L}$.

(2b) *Given any two distinct points $p, q \in R$ there exists a $\psi \in \hat{L}$ such that $\psi p \neq \psi q$ (i.e., $\hat{L}$ separates points).*

Proof: For any two real numbers a, b we have

$$\max \{a, b\} = \tfrac{1}{2} [(a + b) + | a - b |]$$

and

$$\min \{a, b\} = \tfrac{1}{2} [(a + b) - | a - b |]$$

Using (1), this implies that if $\varphi, \psi \in \hat{L}$ then $\max\{\varphi, \psi\} \in \hat{L}$ and $\min\{\varphi, \psi\} \in \hat{L}$.

Let us now assume that the space (R, τ) is completely compact.

For fixed $\varphi \in \hat{C}$ and $p \in R$, given any $q \in R$ we can find, by (2), a $p \in R$ such that $\psi_{p,q}p = \varphi p$ and $\psi_{p,q}q = \varphi q$. For every fixed $\epsilon > 0$ the sets

$$U_q = \{x : x \in R, \mid \varphi x - \psi_{p,q}x \mid < \epsilon\} \qquad (q \in R)$$

are such that $q \in U_q$ and hence they surely form an open covering of R. Since R is completely compact, $R = U_{q_1} \cup \ldots \cup U_{q_n}$. Then

$$\psi_p = \min\{\psi_{p,q_1}, \ldots, \psi_{p,q_n}\}$$

is a map belonging to $\hat{L}$ with $\psi_p p = \varphi p$. For arbitrary $x \in R$ we have $x \in U_{q_\nu}$ for at least one index ν, and hence

$$\psi_p x \leqslant \psi_{p,q_\nu} x \leqslant \varphi x + \epsilon.$$

Since $p \in V_p$, the sets

$$V_p = \{x : x \in R, \mid \varphi x - \psi_p x \mid < \epsilon\} \qquad (p \in R)$$

constitute an open covering of R. As above, $R = V_{p_1} \cup \ldots \cup V_{p_m}$, and

$$\psi = \max\{\psi_{p_1}, \ldots, \psi_{p_m}\}$$

is likewise a map belonging to $\hat{L}$. For any $x \in R$ we have, on the one hand, $\psi x = \psi_{p_\mu} x \leqslant \varphi x + \epsilon$, while on the other hand, $x \in U_{p_{\mu'}}$ for at least one index μ'. Therefore $\psi x \geqslant \psi_{p_{\mu'}} x \geqslant \varphi x - \epsilon$. This implies that $\mid \varphi x - \psi x \mid \leqslant \epsilon$ for all $x \in R$, which in turn implies that $\varphi \in \hat{L}$, and hence $\hat{L}$ is dense in $(\hat{C}, \hat{\tau})$.

If (R, τ) is not completely compact, the above considerations will still hold true on every completely compact subset K of (R, τ). This means that $\hat{L}$ is dense in $\hat{C}$ under the seminorm determined by K. Since this holds for every complete compact subset K, $\hat{L}$ must be dense in $(\hat{C}, \hat{\tau})$.

Properties (2a) and (2b) imply (2): Let p, q be two distinct points of R and let a, b be two real numbers. By (2b), there exists a $\psi \in \hat{L}$ with $\psi p \neq \psi q$. Let $\psi p = a'$ and $\psi q = b'$, say. By (2a) we must have $\hat{c} = c \cdot \hat{1} \in \hat{L}$ for every real number c. If we now put $d = (b - a)/(b' - a')$, then

$$\varphi = d \cdot (\psi - \hat{a}') + \hat{a}$$

is also a map in $\hat{L}$ with $\varphi p = a$ and $\varphi q = b$. ∎

In the above theorem we can drop property (1) from the hypotheses, provided we compensate by imposing more strict algebraic hypotheses on $\hat{L}$; for instance, if we assume that $\hat{L}$ is not only a linear subspace but also a subalgebra (i.e., $\hat{L}$ must be closed under ring multiplication).

39.7 (Stone-Weierstrass Approximation Theorem) *Let $\hat{A}$ be a subalgebra of the topological algebra $(\hat{C}, \tilde{\tau})$ verifying property* (2), *or equivalently properties* (2a) *and* (2b) *of the previous theorem. Then $\hat{A}$ is dense in C.*

Proof: Without loss of generality we can assume that $\hat{A}$ is closed. (Otherwise we could replace $\hat{A}$ by the subalgebra $\bar{\hat{A}}$ which itself verifies the given hypotheses; and if $\bar{\hat{A}}$ is dense, $\hat{A}$ must be dense also.) Since every subalgebra is a linear subspace, by 39.6 we need only show that if $\psi \in \hat{A}$ then $|\psi| \in \hat{A}$ also. The proof of 39.6 also tells us that we may also limit ourselves to the case where (R, τ) is completely compact.

Since R is completely compact, given any $\psi \in \hat{A}$ we can find a positive number a such that $\psi R \subseteq [-a, +a]$. If t denotes a real variable we can define the polynomials in $(t)^2$,

$$\varphi_1(t) = \frac{1}{2a} t^2, \qquad \varphi_{n+1}(t) = \varphi_n(t) + \frac{1}{2a}[t^2 - (\varphi_n(t))^2].$$

None of these polynomials has a nonzero constant term. Hence for each $n \in \mathbf{N}$, $\chi_n = \varphi_n \psi$ is a map belonging to A. For each $p \in R$ we have

$$|\psi p| - \chi_{n+1} p = (|\psi p| - \chi_n p)\left[1 - \frac{1}{2a}(|\psi p| - \chi_n p)\right]$$

and

$$|\psi p| + \chi_{n+1} p = (|\psi p| + \chi_n p)\left[1 + \frac{1}{2a}(|\psi p| + \chi_n p)\right].$$

Since $-a \leqslant \psi p \leqslant +a$, the above equalities imply that

$$-|\psi p| \leqslant \chi_n p \leqslant |\psi p|$$

and, further,

$$0 \leqslant \chi_1 p \leqslant \dots \leqslant \chi_n p \leqslant \chi_{n+1} p \leqslant |\psi p|.$$

Thus the sequence $(\chi_n)_{n \in \mathbf{N}}$ converges pointwise to a map χ. By the above inductive hypothesis we must have

$$\chi p = \chi p + \frac{1}{2a}(|\psi p|^2 - (\chi p)^2),$$

and since $\chi p \geqslant 0$, we also have $\chi p = |\psi p|$. Since $\{\chi_n : n \in \mathbf{N}\}$ is an increasingly filtrating collection of image sets, and since we have just seen that the sequence $(\chi_n)_{n \in \mathbf{N}}$ must converge pointwise to the continuous map $|\psi|$, we also see, by 34.11, that this sequence must converge uniformly to $|\psi|$. Hence $|\psi| \in \bar{\hat{A}} = \hat{A}$. ∎

A special case of this last result is the classical approximation theorem of K. Weierstrass, according to which every continuous real-valued function, defined over a closed, bounded interval of the line, can be uniformly approximated by polynomials.

Notes and Exercises

39A Let (V, τ') be a topological vector space over a nondiscrete topological field (K, τ). Suppose that the field topology has been induced by a valuation $\|\ \|$, We say that a subset M of V is *balanced* or *equilibrated* if $a \in K$ and $\| a \| \leqslant 1$ always imply $a \cdot M \subseteq M$. A subset M of V is said to be *absorbent* when, given any $x \in V$, we can find a positive real number s such that

$$x \in \cap \{a \cdot M : a \in K, \| a \| \geqslant s\}.$$

Exercise

1. If M is equilibrated, so is $\bar{M}$.

2. $\widehat{M} = \cup \{a \cdot M : a \in K, \| a \| \leqslant 1\}$ is the smallest equilibrated subset of V containing M. It is called the *equilibrated-hull* of M.

3. Let $\mathfrak{U}$ be a basis of the neighborhoods filter $\mathfrak{u}_{\tau'}$ of the zero vector. Then $\mathfrak{U} = \{\widehat{U} : U \in \mathfrak{U}\}$ is a basis of $\mathfrak{u}_{\tau'}$.

4. $\mathfrak{u}_{\tau'}$ possesses a basis $\mathfrak{U}$ of closed neighborhoods with the following properties:

(a) The sets belonging to $\mathfrak{U}$ are equilibrated and absorbent.

(b) If $a \in K$, $a \neq 0$ and $U \in \mathfrak{U}$, then $a \cdot U \in \mathfrak{U}$.

(c) Given any $U \in \mathfrak{U}$ there exists a $V \in \mathfrak{U}$ with $U + V \subseteq U$.

Conversely, given any filter basis $\mathfrak{U}$ of V verifying conditions (a)–(c), there then exists one and only one vector-space topology τ' on V such that $\tau'0 = \ulcorner\mathfrak{U}\urcorner$ [9.I, p.6].

39B *Exercise:* Let $\{(V_\iota, \tau_{\iota'}) : \iota \in I\}$ be a collection of topological vector spaces over the topological field (K, τ) and let V be a vector space over K. For each $\iota \in I$ let φ_ι be a linear map from V into V_ι and suppose $\Phi = \{\varphi_\iota : \iota \in I\}$. Then the topology $\Phi^{\tau'}$ on V, transferred by means of Φ^{-1}, is a vector-space topology with respect to (K, τ) [9,I, p. 7].

40. Inductive and Projective Limits

In this section I will always denote an indexing set which is itself an *ordered set*. Whenever, given any two indices $\iota, \lambda \in I$, we can always find a third index μ such that $\mu \leqslant \iota$ and $\mu \leqslant \lambda$, we shall call I *decreasingly filtrated.*

Suppose we have been given a collection of sets $\{R_\iota : \iota \in I\}$. To each couple of indices (ι, λ) where $\iota \geqslant \lambda$ let there be associated a map $\psi_\lambda^\iota : R_\iota \to R_\lambda$ such that the following conditions are satisfied:

(a) *For each $\iota \in I$ the map ψ_ι^ι is the identity map.*

(b) *If $\iota \geqslant \lambda \geqslant \mu$, then $\psi_\mu^\iota = \psi_\mu^\lambda \psi_\lambda^\iota$ (transitivity).*

In the sum set $S = \Sigma\{R_\iota : \iota \in I\}$ (cf. Section 24; the sets R_ι are considered disjoint) we can define an equivalence ρ as follows: To each point $x \in S$ there corresponds a unique index $\iota(x) \in I$ such that $x \in R_{\iota(x)}$. For points $x, y \in S$ we shall say that $(x, y) \in \rho$ if and only of there exists an index $\lambda \in I$ with $\lambda \leqslant \iota(x)$, $\lambda \leqslant \iota(y)$, and $\psi_\lambda^{\iota(x)}(x) = \psi_\lambda^{\iota(y)}(y)$. The quotient set $R = S/\rho$ is called the *inductive limit* of the family of sets $(R_\iota : \iota \in I\}$. Conversely, if we use this terminology it will always be understood that there exist maps ψ_λ^ι associated with the sets R_ι such that properties (a) and (b) are verified.

If α_ι denotes the map of R_ι into S defined on *p.* 160 and if κ denotes the canonical map from S onto S/ρ, then $\psi^\iota = \kappa\alpha_\iota$ is a map from R_ι into the inductive limit R. From $\iota \geqslant \lambda$ one deduces $\psi^\iota = \psi^\lambda \psi_\lambda^\iota$. If $x \in R_\iota$ and $y \in R_\lambda$ then the equality $\psi^\iota x = \psi^\lambda y$ is equivalent to the existence of a $\mu \in I$ such that $\mu \leqslant \iota$, $\mu \leqslant \lambda$ and $\psi_\mu^\iota x = \psi_\mu^\lambda y$. Lastly, we see that R must be equal to

$$\cup \{\psi^\iota R_\iota : \iota \in I\}.$$

40.1 *Let $\{R_\iota : \iota \in I\}$ and $\{R'_\iota : \iota \in I\}$ be two families of sets with respective associated maps ψ_λ^ι, ψ^ι and $\psi_\lambda'^\iota$, ψ'^ι and inductive limits R and R'. Moreover, with each $\iota \in I$ let there be associated a map $\varphi^\iota : R_\iota \to R'_\iota$ in such a way that $\iota \geqslant \lambda$ always implies $\psi_\lambda'^\iota \varphi^\iota = \varphi^\lambda \psi_\lambda^\iota$. Then there exists one and only one map $\varphi : R \to R'$ such that $\psi'^\iota \varphi_\iota = \varphi\psi^\iota$. If for each $\iota \in I$, φ^ι is a map onto R'_ι (a 1-1 map) then φ is also a map onto R'_ι (a 1-1 map). More generally, $\varphi p = \varphi q$ implies the existence of a $\lambda \in I$ and of two points $x, y \in R_\lambda$ such that $\psi^\lambda x = p$, $\psi^\lambda y = q$,* and $\varphi^\lambda x = \varphi^\lambda y$.

Proof: Let p be a point in R and suppose that the points $x \in R_\iota$, $y \in R_\lambda$ satisfy $\psi^\iota x = \psi^\lambda y = p$. Then there exists a $\mu \leqslant \iota, \lambda$ such that $\psi_\mu^\iota x = \psi_\mu^\lambda y$. This implies that

$$\psi_\mu'^\iota \varphi^\iota x = \varphi^\mu \psi_\mu^\iota x = \varphi^\mu \psi_\mu^\lambda y = \psi_\mu'^\lambda \varphi^\lambda y$$

and hence $\psi_\mu'^\iota \varphi^\iota x = \varphi^\mu \psi_\mu^\iota x = \varphi^\mu \psi_\mu^\lambda y = \psi_\mu'^\lambda \varphi^\lambda y$. If to each $p \in R$ we associate an $\iota(p) \in I$ and then an $x_p \in R_{\iota(p)}$ satisfying $\psi^{\iota(p)} x_p = p$, then, independently of how this association has been established, we can define a map $\varphi : R \to R'$ by $\varphi p = \psi'^{\iota(p)} \varphi^{\iota(p)} x_p$ which will manifestly satisfy the condition $\psi'^\iota \varphi^\iota = \varphi\psi^\iota$. Conversely, from this condition and the facts that $x \in R_\iota$ and $\psi^\iota x = p$, it follows that $\varphi p = \varphi\psi^\iota x = \psi'^\iota \varphi^\iota x$; i.e., φ is determined uniquely.

For all $\iota \in I$ let φ^ι now be a map *onto* R'_ι. Given $p' \in R'$ there exists an $\iota \in I$

and an $x' \in R'_\iota$ such that $\psi'^\iota x' = p'$. Furthermore, there exists an $x \in R_\iota$ such that $\varphi^\iota x = x'$. For the point $p = \psi^\iota x \in R$ we then have

$$\varphi p = \varphi \psi^\iota x = \psi'^\iota \varphi^\iota x = \psi'^\iota x' = p'.$$

Hence in this case also φ is a map onto R'.

Given any two points $p, q \in R$ we can find an $\iota \in I$ and points $x_1, y_1 \in R_\iota$ such that $\psi^\iota x_1 = p$ and $\psi^\iota y_1 = q$. From $\varphi p = \varphi q$ we deduce that

$$\psi'^\iota \varphi^\iota x_1 = \varphi \psi^\iota x_1 = \varphi p = \varphi q = \varphi \psi^\iota y_1 = \psi'^\iota \varphi^\iota y_1 .$$

Then there exists a $\lambda \in I$ with $\lambda \leqslant \iota$ and $\psi'^\iota_\lambda \varphi^\iota x_1 = \psi'^\iota_\lambda \varphi^\iota y_1$. For the points $x = \psi^\iota_\lambda x_1$, $y = \psi^\iota_\lambda y_1$ in R_λ we deduce that $\psi^\lambda x = \psi^\lambda \psi^\iota_\lambda x_1 = \psi^\iota x_1 = p$, and likewise $\psi^\lambda y = q$. Moreover,

$$\varphi^\lambda x = \varphi^\lambda \psi^\iota_\lambda x_1 = \psi'^\iota_\lambda \varphi^\iota x_1 = \psi'^\iota_\lambda \varphi^\iota y_1 = \varphi^\lambda \psi^\iota_\lambda y_1 = \varphi^\lambda y.$$

If all the φ^ι are 1-1 maps, we therefore have $x = y$ and $p = \psi^\lambda x = \psi^\lambda y = q$; i.e., φ is also 1-1. ∎

The map φ of the above theorem is also known as the inductive limit of the maps φ^ι. Conversely, if we use this terminology, it will always be understood that the maps φ^ι satisfy the conditions of the above theorem. We obtain two important special cases when all the sets R_ι or R'_ι are identical and when the associated maps ψ^ι_λ or ψ'^ι_λ, respectively, are identity maps.

40.2 *Let I be decreasingly filtrated and for* $\nu = 1, \ldots, n (n \in \mathbf{N})$ let $\{{}_\nu R_\iota : \iota \in I\}$ *be a collection of sets with associated maps* ${}_\nu\psi^\iota_\lambda$, ${}_\nu\psi^\iota$ *and inductive limit* ${}_\nu R$. *Moreover, suppose* $R_\iota = {}_1R_\iota \times \ldots \times {}_nR_\iota$ *and* $\psi^\iota_\lambda = {}_1\psi^\iota_\lambda \times \ldots \times {}_n\psi^\iota_\lambda$. *Then the maps* $\psi^\iota_\lambda : R_\iota \to R_\lambda$ *satisfy conditions* (a) *and* (b). *Hence we are assured of the existence of the inductive limit R of the family* $\{R_\iota : \iota \in I\}$, *and there exists a unique* 1-1 *map of R onto,* ${}_1R \times \ldots \times {}_nR$ *with* $\varphi\psi^\iota = {}_1\psi^\iota \times \ldots \times {}_n\psi^\iota$.

Proof: The first contention follows from

$$\psi^\lambda_\mu \psi^\iota_\lambda = ({}_1\psi^\lambda_\mu \, {}_1\psi^\iota_\lambda) \times \ldots \times ({}_n\psi^\lambda_\mu \, {}_n\psi^\iota_\lambda).$$

For $\iota \geqslant \lambda$ the projection maps ${}_\nu\pi^\iota : R_\iota \to {}_\nu R_\iota$ satisfy the condition

$${}_\nu\psi^\iota_\lambda \, {}_\nu\pi^\iota = {}_\nu\pi^\lambda \varphi^\iota_\lambda .$$

Hence, by 40.1, for each ν the inductive limit ${}_\nu\pi$ of the ${}_\nu\pi^\iota$ does exist. This map ${}_\nu\pi$ is a map from R onto ${}_\nu R$, and ${}_\nu\psi^\iota \, {}_\nu\pi^\iota = {}_\nu\pi\psi^\iota$. By $\varphi = {}_1\pi \times \ldots \times {}_n\pi$ we can define a map $\varphi : R \to {}_1R \times \ldots \times {}_nR$. Since ${}_1\pi^\iota \times \ldots \times {}_n\pi^\iota$ is the identity on R_ι we have

$$\begin{aligned} \varphi\psi^\iota &= ({}_1\pi\psi^\iota) \times \ldots \times ({}_n\pi\psi^\iota) = ({}_1\psi^\iota \times \ldots \times {}_n\psi^\iota)({}_1\pi^\iota \times \ldots \times {}_n\pi^\iota) \\ &= {}_1\psi^\iota \times \ldots \times {}_n\psi^\iota. \end{aligned}$$

If $p_1 \times \dots \times p_n$ is a point in ${}_1R \times \dots \times {}_nR$, then there exists an $\iota \in I$ and points $x_\nu \in {}_\nu R_\iota$ with ${}_\nu\psi^\iota x_\nu = p_\nu$ for $\nu = 1, \dots, n$. (This stems from the facts that I is descendingly filtrated and that we are dealing with the product of only a finite number of sets—otherwise we can not usually find such a common index ι.) For the point $p = \psi^\iota(x_1 \times \dots \times x_n)$ of R we then have

$$\varphi p = \varphi\psi^\iota(x_1 \times \dots \times x_n) = {}_1\psi^\iota x_1 \times \dots \times {}_n\psi^\iota x_n = p_1 \times \dots \times p_n .$$

Hence φ is a map from R onto ${}_1R \times \dots \times {}_nR$.

Lastly, by 40.1 we deduce from $\varphi p = \varphi q$ the existence of a common index $\lambda \in I$ and of points $x, y \in R_\lambda$ such that $\psi^\lambda x = p$, $\psi^\lambda y = q$, and ${}_\nu\pi^\lambda x = {}_\nu\pi^\lambda y$. This implies $x = y$ and therefore $p = q$; i.e., φ is also a 1-1 map.

Conversely, if for the map $\varphi : R \to {}_1R \times \dots \times {}_nR$ the conditions

$$\varphi\psi^\iota = {}_1\psi^\iota \times \dots \times {}_n\psi^\iota$$

hold, then we can deduce from $x \in R_\iota$ and $p = \psi^\iota x$ that

$$\varphi p = ({}_1\psi^\iota\, {}_1\pi^\iota x) \times \dots \times ({}_n\psi^\iota\, {}_n\pi^\iota x) = ({}_1\pi \times \dots \times {}_n\pi)\, \psi^\iota x = ({}_1\pi \times \dots \times {}_n\pi)\, p,$$

and hence $\varphi = {}_1\pi \times \dots \times {}_n\pi$. Consequently, φ is also uniquely determined. ∎

For each index $\iota \in I$ let τ_ι be a topology on R_ι, and let the maps $\psi^\iota_\lambda : R_\iota \to R_\lambda$ be continuous maps of (R_ι, τ_ι) into $(R_\lambda, \tau_\lambda)$. Then we shall call the topological space $(R, \tau) = \Sigma(R_\iota, \tau_\iota)/\rho$ the inductive limit of the topological spaces (R_ι, τ_ι) and we shall call its topology τ the inductive limit of the topologies τ_ι. This terminology allows us to state the following result.

40.3 (1) *The maps $\psi^\iota : R_\iota \to R$ are continuous maps from (R_ι, τ_ι) into (R, τ).*

(2) *τ is the finest topology on R under which all the ψ^ι are continuous.*

(3) *If φ is the inductive limit of the continuous maps $\varphi^\iota : (R_\iota, \tau_\iota) \to (R'_\iota, \tau'_\iota)$, then φ is also a continuous map from the inductive limit (R, τ) into the inductive limit (R', τ').*

Proof

(1) Since $\psi^\iota = \kappa\alpha_\iota$, our first contention follows from 21.7 and 24.1.

(2) Let τ^* be a topology on R for which all maps ψ^ι are continuous. Since the α_ι are continuous, open maps into the sum space $\Sigma(R_\iota, \tau_\iota)$ (24.1), the continuity of the ψ^ι implies that κ is also a continuous map of $\Sigma(R_\iota, \tau_\iota)$ onto (R, τ^*). This implies that $\tau^* \geqslant (\Sigma\tau_\iota)_\kappa = \tau$.

(3) By 24.3, $\varphi^* = \Sigma\{\varphi^\iota : \iota \in I\}$ is a continuous map of $\Sigma(R_\iota, \tau_\iota)$ into $\Sigma(R'_\iota, \tau'_\iota)$. Since $\varphi\kappa\alpha_\iota = \varphi\psi^\iota = \psi'^\iota\varphi^\iota = \kappa'\alpha'_\iota\varphi^\iota$, we have $\varphi\kappa = \kappa'\varphi^*$. Hence $\varphi\kappa$, and therefore also φ, are continuous (21.7). ∎

40.4 *If the spaces (R_ι, τ_ι) are locally connected, so is their inductive limit (R, τ). If the spaces (R_ι, τ_ι) are connected and I is descendingly filtrated, then (R, τ) is connected.*

Proof: The preservation of locally connectedness follows immediately from 21.7 and 24.5. If the spaces (R_ι, τ_ι) are connected, so are their images $\psi^\iota R_\iota$ (19.3). Let $\{I_1, I_2\}$ be a decomposition of the decreasingly filtrated index set I, and suppose $\iota_1 \in I_1$, $\iota_2 \in I_2$. Then there exists a $\lambda \in I$ with $\lambda \leqslant \iota_1$ and $\lambda \leqslant \iota_2$. Suppose that $\lambda \in I_1$, say. Since $\psi_\lambda^{\iota_2} R_{\iota_2} \subseteq R_\lambda$, we must have $\psi^{\iota_2} R_{\iota_2} \subseteq \psi^\lambda R_\lambda$. Hence the intersection of the two sets

$$V_1 = \cup \{\psi^\iota R_\iota : \iota \in I_1\} \qquad \text{and} \qquad V_2 = \cup \{\psi^\iota R_\iota : \iota \in I_2\}$$

is nonempty, and hence, by 14.5, $R = \cup \{\psi^\iota R_\iota : \iota \in I\}$ must be connected. ∎

40.5 *Let (R, τ) be the inductive limit of the spaces (R_ι, τ_ι). Then M is an open (closed) subset of (R, τ) if and only if $(\psi^\iota)^{-1}M$ is an open (closed) subset of (R_ι, τ_ι) for all $\iota \in I$.*

Proof: Because of the continuity of the ψ^ι (40.3) our condition is necessary. Conversely, if all the $(\psi^\iota)^{-1}M$ are open, then $\kappa^{-1}M = \Sigma (\psi^\iota)^{-1}M$ must also be open (24.2). From 21.1 we see that M must also be open. Similar conclusions hold for closed subsets. ∎

As in 40.2, let I be decreasingly filtrated, let $({}_\nu R, {}_\nu\tau)$ be the inductive limit of the space $({}_\nu R_\iota, {}_\nu\tau_\iota)$, and let (R, τ) be the inductive limit of the product space. $({}_1R_\iota, {}_1\tau_\iota) \times \ldots \times ({}_nR_\iota, {}_n\tau_\iota)$. Since the projection maps ${}_\nu\pi^\iota$ are continuous, their inductive limit must also be continuous (40.3). Hence the map φ of 40.2 is a continuous 1-1 map from (R, τ) onto the product space $({}_1R, {}_1\tau) \times \ldots \times ({}_nR, {}_n\tau)$. However, this map is generally not a homeomorphism (40A):

Let us again assume that I is decreasingly filtrated. Let the sets R_ι now be groups and the maps ψ_λ^ι homomorphisms of these groups. For the maps $\mu^\iota : R_\iota \times R_\iota \to R_\iota$ which define multiplication in these groups, we then have $\psi_\lambda^\iota \mu^\iota = \mu^\lambda(\psi_\lambda^\iota \times \psi_\lambda^\iota)$. By 40.1, the μ^ι have an inductive limit μ which is a map from the inductive limit R^* of the sets $R_\iota \times R_\iota$ onto the inductive limit R of the R_ι. By 40.2, there also exists a 1-1 map φ from R^* onto $R \times R$. For any two elements $a, b \in R$ we can therefore define a product in R by $a \cdot b = \mu\varphi^{-1}(a, b)$. We can obtain the element $a \cdot b$ as follows: There always exists an index ι and elements $x, y \in R_\iota$ such that $\psi^\iota x = a$ and $\psi^\iota y = b$. Hence $a \cdot b = \psi^\iota(x \cdot y)$. By simple calculations, we can also establish the following result.

40.6 *The inductive limit R of the groups R_ι is also a group with respective to the above-defined product. The inverse in R is the inductive limit of the inverses in the groups R_ι. The map ψ^ι is a homomorphism of the group R_ι into the group R.*

If for each index ι, (R_ι, τ_ι) is a topological group and if the maps ψ_λ^ι are continuous homomorphisms, then, by what precedes, it must follow that the inductive limit (R, τ) is both a group and a topological space. Moreover, the maps ψ^ι are continuous homomorphisms. In general, however, (R, τ) is not a topological group (40A): But since the trivial topology is always a group topology on the group R and since the inf of group topologies is again a group topology (36E), there must exist a finest group topology τ_g on R such that $\tau \leqslant \tau_g$. The topological group (R, τ_g) is then known as the inductive limit of the topological groups (R_ι, τ_ι). We arrive at similar conclusions for rings and for vector spaces over a fixed field.

There exists a second type of formation of limits, proceeding as follows: Let $\{R_\iota : \iota \in I\}$ again be a family of sets. To each couple of indices (ι, λ) with $\iota \geqslant \lambda$ let us associate a map $\chi_\iota^\lambda : R_\lambda \to R_\iota$ verifying the following conditions:

(a′) *For each* $\iota \in I$, χ_ι^ι *is the identity.*

(b′) *If* $\iota \geqslant \lambda \geqslant \mu$, *then* $\chi_\iota^\mu = \chi_\iota^\lambda \chi_\lambda^\mu$.

We define the *projective limit* R of the sets R_ι to be the set of all points $x = \times (x_\iota : \iota \in I\}$ in the product set $\times \{R_\iota : \iota \in I\}$ for which $\iota \geqslant \lambda$ always implies $\pi_\iota x = x_\iota = \chi_\iota^\lambda x_\lambda$. Then $\chi_\iota x = x_\iota = \pi_\iota x$ defines a map $\chi_\iota : R \to R_\iota$. From $\iota \geqslant \lambda$ one deduces $\chi_\iota = \chi_\iota^\lambda \chi_\lambda$. If I is decreasingly filtrated and if for any two points $x, y \in R$, for any fixed index $\iota \in I$ and all $\lambda \leqslant \iota$ we always have $\chi_\lambda x = \chi_\lambda y$, then x and y must be equal: For, given any $\mu \in I$, we can find an $\lambda \in I$ such that $\lambda \leqslant \iota$ and $\lambda \leqslant \mu$, and then we have

$$\chi_\mu x = \chi_\mu^\lambda \chi_\lambda x = \chi_\mu^\lambda \chi_\lambda y = \chi_\mu y.$$

Even if $R_\iota \neq \emptyset$ for all $\iota \in I$, there may exist no point x in the product-space $\times R_\iota$ for which $\pi_\lambda x = \chi_\iota^\lambda \pi_\lambda x$. Even if all maps χ_ι^λ are onto R_ι, the projective limit may be empty (cf. 40.9).

40.7 *Let* $(R_\iota : \iota \in I\}$ *and* $(R'_\iota : \iota \in I\}$ *be two collections of sets with respective associated maps* χ_ι^λ, χ_ι *and* $\chi_\iota'^\lambda$, χ'_ι *and projective limits* R *and* R'. *To each* $\iota \in I$ *let a map* $\varphi^\iota : R_\iota \to R'_\iota$ *be associated, such that* $\iota \geqslant \lambda$ *will then always imply* $\varphi^\iota \chi_\iota^\lambda = \chi_\iota'^\lambda \varphi^\lambda$. *Then* $\varphi = \times \{\varphi^\iota \chi_\iota : \iota \in I\}$ *is a map from* R *into* R' *and* $\varphi^\iota \chi_\iota = \chi'_\iota \varphi$ *for all* $\iota \in I$. *This property characterizes* φ. *If for each* $\iota \in I$, φ^ι *is a* 1-1 *map, then* φ *also is* 1-1.

Proof: From $x \in R$, $x' = \varphi x$, and $\iota \geqslant \lambda$ it follows that

$$x'_\iota = \varphi^\iota \chi_\iota x = \varphi^\iota \chi_\iota^\lambda \chi_\lambda x = \chi_\iota'^\lambda \varphi^\lambda \chi_\lambda x = \chi_\iota'^\lambda x'_\lambda .$$

Hence $x' \in R'$, i.e., φ is a map into R'. In addition, $\varphi^\iota \chi_\iota x = x'_\iota = \chi'_\iota \varphi x$, and hence $\varphi^\iota \chi_\iota = \chi'_\iota \varphi$. Conversely, this last equality implies $\varphi x = \times \chi'_\iota \varphi x = \times \varphi^\iota \chi_\iota x$, and φ is therefore uniquely determined by this condition. From $\varphi x = \varphi y$ it also follows that $\varphi^\iota \chi_\iota x = \chi'_\iota \varphi x = \chi'_\iota \varphi y = \varphi^\iota \chi_\iota y$ for all $\iota \in I$. If the φ^ι are all

1-1 maps, we have $\chi_\iota x = \chi_\iota y$ for all $\iota \in I$ and hence $x = y$. This tells us that φ is also a 1-1 map. ∎

We call the map φ the projective limit of the maps φ^ι. Conversely, whenever we use this terminology it will be understood that the φ^ι obey the conditions imposed in the above theorem. As in 40.1, the above theorem also contains the special case where $\varphi^\iota : R \to R'_\iota$ and $\varphi^\iota : R_\iota \to R'$. If each φ_ι is a map onto R'_ι, it does not necessarily follow that the projective limit φ is a map onto R'; even if, in addition, χ'_λ, χ''_λ are mappings onto R_ι and R'_ι, respectively (40B).

If $\{(R_\iota, \tau_\iota) : \iota \in I\}$ is a collection of topological spaces and if the χ^λ_ι are continuous maps from $(R_\lambda, \tau_\lambda)$ into (R_ι, τ_ι), then the relative topology $\tau = (\times\ \tau_\iota)_R$ of the projective limit R of the R_ι is known as the projective limit of the topologies τ_ι, while (R, τ) is known as the projective limit of the topological spaces (R_ι, τ_ι).

40.8 (1) *The map* χ_ι *is a continuous map from the projective limit* (R, τ) *into* (R_ι, τ_ι).

(2) τ *is the coarsest topology on* R *rendering all the* χ_ι *continuous.*

(3) *The projective limit of continuous maps is a continuous map on the projective limit.*

(4) *If for each* $\iota \in I$, τ_ι *is a* T_2*-topology, then* R *is closed in the product space* $\times\ (R_\iota, \tau_\iota)$.

Proof

(1) Since χ_λ is the restriction to R of the projection map $\pi_\lambda : \times\ R_\iota \to R_\lambda$, χ_λ must be continuous (22.3).

(2) If τ^* is a topology on R with respect to which all χ_ι are continuous, then for any point $x \in R$, we have

$$\tau^* x \leqq \bigwedge \{\chi_\iota^{-1}\tau_\iota\chi_\iota x : \iota \in I\} = \overset{\sqcap}{R} \wedge \bigwedge \{\pi_\iota^{-1}\tau_\iota\pi_\iota x : \iota \in I\} = \overset{\sqcap}{R} \wedge (\times\ \tau_\iota)\, x = \tau x.$$

(3) The continuity of the maps φ^ι and χ_ι implies the continuity of the projective limit $\varphi = \times\ \{\varphi^\iota \chi_\iota : \iota \in I\}$ (22.9).

(4) For any filter $\mathfrak{a}$ of R and any two indices ι, λ with $\iota \geqslant \lambda$ we have $\pi_\iota \mathfrak{a} = \chi^\lambda_\iota \pi_\lambda \mathfrak{a}$. If the point $\chi \in \times\ R_\iota$ belongs to the closure of R, then $\mathfrak{a} = \overset{\sqcap}{R} \wedge (\times)\, \tau_\iota \chi \neq \mathfrak{o}$. By the continuity of the projection-maps π_ι and of the maps χ^λ_ι we have, on the one hand, $\pi_\iota \mathfrak{a} \leqslant \tau_\iota \pi_\iota \chi$, and on the other hand,

$$\pi_\iota \mathfrak{a} = \chi^\lambda_\iota \pi_\lambda \mathfrak{a} \leqq \chi^\lambda_\iota \tau_\lambda \pi_\lambda x \leqq \tau_\iota \chi^\lambda_\iota \pi_\lambda x,$$

whence $\tau_\iota \pi_\iota x \wedge \tau_\iota \chi^\lambda_\iota \pi_\lambda x \neq \mathfrak{o}$. If all the topologies τ_ι satisfy $(\mathbf{T}_2)$, then $\pi_\iota x = \chi^\lambda_\iota \pi_\lambda x$ and therefore $x \in R$. ∎

40.9 *Let (R, τ) be the projective limit of the spaces (R_ι, τ_ι). If for each $\iota \in I$, τ_ι is a T_1-, T_2-, T_3- or T_{3a}-topology, then τ is of the same nature. If all the (R_ι, τ_ι) are completely compact, T_2-spaces, then (R, τ) is completely compact.*

If I is descendingly filtrated and all the (R_ι, τ_ι) are completely compact, nonempty T_2-spaces, then their projective limit (R, τ) is also nonempty. If, moreover, the spaces (R_ι, τ_ι) are connected, then (R, τ) is itself connected.

Proof: If the spaces (R_ι, τ_ι) are all T_ι-spaces ($\iota = 1, 2, 3, 3a$) then so is the product space $\times (R_\iota, \tau_\iota)$ (22.6). The same conclusion must also hold for the subspace R (9.4, 18.8). If all (R_ι, τ_ι) are completely compact T_2-spaces, then $\times (R_\iota, \tau_\iota)$ is a closed subset and hence completely compact. By 40.8, the same reasoning and conclusion also hold for R.

Let us now assume that all the (R_ι, τ_ι) are completely compact, nonempty T_2-spaces, and let I be decreasingly filtrated. For each $\lambda \in I$ and for any finite set of indices $\iota_1, \ldots, \iota_n \in I$ with $\lambda \leqslant \iota_1, \ldots, \iota_n$, let

$$Z(\lambda; \iota_1 \ldots, \iota_n) = \{x : x \in \times R_\iota, \chi^\lambda_{\iota_\nu} \pi_\lambda x = \pi_{\iota_\nu} x \quad \text{for} \quad \nu = 1, \ldots, n\}.$$

The sets $Z(\lambda; \iota_1, \ldots, \iota_n)$ are manifestly nonempty, while the collection $\mathfrak{Z}$ of all such subsets of $\times R_\iota$ forms a filter basis: For if $\mu \leqslant \lambda, \lambda'$ and

$$x \in Z(\mu; \lambda, \lambda', \iota_1, \ldots, \iota_n, \iota_1', \ldots, \iota_{n'}')$$

then we have

$$\chi^\lambda_{\iota_\nu} \pi_\lambda x = \chi^\lambda_{\iota_\nu} \chi^\mu_\lambda \pi_\mu x = \chi^\mu_{\iota_\nu} \pi_\mu x = \pi_{\iota_\nu} x ,$$

and likewise $\chi^{\lambda'}_{\iota'_\nu} \pi_{\lambda'} x = \pi_{\iota'_\nu}$. This implies that

$$Z(\mu; \lambda, \lambda', \iota_1, \ldots, \iota_n, \iota_1', \ldots, \iota_{n'}') \subseteq Z(\lambda; \iota_1, \ldots, \iota_n) \cap Z(\lambda'; \iota_1', \ldots, \iota_n').$$

As in 40.8 (4), we can deduce that the member sets of $\mathfrak{Z}$ are closed. We also see immediately that $\cap \{Z : Z \in \mathfrak{Z}\} = R$. Since $\times R_\iota$ is completely compact, we immediately see by 11.19 that $R \neq \emptyset$.

Let us now assume that the spaces (R_ι, τ_ι) are also connected. The map $\chi^\lambda_\lambda \times \chi^\lambda_{\iota_1} \times \cdots \times \chi^\lambda_{\iota_n}$ is continuous, mapping R_λ onto a connected subset T of the subspace $R_\lambda \times R_{\iota_1} \times \cdots \times R_{\iota_n}$ of the space $\times (R_\iota, \tau_\iota)$ (19.3). The subspace $T \times \times \{R_\mu : \mu \in I, \mu \neq \lambda, \iota_1, \ldots, \iota_n\}$ is homeomorphic to $Z\{\lambda; \iota_1, \ldots, \iota_n)$ and is therefore itself connected (22.6). Since $\times (R_\iota, \tau_\iota)$ is a completely compact T_2-space and hence also a T_4-space (11.17), we see, using 14.6, that

$$R, \tau) = \cap \{Z : Z \in \mathfrak{Z}\}$$

must be connected. ∎

Suppose the sets R_ι or the spaces (R_ι, τ_ι) are groups, or topological groups, as the case may be. If we define a product for elements $x, y \in \times R_\iota$ by $x \cdot y = \times (x_\iota \cdot y_\iota)$, then $\times R_\iota$ or $\times (R_\iota, \tau_\iota)$ becomes a group, or topological group, respectively (cf. p. 245). If, moreover, the maps χ_ι^λ are homomorphisms, then the following result holds for the product of two elements x, y of the projective limit R of the R_ι,

$$\chi_\iota^\lambda \pi_\lambda(x \cdot y) = \pi_\iota^\lambda(x_\lambda \cdot y_\lambda) = (\chi_\iota^\lambda x_\lambda) \cdot (\chi_\iota^\lambda y_\lambda) = x_\iota \cdot y_\iota = \pi_\iota(x \cdot y).$$

Hence R is a subgroup of $\times R_\iota$. Consequently:

40.10 *The projective limit of (topological) groups is again a (topological) group. The maps χ_ι are (continuous) homomorphisms. The projective limit of (continuous) homomorphisms is itself a (continuous) homomorphism.*

Using proofs similar to the above proof, we obtain results analogous to those above for rings and for vector spaces over a fixed field.

Notes and Exercises

40A Let $(p_n)_{n \in \mathbf{N}}$ be the sequence of prime numbers. For each $x \in \mathbf{N}$ let τ'_ν be the topology induced on the set of rationals by the p_ν-adic valuation (38B). Suppose further that $\tau_n = \bigwedge \{\tau'_\nu : \nu \geqslant n\}$.

Exercise

1. If we view the set $\mathbf{Q}$ of rational numbers as an additive group, then $(\mathbf{Q}, \tau_n)$ is a T_1-topological group for each $n \in \mathbf{N}$.

2. For any two natural numbers m, n with $m \leqslant n$ let ψ_n^m be the identity on $\mathbf{Q}$. Then the inductive limit τ of the topologies τ_n is identical to the sup-topology $\bigvee \{\tau_n : n \in \mathbf{N}\}$. This implies that τ is a T_1-topology.

3. This topology τ is, nevertheless, not a group topology. The inductive limit of the topological groups $(\mathbf{Q}, \tau_n)$ is the additive group $\mathbf{Q}$ under the trivial topology.

This implies further that the inductive limit of the product spaces $(\mathbf{Q}, \tau_n) \times (\mathbf{Q}, \tau_n)$ cannot be homeomorphic to the product space $(\mathbf{Q}, \tau) \times (\mathbf{Q}, \tau)$: By 40.3, the inductive limit μ of the group operations (cf. p. 272) is a continuous map from the inductive limit of the product spaces $(\mathbf{Q}, \tau_n) \times (\mathbf{Q}, \tau_n)$ onto $(\mathbf{Q}, \tau)$. If $(\mathbf{Q}, \tau_n) \times (\mathbf{Q}, \tau_n)$ were homeomorphic to $(\mathbf{Q}, \tau) \times (\mathbf{Q}, \tau)$ we would have as a consequence that the group operation in $(\mathbf{Q}, \tau)$ is continuous and hence that τ is a group topology—contradiction: (the operation of taking inverses in the inductive limit of topological groups is also always continuous with respect to the inductive-limit topology).

40B Let I be the set of all negative integers, ordered as usual. For each $\iota \in I$ let $R_\iota = \mathbf{N}$ and $R'_\iota = \{1, ..., \iota\}$. For any two numbers $\iota, \lambda \in I$ with $\iota \geqslant \lambda$ let χ_ι^λ be the identity on $\mathbf{N}$ and $\chi_\iota'^\lambda$ be defined by

$$\chi_\iota'^\lambda x' = \begin{cases} x' & \\ & \text{for} \\ -\iota & \end{cases} \quad \begin{matrix} 1 \leqslant x' \leqslant -\iota \\ \\ -\iota < x' \leqslant -\lambda. \end{matrix}$$

Lastly, let the map $\varphi^\iota : R_\iota \to R'_\iota$ be defined by

$$\varphi^\iota x = \begin{cases} x & \\ & \text{for} \\ -\iota & \end{cases} \quad \begin{matrix} 1 \leqslant x \leqslant -\iota \\ \\ -\iota < x \end{matrix}$$

Exercise: The maps χ_ι^λ, $\chi_\iota'^\lambda$ and φ^ι are onto R_ι and R'_ι, respectively. The maps φ^ι possess a projective limit φ. The point $\times \{-\iota : \iota \in I\}$ belongs to the projective limit of the R'_ι yet it is not attained as an image under the map φ.

40C *Exercise:* The inductive (projective) limit (R, τ) of the topological spaces (R_ι, τ_ι) is characterized, up to homeomorphism, by the following property:

Given any other space (R', τ') and any collection of continuous maps $\varphi^\iota : (R_\iota, \tau_\iota) \to (R', \tau')$ with $\varphi^\iota = \varphi^\lambda \psi_\lambda^\iota (\varphi^\iota : (R', \tau') \to (R_\iota, \tau_\iota)$ with $\varphi^\iota \chi_\iota^\lambda = \varphi^\lambda)$, then we can find a continuous map $\varphi^* : (R, \tau) \to (R', \tau')$ with

$$\varphi^\iota = \varphi^* \psi^\iota (\varphi^* : (R', \tau') \to (R', \tau) \text{ with } \varphi^\iota = \chi_\iota \varphi^*).$$

Similar theorems hold for topological groups, etc.

NOTATIONS EMPLOYED

The numbers in parentheses refer to the page where the corresponding symbol was first defined.

$D(R)$	Diagonal (210)
R	Reference set, space (4)
$U_\epsilon(p)$, $U_\epsilon^\delta(p)$	ϵ-Neighborhood of p under the (pseudo) metric δ
W_ϵ, W_ϵ^δ	ϵ-Uniform neighborhood under the (pseudo) metric δ
$\mathbf{N}$	Natural numbers (zero excluded)
$\mathbf{Q}$	Rational numbers
$\mathbf{R}$	Real numbers, real line
$\bar{\mathbf{R}}$	Real line extended by $\pm\infty$
$\mathfrak{o}$	Zero filter, null filter
$\mathfrak{r}(\mathfrak{a})$	Round filter associated with filter $\mathfrak{a}$
$\mathfrak{u}_\tau$	Neighborhoods filter of the neutral element of a topological group
$\mathfrak{w}$	Filter of uniform neighborhoods, uniform structure
$\mathfrak{w}_\delta$	Uniform structure belonging to the pseudometric δ
${}_\varphi\mathfrak{w}$	Uniform structure transferred by φ^{-1}
$\Delta(\mathfrak{w})$	Set of pseudometrics δ such that $\mathfrak{w}_\delta \geqslant \mathfrak{w}$
I	Indexing set
α_κ	Injection map into a sum set
δ	(Pseudo-) metric
$\epsilon_\mathfrak{p}$	cf. (156)
κ	Canonical map onto quotient set
π_ι	Projection map in a product set
σ	Trace map
τ	Topology
τ_M	Relative topology on the subspace M
τ_δ	Topology induced by the pseudometric δ
$\tau_\mathfrak{w}$	Topology induced by the uniform structure $\mathfrak{w}$
${}_\varphi\tau$	Topology transferred by means of φ^{-1}
τ_φ	Quotient topology associated with map φ
$\hat{\tau}_\mathfrak{K}$	Topology of compact convergence

$\emptyset$	Empty set
$\in$	Relation of "belonging," between element and set
$\complement$, $\complement R$	Complement with respect to set R
$\subseteq$	Relation of inclusion
$\subset$	Proper inclusion
$\cup$, $\bigcup$	Set-theoretic union
$\cap$, $\bigcap$	Set-theoretic intersection
$\leqslant$, $<$	Order relation
$\wedge$, $\bigwedge$	Order-theoretic inf (infinum)
$\vee$, $\bigvee$	Order-theoretic sup (supremum)
$\ll$	Refinement relation between systems of sets
$\lvert a \rvert$	Absolute value of number a
$\lvert M \rvert$	Cardinality of set M
$\lVert x \rVert$	Norm of vector x
$\ulcorner\mathfrak{A}\urcorner$	Filter generated by filter basis $\mathfrak{A}$
$\ulcorner M\urcorner, \ulcorner p\urcorner$	Principal filter generated by set M, or by point p, respectively
$\mathrm{Rk}\,(\mathfrak{a})$	Rank of the filter $\mathfrak{a}$
$Pz(\mathfrak{a})$	Power of the filter $\mathfrak{a}$
$\mathrm{Tr}\,(\mathfrak{a})$, $\mathrm{Tr}_M\,(\mathfrak{a})$	Trace, on the set M, of the filter $\mathfrak{a}$
$\bar{M}$, $h_\tau(M)$	Closure of set M (in the topology τ)
$\mathring{M}$, $k_\tau(M)$	Interior of set M (in the topology τ)
(R, τ)	Topological space
(R, δ)	(Pseudo-) metric space
$(R, \mathfrak{w})$	Uniform space
$\mathrm{Bd}\,(R, \tau)$	Basis degree of the space (R, τ)
$\mathrm{Sd}\,(R, \tau)$	Separability degree of the space (R, τ)
$\beta(R, \tau)$	Stone-Čech compactification of (R, τ)
$\omega(R, \tau)$	Wallman compactification of (R, τ)
R/ρ, $(R, \tau)/\rho$	Quotient set, respectively, quotient space, modulo the equivalence ρ
$\times R_\iota$, $\times (R_\iota, \tau_\iota)$ $R_1 \times \ldots \times R_n$ $(R_1, \tau_1) \times \ldots \times (R_n, \tau_n)$	Product set, respectively, product space
$\times p_\iota$, $p_1 \times \ldots \times p_n$	Point with the coordinates p_ι, respectively, p_ν in the product set $\times R_\iota$, respectively, $R_1 \times \ldots \times R_n$

$\times\ \varphi_\iota, \varphi_1 \times \dots \times \varphi_n$	Product of maps
$\sum R$, $\sum (R_\iota, \tau_\iota)$	Sum set, respectively, sum space
$\sum \varphi^\iota$	Sum of maps
$\varphi : R \to R'$	Map from R into R'
$\varphi : (R, \tau) \to (R', \tau')$	Continuous map from (R, τ) into (R', τ')
$\varphi : (R, \mathfrak{w}) \to (R', \mathfrak{w}')$	Uniformly continuous map of the uniform space $(R, \mathfrak{w})$ into the uniform space $(R', \mathfrak{w}')$
$[a, b]$, $[a, b[$, $]a, b]$, $]a, b[$, $[\leftarrow, a]$, $[\leftarrow, a[$ $]a, \rightarrow]$, $[a, \rightarrow]$	Intervals

BIBLIOGRAPHY

[1] Alexandroff, P. S., "Einführung in die Mengenlehre und die Theorie der reellen Funktionen." Deut. Verlag Wiss., 1956.

[2] Alexandroff, P. and Hopf, H., "Topologie I." Springer, Berlin, 1935.

[3] Appert, A., Propriétés des espaces abstraits les plus généraux, 2 Vols. *Actualities sci. et ind.* Nos. 145, 146 (1934).

[4] Appert, A., and Fan, K., Espaces topologiques intermédiaires. *Actualities sci. et ind.* No. 1121 (1951).

[5] Banach, S., "Théorie des opérations linéaires." Warsaw, 1932.

[6] Birkhoff, G., "Lattice theory," Math. Soc. Coll. Publ. XXV). Am. Math. Soc., New York, 1948.

[7] Bourbaki, N., Théorie des ensembles. *Actualities sci. et ind.* Nos. 846/1141, 1212). 0000.

[8] Bourbaki, N., Topologie générale. *Actualities sci. et ind.* Nos. 858/1142, 916/1143, 1029, 1045, 1084.

[9] Bourbaki, N., Espaces vectoriels topologiques. *Actualities sci. et ind.* No. 1189 (1953).

[10] Breuer, J., "Introduction to the Theory of Sets." Prentice-Hall, Englewood Cliffs, New Jersey, 1959.

[11] Chevalley, C., "Theory of Lie Groups, I." Princeton Univ. Press, Princeton, New Jersey, 1946.

[12] Fraenkel, A., "Zehn Vorlesungen über die Grundlegung der Mengenlehre." Teubner, Leipzig and Berlin, 1927.

[13] Fraenkel, A., "Einleitung in die Mengenlehre," 3rd ed., Springer, Berlin, 1932.

[14] Fraenkel, A., "Abstract Set Theory." North-Holland Publ. Co., Amsterdam, 1953.

[15] Fraenkel, A., and Bar-Hillel, Y., "Foundations of Set Theory." North-Holland Publ. Co., Amsterdam, 1958.

[16] Fréchet, M., "Les espaces abstraits." Gauthier-Villars, Paris, 1928.

[17] Grelling, K., "Mengenlehre," Math. Phys. Bibl. 58. Teubner, Leipzig and Berlin, 1924.

[18] Hall, D. W., and Spencer, G. L., "Topology." Wiley, New York, 1955.

[19] Hausdorff, F., "Grundzüge der Mengenlehre." Veit, Leipzig, 1914; Chelsea, New York, 1949.

[20] Hausdorff, F., "Mengenlehre." de Gruyter, Berlin, 1935.

[21] Hermes, H., "Einführung in die Verbandstheorie." Springer, Berlin, 1955.

[22] Hurewicz, W., and Wallman, H., "Dimension Theory." Princeton Univ. Press, Princeton, New Jersey, 1948.

[23] Kamke, E., "Mengenlehre," 3rd ed. Sammlung Göschen 999/999a). de Gruyter, Berlin, 1955.

[24] Kelley, J. L., "General Topology." van Nostrand, Princeton, New Jersey, 1955.

[25] Kerekjarto, B. v., "Vorlesungen über Topologie I." Springer, Berlin, 1923.

[26] Kuratowski, C., "Topologie I, II," Monogr. Mat., Vols. 11, 21. Sem. Math. Univ., Warsaw and Wroclaw, 1948, 1950.

[27] Kuratowski, K., "Einführung in die Mengenlehre und Topologie" (Polish). Warsaw, 1955.

[28] Kuratowski, K., and Mostowski, A., "Teoria Mnogosci (Mengenlehre)," (Monogr. Math., Vol. 27. Warsaw and Wroclaw, 1952.

[29] Menger, K., "Dimensionstheorie." Teubner, Leipzig and Berlin, 1928.

[30] Menger, K., "Kurventheorie." Teubner, Leipzig and Berlin, 1932.

[31] Moore, R. L., "Foundation of Point Set Theory," Am. Math. Coll. Publ. XIII. Am. Math. Soc., New York, 1932.

[32] Nakano, H., "Topology and Linear Topological Spaces." Maruzen, Tokyo, 1951.

[33] Newman, M. H. A., "Elements of the Topology of Plane Sets of Points." Cambridge Univ. Press, London, 1939.

[34] Nöbeling, G., "Grundlagen der analytischen Topologie." Springer, Berlin, 1954.

[35] Patterson, E. M., "Topology." Oliver and Boyd, Edinburgh and London, 1956.

[36] Pontrjagin, L. S., "Topological Groups." Princeton Univ. Press, Princeton, New Jersey, 1946.

[37] Pontrjagin, L. S., "Topologische Gruppen." Teubner, Leipzig, 1957.

[38] Schoenflies, A., "Entwicklung der Mengenlehre und ihre Anwendungen I." Teubner, Leipzig and Berlin, 1913.

[39] Sierpinski, W., "Leçons sur les nombres transfinis." Gauthier-Villars, Paris, 1950.

[40] Sierpinski, W., "Algèbre des ensembles," Monogr. Math., Vol. 23). Warsaw and Wroclaw, 1951.

[41] Sierpinski, W., "General Topology," Math. Exp. 7. Toronto Univ. Press, Toronto, 1952.

[42] Vaidyanathaswamy, R., "Treatise on Set Topology I." Ind. Math. Soc., Madras, 1947.

[43] Wang, H., and Naughton, Mc. R., "Les systèmes axiomatiques de la théorie des ensembles," Gauthier-Villars, Paris, 1953.

[44] Weil, A., Sur les espaces à structure uniforme et sur la topologie générale, *Actualities sci. et ind.* No. 551 (1937).

[45] Whyburn, G. T., "Analytic Topology," Am. Math. Soc. Coll. Publ. XXVIII. Am. Math. Soc., New York, 1942.

Subject Index

Printed in Dunstable, United Kingdom

91336875R00167